recent advances in phytochemistry

volume 32

Phytochemical Signals and Plant–Microbe Interactions

RECENT ADVANCES IN PHYTOCHEMISTRY

Proceedings of the Phytochemical Society of North America
General Editor: John T. Romeo, University of South Florida, Tampa, Florida

Recent Volumes in the Series:

Volume 23 **Plant Nitrogen Metabolism**
Proceedings of the Twenty-eighth Annual Meeting of the Phytochemical Society of North America, Iowa City, Iowa, June, 1988

Volume 24 **Biochemistry of the Mevalonic Acid Pathway to Terpenoids**
Proceedings of the Twenty-ninth Annual Meeting of the Phytochemical Society of North America, Vancouver, British Columbia, Canada, June, 1989

Volume 25 **Modern Phytochemical Methods**
Proceedings of the Thirtieth Annual Meeting of the Phytochemical Society of North America, Quebec City, Quebec, Canada, August, 1990

Volume 26 **Phenolic Metabolism in Plants**
Proceedings of the Thirty-first Annual Meeting of the Phytochemical Society of North America, Fort Collins, Colorado, June, 1991

Volume 27 **Phytochemical Potential of Tropical Plants**
Proceedings of the Thirty-second Annual Meeting of the Phytochemical Society of North America, Miami Beach, Florida, August, 1992

Volume 28 **Genetic Engineering of Plant Secondary Metabolism**
Proceedings of the Thirty-third Annual Meeting of the Phytochemical Society of North America, Pacific Grove, California, June – July, 1993

Volume 29 **Phytochemistry of Medicinal Plants**
Proceedings of the Thirty-fourth Annual Meeting of the Phytochemical Society of North America, Mexico City, Mexico, August, 1994

Volume 30 **Phytochemical Diversity and Redundancy in Ecological Interactions**
Proceedings of the Thirty-fifth Annual Meeting of the Phytochemical Society of North America, Sault Ste. Marie, Ontario, Canada, August, 1995

Volume 31 **Functionality of Food Phytochemicals**
Proceedings of the Thirty-sixth Annual Meeting of the Phytochemical Society of North America, New Orleans, Louisiana, August, 1996

Volume 32 **Phytochemical Signals and Plant–Microbe Interactions**
Proceedings of a joint meeting of the Phytochemical Society of North America and the Phytochemical Society of Europe, Noordwijkerhout, The Netherlands, April, 1997

A Continuation Order Plan is available for this series. A continuation order will bring delivery of each new volume immediately upon publication. Volumes are billed only upon actual shipment. For further information please contact the publisher.

recent advances in phytochemistry

volume 32

Phytochemical Signals and Plant–Microbe Interactions

Edited by

John T. Romeo
University of South Florida
Tampa, Florida

Kelsey R. Downum
Florida International University
Miami, Florida

and

Rob Verpoorte
Leiden University
Leiden, The Netherlands

PLENUM PRESS • NEW YORK AND LONDON

Library of Congress Cataloging-in-Publication Data

Phytochemical signals and plant-microbe interactions / edited by John
T. Romeo, Kelsey R. Downum, and Rob Verpoorte.
      p.   cm. -- (Recent advances in phytochemistry ; v. 32)
   "Proceedings of a joint meeting of the Phytochemical Society of
 North America and the Phytochemical Society of Europe, held April
 20-23, 1997, in Noordwijkerhout, The Netherlands"--T.p. verso.
   Includes bibliographical references and index.
   ISBN 0-306-45917-5
   1. Plant-pathogen relationships--Congresses. I. Romeo, John T.
 II. Downum, Kelsey R., 1952-   . III. Verpoorte, R. IV. Series.
 QK861.R38 vol. 32
 [SB732.7]
 572'.2 s--dc21
 [571.9'82]   98-19421
                                                                CIP

Cover photograph: Tumors produced by *Agrobacterium tumefaciens* on the stem of the tobacco plant *Nicotiana glaucà*. (Photo provided by A. J. G. Regensburg–Tuïnk, Leiden, The Netherlands.)

Proceedings of a joint meeting of the Phytochemical Society of North America and the Phytochemical Society of Europe, held April 20–23, 1997, in Noordwijkerhout, The Netherlands

ISBN 0-306-45917-5

© 1998 Plenum Press, New York
A Division of Plenum Publishing Corporation
233 Spring Street, New York, N.Y. 10013

http://www.plenum.com

10 9 8 7 6 5 4 3 2 1

All rights reserved

No part of this book may be reproduced, stored in a retrieval system, or transmitted in any form or by any means, electronic, mechanical, photocopying, microfilming, recording, or otherwise, without written permission from the Publisher

Printed in the United States of America

PREFACE

The papers assembled in this volume were originally presented at the joint meeting of the Phytochemical Society of North America and Phytochemical Society of Europe held in Noordwijkerhout, The Netherlands, April 20–23, 1997. The meeting was organized by an international panel of scientists from both societies. The symposium from which the related contributions on phytochemical signaling and plant–microbe interactions were taken was entitled "Communication of Plants with the Environment." The chapters included in this volume cover traditional areas in plant chemical ecology, as well as address fundamental issues on the involvement of phytochemicals in intra- and interspecific signaling, applications of the knowledge gained from such studies, and evolutionary origins. The term microbes is used here in the broadest sense to encompass bacteria, fungi, and nematodes.

An array of plant and fungal metabolites ranging from simple phenolics, salicylic acid, hydroxamic acids, flavonoids, polysaccharides, fatty acid derived octadecanoids, to trichothecenes and perylenequinones are discussed. A number of important themes emerge: i) the multifunctional roles of many phytochemicals, ii) the power of molecular techniques for studying biosynthetic pathways and gene function, iii) the central role of natural products in pathogenesis and disease resistance, and iv) the identification of promising areas for future research and development of applications.

The opening chapter by Osbourn *et al.* discusses the emerging role and significance of saponins as antifungal agents, compounds already well-documented as important mediators of plant–insect interactions. McCormick *et al.* treat mycotoxins. They employed selective gene disruption to generate fungal mutants that were used to elucidate the biosynthetic pathway of trichothecene toxins and to study their role in pathogenesis. Less known but equally intriguing are fungal pathogens that utilize active oxygen in pathogenesis of their host. Daub *et al.* address this issue. They used molecular techniques to study perylenequinone toxins of the fungal genus *Cercopora*. The study of gene function is leading to a fundamental understanding of the basis of cellular resistance to active oxygen and may provide novel control strategies for important fungal pathogens of plants.

Our detailed understanding of the involvement of phytochemicals in many plant disease processes underscores the primitive state of our knowledge of other plant–fungal interactions. Tree–fungal ectomycorrhizal symbiotic relationships are discussed by Koide *et al.* A putative role of phenolics in mediating recognition and succession phenomena is proposed, an idea ripe for exploration.

Friebe *et al.* reinforce the concept of multiple roles played by natural products in a number of ecological systems. Hydroxamic acids, well known allelopathic and antiherbivore agents, are specifically addressed in studies of recognition phenomena and chemotaxis between a specific plant and nematode. The inherent variety of chemical signals in plant–nematode interactions is developed in the paper by Gheysen. Adaptation of nematodes to plant parasitism involves the use of chemical signals as hatching factors, host locating agents, settling factors, and pheromones. Plant responses involve not only the activation of plant defenses but also activation of phytohormones.

The ongoing debate over the role of salicylic acid as a signaling molecule in plant disease resistance is emphasized in the chapter by Klessig *et al.* Answers are being provided by genetic analyses of mutants and their effector proteins. Multiple modes of action of salicylic acid for controlling different plant processes and for facilitating disease/defense responses are being discovered. In the future, cross-talk between various signaling pathways will have to be examined. Other signals, plant flavonoids and bacterial polysaccharides, both playing a role in plant-rhizobial symbiosis, are discussed by van Workum and Kijne. These infochemicals are involved in cell-to-cell communication leading to host plant specificity in rhizobial nodulation. Additional proposed roles of flavonoids beyond plant-rhizobial symbiosis as regulators of plant development and their effect on auxin transport are developed in the paper by Spaink.

Octadecanoid plant signals, jasmonic acid and its precursors, produced in response to various biotic and abiotic stresses are examined by Weiler *et al.* An elegant evolutionary hypothesis is proposed that suggests herbivores, pathogens, and mechanical forces all induce the biosynthesis of different cyclic oxylipins. These activate genes that induce the synthesis of compounds as diverse as protease inhibitors, low molecular weight phytoalexins, and structural polymers. These multifaceted signal transducers are hypothesized to have evolved originally in response to oxidative damage to membrane lipids.

The final chapters address the application of molecular biological tools for modifying the chemistry of plants for man's use. Bundock and Hookykas emphasize the potential use of *Agrobacterium* for transferring disease and chemical resistance traits to crop plants, and for manipulating secondary metabolic pathways with tailor-made microbes. Beeching *et al.* offer an alternative approach, the modification of the phenylpropanoid defense cascade to prevent or slow post-harvest physiological deterioration in Cassava, a vital crop of the third world.

PREFACE

We thank the local meeting organizers from Leiden University for their hospitality and efficiency. JTR expresses appreciation to Dawn McGowan for her technical expertise and her commitment to the book.

John T. Romeo
University of South Florida

Kelsey R. Downum
Florida International University

Rob Verpoorte
Leiden University

CONTENTS

1. Saponins and Plant Defense 1
   Anne E. Osbourn, Jos P. Wubben, Rachel E. Melton,
   Jonathan P. Carter, and Michael J. Daniels

2. Role of Toxins in Plant Microbial Interactions 17
   Susan P. McCormick, Thomas M. Hohn, Anne E. Desjardins,
   Robert H. Proctor, and Nancy J. Alexander

3. Active Oxygen in Fungal Pathogenesis of Plants: The Role of
   Cercosporin in *Cercospora* Diseases 31
   Margaret E. Daub, Marilyn Ehrenshaft, Anne E. Jenns, and
   Kuang-Ren Chung

4. Tree–Fungus Interactions in Ectomycorrhizal Symbiosis 57
   Roger T. Koide, Laura Suomi, and Robert Berghage

5. Allelochemicals in Root Exudates of Maize: Effects on Root Lesion
   Nematode *Pratylenchus zeae* 71
   Annette Friebe, Wilma Klever, Richard Sikora, and
   Heide Schnabl

6. Chemical Signals in the Plant–Nematode Interaction: A Complex
   System? 95
   Godelieve Gheysen

7. Salicylic Acid-Mediated Signal Transduction in Plant Disease
   Resistance 119
   Daniel F. Klessig, Jörg Durner, Jyoti Shah, and Yinong Yang

8. Biosynthesis of Rhizobial Exopolysaccharides and Their Role in
   the Root Nodule Symbiosis of Leguminous Plants 139
   Wilbert A. T. van Workum and Jan W. Kijne

9. Flavonoids as Regulators of Plant Development: New Insights from Studies of Plant–Rhizobia Interactions 167
   Herman P. Spaink

10. Fatty Acid-Derived Signaling Molecules in the Interaction of Plants with Their Environment 179
    Elmar W. Weiler, Dietmar Laudert, Florian Schaller, Boguslava Stelmach, and Peter Hennig

11. Interactions between *Agrobacterium tumefaciens* and Plant Cells 207
    Paul Bundock and Paul Hooykaas

12. Wound and Defense Responses in Cassava as Related to Post-Harvest Physiological Deterioration 231
    John R. Beeching, Yuanhuai Han, Rocío Gómez-Vásquez, Robert C. Day, and Richard M. Cooper

Index 249

Chapter One

SAPONINS AND PLANT DEFENSE

Anne E. Osbourn, Jos P. Wubben, Rachel E. Melton,
Jonathan P. Carter, and Michael J. Daniels

Sainsbury Laboratory
John Innes Centre
Colney Lane, Norwich NR4 7UH, United Kingdom

Introduction 1
Oat Saponins 2
   Avenacins and Avenacosides 2
   Avenacins and Disease Resistance 4
      Fungal Detoxification of Avenacins 4
      Avenacinase-Like Proteins in the Genus *Gaeumannomyces* 4
      Natural and Induced Variation in Avenacin Content in Oats 5
   Detoxification of Oat Leaf Saponins by *Septoria avenae* 6
Tomato Saponins 9
   Enzymatic Detoxification of α-Tomatine by Fungi 9
   Heterologous Expression of *Septoria lycopersici* Tomatinase in the
      Biotroph, *Cladosporium fulvum* 10
Relatedness of Saponin Degrading Enzymes 10
Summary 13

INTRODUCTION

    Saponins are an important group of glycosylated natural products which are widely distributed in the plant kingdom.[1] They have attracted considerable attention because of their effects (both beneficial and detrimental) on human health, and because of their diverse range of other biological properties.[1-5] Saponins are also of considerable interest to plant pathologists because these molecules often exhibit potent antifungal activity, suggestive of a role in the

Phytochemical Signals and Plant–Microbe Interactions, edited by Romeo *et al.*
Plenum Press, New York, 1998.

protection of plants against attack by phytopathogenic fungi.[6,7] Because saponins are often present in relatively high levels in healthy plants, they may be regarded as pre-formed antimicrobial agents, in contrast to phytoalexins, which are synthesized in response to pathogen attack.

Saponins can be divided into three major groups depending on the structure of the aglycone (known as the sapogenin), which may be a steroidal glycoalkaloid, a steroid, or a triterpenoid.[1-4] Sapogenins belonging to all three groups usually have a sugar chain attached via an ether linkage at the hydroxyl group at the C-3 position. Saponins with a single sugar chain are referred to as monodesmosidic. Bisdesmosidic saponins have a second sugar moiety attached at C-28 (triterpenoid saponins) or at C-26 (steroidal saponins). Bisdesmosidic saponins lack many of the properties and biological activities of monodesmodic saponins,[1-4] but removal of the sugar at the C-26 or C-28 position gives the amphipathic monodesmosidic form, which is usually biologically active. For a number of bisdesmosidic saponins this conversion has been shown to occur *in planta* in response to tissue damage, and is carried out by specific plant glycosyl hydrolases.[8-13]

The toxicity of saponins to fungi is likely due primarily to their ability to complex with sterols, causing loss of membrane integrity.[1,3] Fungi which are successful pathogens of saponin-containing plants appear to have two major strategies for protecting themselves from host saponins.[6,7] The first involves intrinsic resistance at the membrane level. The second involves active detoxification by specific glycosyl hydrolase enzymes which remove sugars from the sugar chain at the C-3 position.

Despite their widespread distribution across a diverse range of plant species, detailed investigations of the potential role of saponins as determinants of disease resistance have been restricted mainly to oats (*Avena*) and tomato (*Lycopersicon*).[6] This is probably because the saponin composition of these plants is relatively simple and the structures and antifungal properties of oat and tomato saponins are well-documented. This chapter is concerned with recent progress in assessing the contribution of oat and tomato saponins to plant defense against fungal attack. Approaches currently being taken involve the generation of fungal mutants which are defective in saponin detoxification, heterologous expression of saponin-detoxifying enzymes in other fungi, and the isolation of plant mutants which lack saponins.

OAT SAPONINS

Avenacins and Avenacosides

Oats are unusual because they contain members of two different groups of saponins, the triterpenoids and the steroids (represented by the avenacins and avenacosides, respectively).[1] The distribution of the two types within the plant is mutually exclusive, the avenacosides being located in the leaves and shoots,[14,15] and the avenacins in the root.[16,17] The occurrence of these saponins is restricted primarily to the genus *Avena*, although the avenacins are also known to occur in the closely related species *Arrhenatherum elatius*.[18–20]

The avenacin family consists of four closely related saponins, avenacins A-1 and B-1 (which are esterified with N-methyl anthranilic acid and hence are autofluorescent under ultra-violet illumination), and A-2 and B-2 (which are esterified with benzoic acid and are only weakly autofluorescent).[21,22] The chemical structure of the major oat root saponin, avenacin A-1, is illustrated in Figure 1. All four compounds have the same trisaccharide moiety attached to the C-3 carbon atom, consisting of β,1-2- and β,1-4-linked terminal D-glucose molecules attached via L-arabinose to the aglycone. The avenacins are monodesmosidic and are present in healthy oat roots as biologically active molecules.

In contrast, the steroidal oat saponins, avenacosides A and B, are bisdesmosidic,[14,15] and are converted into the biologically active monodesmosidic 26-desglucoavenacosides (26-DGAs) by a plant enzyme in response to wounding or pathogen attack.[9,10,12,13] This enzyme is specific for the C-26 glucose and does not

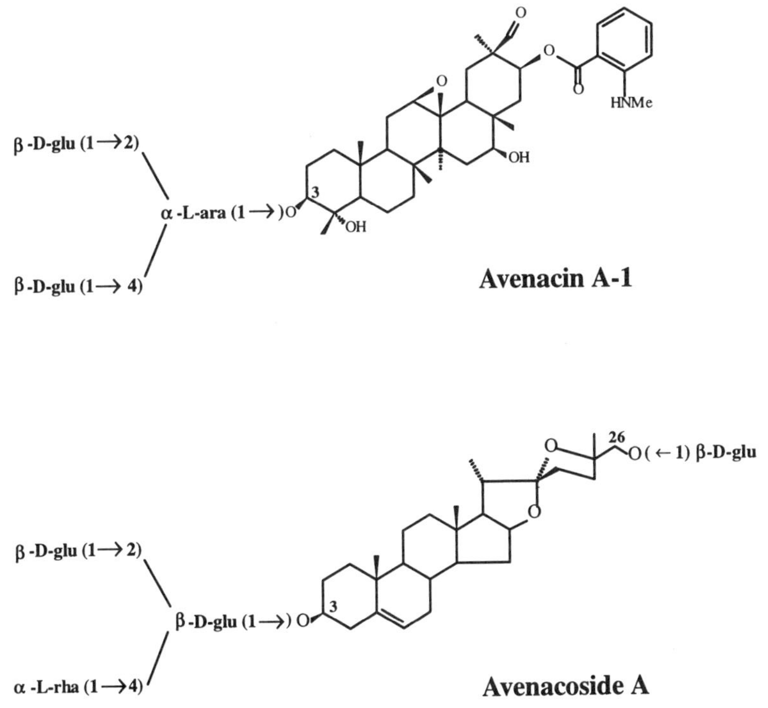

Figure 1. Structures of the two groups of oat saponins. The monodesmosidic triterpenoid avenacin A-1 is the major saponin in oat roots. Avenacoside A a bisdesmosidic steroidal saponin found in oat leaves.

remove sugars from the C-3 sugar chain. The chemical structure of avenacoside A is illustrated in Figure 1. Avenacoside B differs from avenacoside A only in that it has an additional β,1-3-linked D-glucose molecule attached to the C-3 sugar chain.

Avenacins and Disease Resistance

Fungal Detoxification of Avenacins. The major oat root saponin, avenacin A-1, has been shown to occur exclusively in the root epidermis,[20] and so is likely to present an effective barrier to attack by soil fungi. The contribution of avenacin A-1 to disease resistance has been particularly well studied for interactions between the root pathogen *Gaeumannomyces graminis* and oats. *G. graminis* var. *tritici*, which infects the roots of wheat and barley and causes the disease known as 'take-all,' is relatively sensitive to avenacin A-1 *in vitro* and is unable to infect oats. Consequently, the resistance of oats to this pathogen has been attributed to the presence of avenacin A-1 in oat roots.[17–20,23–26] Another form of *G. graminis* (*G. graminis* var. *avenae*) can infect oats in addition to wheat and barley,[27] and is much less sensitive to avenacin A-1.[23–25] Isolates of *G. graminis* var. *avenae* produce an enzyme known as avenacinase which detoxifies avenacin A-1 by the removal of β,1-2- and β,1-4-linked terminal D-glucose molecules from the sugar chain at the C-3 position.[23–25] Isolates of *G. graminis* var. *tritici* show little or no detoxifying activity (although they do produce proteins which are related to avenacinase; see below). Thus, resistance of *G. graminis* var. *avenae* to avenacin A-1 is correlated with the ability to actively detoxify the saponin, suggesting that avenacinase may be a determinant of host specificity. This has been confirmed by genetic tests in which mutants of *G. graminis* var. *avenae* specifically defective in avenacinase production were generated by targeted gene disruption.[26] These mutants were completely non-pathogenic to oats, indicating that avenacinase is essential in order for *G. graminis* var. *avenae* to infect this host. However, the enzyme is clearly not a general pathogenicity determinant because the mutants were still fully pathogenic to the alternative host, wheat, which does not contain avenacins.

Avenacinase activity is not restricted to *G. graminis* var. *avenae*, since the oat pathogen *Fusarium avenaceum* is also known to detoxify avenacin A-1 by the removal of sugars.[24] A recent survey of over 200 filamentous fungi isolated from field-grown wheat and oat plants revealed that (at least for the sampling sites included in this study) fungi from oat roots are generally more resistant to avenacin A-1 than those from wheat, indicating that avenacin A-1 may influence the composition of the fungal populations present in the rhizosphere. Interestingly, many of these avenacin-resistant fungi are also able to degrade avenacin A-1 by various mechanisms (Carter, Daniels, and Osbourn, unpublished).

Avenacinase-like Proteins in the Genus Gaeumannomyces. DNA sequences that cross-hybridize to the *G. graminis* var. *avenae* avenacinase gene are also found in many isolates of *G. graminis* var. *tritici* and in other related *Gaeumannomyces* species, all of which are unable to infect oats.[28] These fungi have little or no avenacin A-1-detoxifying activity under standard assay conditions, but they do have a weak ability to deglucosylate avenacin A-1 which is detectable only after prolonged incubation with the substrate. This enzyme activity resides in "avenacinase-like proteins" or ALPs, which have similar physicochemical properties to avenacinase and which are specifically recognized by polyclonal anti-avenacinase antisera. Purification of the *G. graminis* var. *tritici* ALP indicated a specific activity for avenacin A-1 which was 25 times lower than that of avenacinase from *G. graminis* var. *avenae*.[6] Sequence analysis of the cross-hybridizing DNA from *G. graminis* vars. *tritici* and *graminis*[28] indicates the existence of open reading frames which would encode proteins closely related to avenacinase (96% and 91% amino acid similarity, respectively).[28] Given that these open reading frames appear to represent single copy genes in both *G. graminis* var. *tritici* and var. *graminis*, it is likely that the cloned DNA sequences encode the ALPs which have been biochemically characterized. The function of the ALPs is unknown. While they do not appear to be able to deglucosylate avenacin A-1 effectively, they may be important for detoxification of other saponins encountered during infection of alternative hosts such as wild grasses, the saponin contents of which are unknown. Alternatively, these enzymes also have hydrolytic activity towards standard β-glucosyl hydrolase substrates, and may be required for nutrition during saprophytic growth.[6,28]

Natural and Induced Variation in Avenacin Content in Oats. Studies of naturally occurring or artificially-induced oat variants which lack avenacin A-1 offer a direct route towards assessing the role of this saponin in protecting plants against fungal attack.[7] Searches for natural variation in avenacin content have identified one diploid oat species (*Avena longiglumis*) which lacks avenacin A-1.[20] Significantly, this species is susceptible to infection by *G. graminis* var. *tritici*. Unfortunately, *A. longiglumis* does not hybridize readily with other diploid oat species which produce avenacins, making it difficult to test whether the ability to synthesize avenacins and resistance to *G. graminis* var. *tritici* are connected. Recently, mutants of the diploid oat species, *Avena strigosa*, lacking avenacin A-1 have been isolated following sodium azide mutagenesis (Osbourn, Melton, and Daniels, unpublished). These mutants show increased susceptibility to isolates of *G. graminis* var. *avenae* (Fig. 2A), and also to infection by other root-infecting fungi such as the *Fusarium* species *F. avenaceum*, *F. culmorum*, and *F. graminearum*. While the *A. strigosa* wild type is immune to infection by *G. graminis* var. *tritici*, roots of avenacin A-1-minus mutants give discrete black lesions (Fig. 2B). The extent of fungal growth within these *G. graminis* var. *tritici*-inoculated oat mutants and the nature of the host response are currently being investigated by cytological studies of the infection process. Overall, these observations are consistent with a role for avenacin A-1 as a determinant of

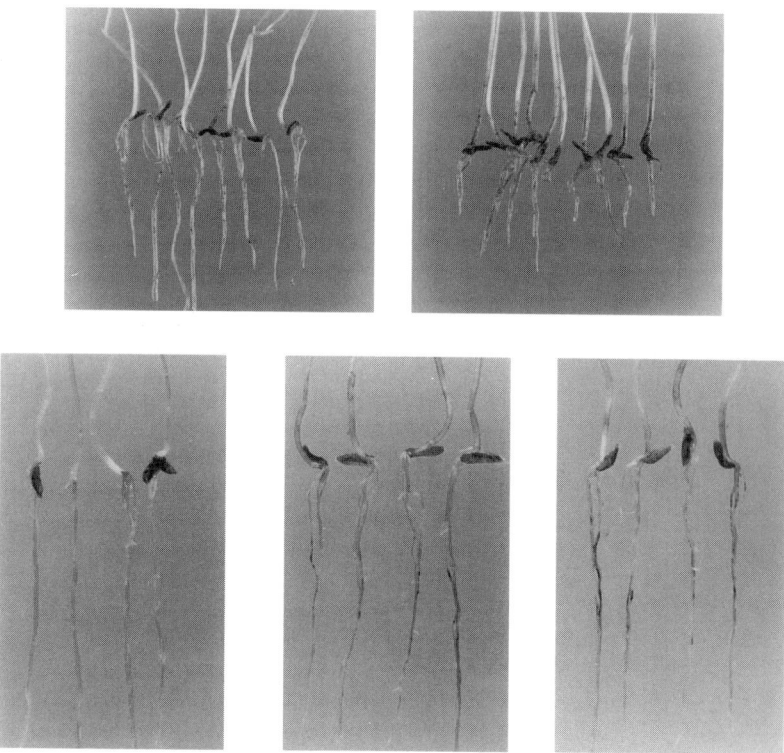

Figure 2. Inoculation of wild-type and avenacin A-1-minus mutants of *Avena strigosa* with *G. graminis*. (A) Wild type *A. strigosa* (left hand panel) and the avenacin A-1-minus mutant #1243 (right hand panel) inoculated with *G. graminis* var. *avenae* isolate A3. (B) Wild type *A. strigosa* (left hand panel) and avenacin-minus mutants #109 and #610 (central and right hand panels respectively) inoculated with *G. graminis* var. *tritici* isolate T5. Disease symptoms are visible as dark lesions on the roots and, in cases of severe disease, as browning of the leaf sheath.

resistance of oats to a range of phytopathogenic fungi, and preliminary genetic analysis indicates that the absence of this saponin and increased disease susceptibility are indeed causally related.

Detoxification of Oat Leaf Saponins by *Septoria avenae*

The foliar pathogen of oats, *Septoria avenae*, is able to hydrolyze 26-DGAs A and B to the common aglycone, nuatigenin.[29] This enzyme activity has been characterized in detail and involves the sequential hydrolysis of sugars from the C-3 chain, beginning with the terminal L-rhamnose (Fig. 3A). Removal of the L-rhamnose alone is sufficient to give a substantial reduction in toxicity (Fig. 3B).[29]

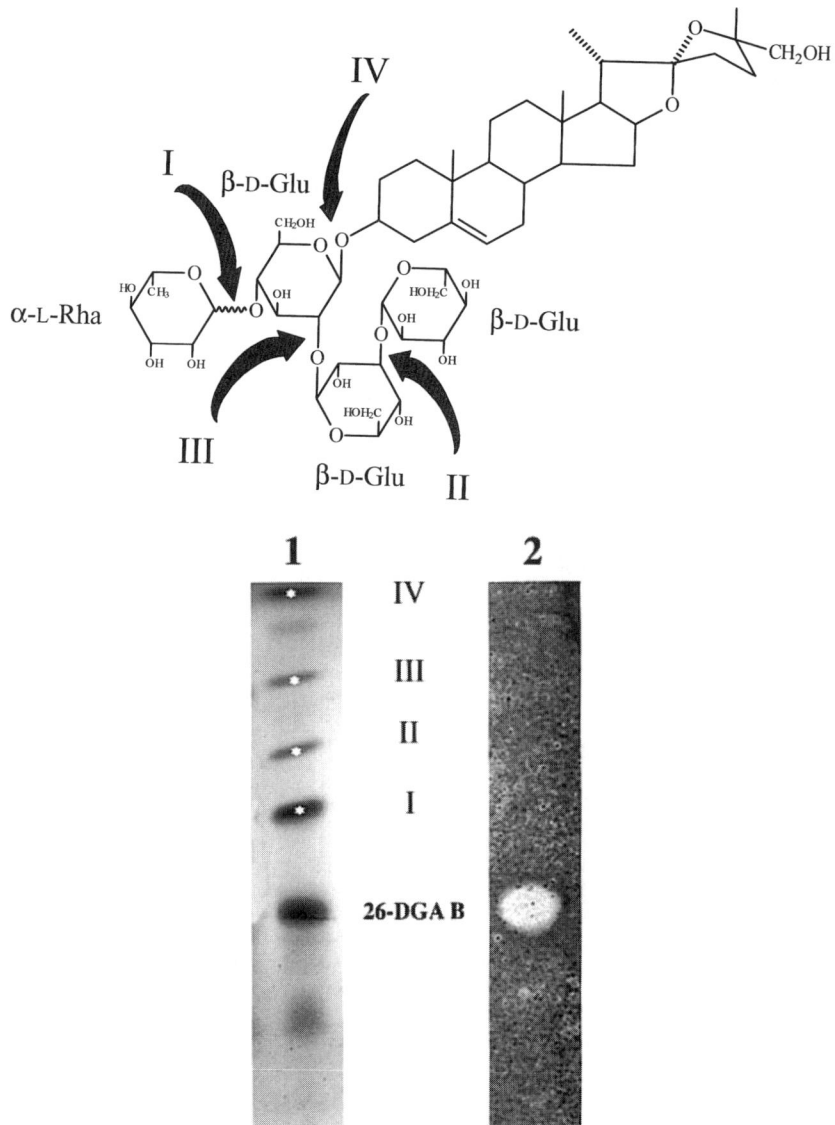

Figure 3. Hydrolysis of the foliar oat saponin 26-desglucoavenacoside B by *Septoria avenae*. (A) *S. avenae* removes sugars from the C-3 sugar chain of 26-DGAs A and B and avenacosides A and B. The process involves sequential hydrolysis, beginning with the terminal α-L-rhamnose (step I), followed by the remaining D-glucose molecules (steps II–IV), as illustrated here for 26-DGA B. (B) Separation of the products of hydrolysis of 26-DGA B by thin layer chromatography (lane 1); lane 2, a comparable thin layer chromatography plate which has been sprayed with spores of an avenacoside-sensitive *Septoria* isolate. Fungal growth is inhibited by 26 DGA B but not by the hydrolysis products.

An enzyme (avenacosidase) capable of carrying out all the stages of this hydrolysis was purified from the culture filtrate of *S. avenae* (Fig. 4).[29] The properties of this enzyme, in terms of its molecular mass and pI, are similar to those of avenacinase from *G. graminis* var. *avenae*,[25,26] and amino acid sequence analysis of peptide fragments derived from the enzyme by proteolytic cleavage confirmed that avenacosidase is indeed related to avenacinase (Wubben, Daniels, and Osbourn, unpublished). However, *S. avenae* avenacosidase is unable to hydrolyze the oat root saponin, avenacin A-1.

The gene encoding this avenacosidase activity has been isolated using PCR primers derived from the experimentally determined amino acid sequence of the purified protein. The gene is predicted to encode a protein which shares 62% amino acid similarity with *G. graminis* var. *avenae* avenacinase. Experiments are currently in progress to generate an avenacosidase-minus mutant of *S. avenae* by targeted gene disruption, following the approach used for *G. graminis* var. *avenae* avenacinase.[26] To date, a single mutant of *S. avenae* which has undergone

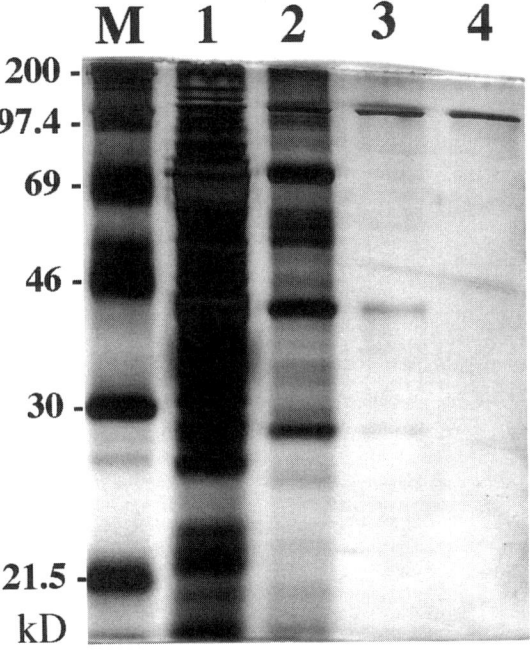

Figure 4. Purification of avenacoside activity from *Septoria avenae*. Silver-stained sodium dodecyl sulphate-polyacrylamide gel showing fractions containing avenacoside-hydrolyzing activity from the *S. avenae* isolate WAC 1293 at different stages of purification. Lane M, molecular weight markers; lane 1, dialyzed sample after ammonium sulphate precipitation; lane 2, after free-flow isoelectric focusing; lane 3, after DEAE anion-exchange high-performance liquid chromatography; and lane 4, after size-exclusion high-performance liquid chromatography.

a disruption event in the avenacosidase gene has been obtained. While it is clear that the anticipated molecular event has taken place and that the avenacosidase gene has indeed been mutated, the mutant retains residual avenacosidase activity and shows little or no reduction in pathogenicity to oats. Fractionation of the residual enzyme activity indicates the existence of at least one other avenacosidase enzyme with similar properties to the one which has been purified. A stringent test of the contribution of avenacosidase activity to pathogenicity of *S. avenae* to oats will therefore require the generation of multiple mutants.

TOMATO SAPONINS

Enzymatic Detoxification of α-Tomatine by Fungi

Tomato plants contain the steroidal glycoalkaloid, α-tomatine, which is a monodesmosidic saponin with a tetrasaccharide group (known as β-lycotetraose) attached to the C-3 carbon (Fig. 5).[5] β-lycotetraose consists of two molecules of D-glucose and one each of D-galactose and D-xylose. The presence of an intact sugar chain at the C-3 position is essential for full toxicity.[30]

A number of fungi which infect tomato produce enzymes which detoxify α-tomatine by the removal of sugars.[6,7] These enzymes are collectively referred to as tomatinases, although their mechanisms of action differ. The tomatinase of the tomato leaf spot fungus, *Septoria lycopersici*, acts in a similar way to avenacinase, since both enzymes hydrolyze β,1-2-linked terminal D-glucose residues from their saponin substrates.[31–33] Indeed, the predicted amino acid sequence of tomatinase reveals that this enzyme clearly belongs to the same family of β-glycosyl hydrolases as *G. graminis* var. *avenae* avenacinase[32,33] and *S. avenae* avenacosidase (Wubben, Daniels, and Osbourn, unpublished). Attempts to generate tomatinase-minus mutants of *S. lycopersici* by targeted gene disruption are in progress in order to assess the contribution of tomatinase to the *S. lycopersici* infection process.

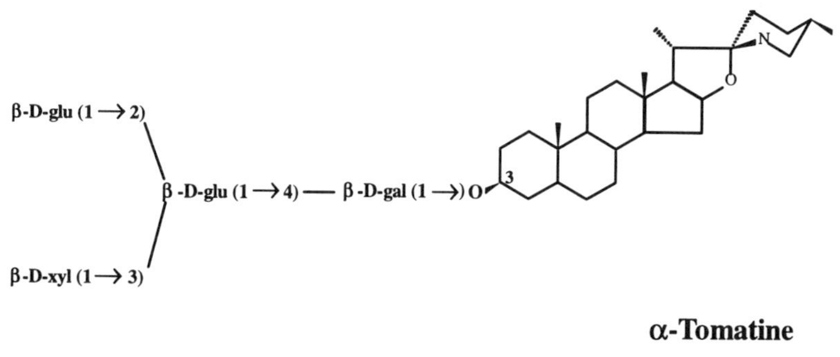

β-D-glu (1 → 2)
β-D-glu (1 → 4) — β-D-gal (1 →) O
β-D-xyl (1 → 3)

α-Tomatine

Figure 5. Structure of the tomato steroidal glycoalkaloid saponin α-tomatine. The tetrasaccharide group attached at carbon 3 is known as β-lycotetraose, and the aglycone as tomatidine.

Heterologous Expression of *Septoria lycopersici* Tomatinase in the Biotroph, *Cladosporium fulvum*

The tomato pathogen, *Cladosporium fulvum*, is highly sensitive to 100 μM α-tomatine *in vitro*[34] and does not appear to be able to enzymatically degrade the saponin. The α-tomatine content of green leaf tissue of tomato is estimated to be approximately 1mM,[5] and hence may be anticipated to inhibit *C. fulvum*. However, α-tomatine is believed to be located primarily in the vacuoles of the plant cells, and since *C. fulvum* grows in the intercellular spaces of tomato leaves, the fungus may simply avoid the release of α-tomatine during infection.[7] In fact, heterologous expression of *S. lycopersici* tomatinase in *C. fulvum* results in increased pathogenicity to compatible tomato lines, suggesting that α-tomatine may play some role in restricting the growth of *C. fulvum* (Melton, Flegg, Oliver, and Osbourn, unpublished). Perhaps more significantly, tomatinase-expressing *C. fulvum* transformants show increased ability to grow on incompatible tomato lines relative to control strains (although they do not go on to sporulate). Thus, α-tomatine may play a secondary role during incompatible *C. fulvum*/tomato interactions when release of the saponin from damaged plant cells may contribute to killing or containing the invading fungus, as originally suggested by Dow and Callow.[34] The isolation of α-tomatine-minus mutants of tomato will allow a direct test of the role of this saponin in determining the outcome of interactions between *C. fulvum* and other fungal pathogens with tomato.

RELATEDNESS OF SAPONIN DEGRADING ENZYMES

The saponin detoxifying enzymes described in this chapter each act on structurally distinct substrates representing each of the three major groups of saponins. *G. graminis* var. *avenae* avenacinase degrades the triterpenoid saponin, avenacin A-1; *S. avenae* avenacosidase, the steroidal avenacosides; and *S. lycopersici*, the steroidal glycoalkaloid, α-tomatine (Table 1).

These enzymes show clear differences in their substrate specificities. Avenacinase has a relative rate of hydrolysis towards α-tomatine of 2% of that towards avenacin A-1, while *S. lycopersici* tomatinase has only weak activity towards avenacin A-1 (less than 0.01% of its activity towards α-tomatine).[32] *S. avenae* avenacosidase has little or no activity towards the non-host saponins avenacin A-1 and α-tomatine. Thus, the substrate specificities of the three enzymes reflect the host preferences of the fungi of origin. Despite these differences, these enzymes are all clearly related, and show substantial similarities at the amino acid level (Table 1).

G. graminis var. *avenae* avenacinase and *S. lycopersici* tomatinase both hydrolyze terminal D-glucose molecules from their respective sub-

SAPONINS AND PLANT DEFENSE

Table 1. Related saponin detoxifying enzymes

Fungus	Saponin	Class	Enzyme	Sugars removed	% amino acid similarity with avenacinase
Gaeumannomyces graminis	Avenacins (Oat roots)	Triterpenoid	Avenacinase	D-glu	100
Septoria avenae	Avenacosides (Oat leaves)	Steroid	Avenacosidase	D-glu	L-rha
Septoria lycopersici	α-Tomatine (Tomato leaves)	Steroidal glycoalkaloid	Tomatinase	D-glu	68

strates,[23–26,31–33] while *S. avenae* avenacosidase hydrolyzes L-rhamnose and D-glucose.[29] Comparison with other sequences in the databases indicates that, as might be expected from their biochemical function, these enzymes are related to a family of β-glycosyl hydrolases (defined as family 3 by Henrissat).[6,35,36] The effects (if any) of these other enzymes on saponins have not been described.[6] They include fungal and bacterial cellobiose-degrading enzymes,[6,35,36] and also the *Histoplasma* H-antigen[37] and CBG1 from *Agrobacterium tumefaciens* (which is associated with virulence to gymnosperms).[38] From the phenogram presented (Fig. 6), it can be seen that the protein sequence of *S. lycopersici* tomatinase is more similar to the *G. graminis* avenacinase and ALPs than to other members of the family 3 β-glycosyl hydrolases. *S. avenae* avenacosidase, however, does not branch with these enzymes, but is grouped with *BGL1* from *Aspergillus aculatus* and *Histoplasma capsulatum* H-antigen. Characterization of more saponin-detoxifying enzymes belonging to family 3, and more detailed analysis of the substrate specificities of family 3 β-glycosyl hydrolases in general, are required in order to understand the significance of these subdivisions. Currently, there is no structure available for this major family of enzymes, although the recent crystallization of *S. lycopersici* tomatinase now offers an opportunity to investigate the relationship between substrate specificity and enzyme structure (Bamford, Osbourn, and Hemmings, unpublished).

It is unlikely that all saponin detoxifying enzymes produced by plant pathogenic fungi will be related. For instance, the tomatinase enzyme produced by *Fusarium oxysporum* f.sp. *lycopersici* has a different mechanism of action from that of the *S. lycopersici* enzyme, and releases the intact β-lycotetraose group to give the aglycone, tomatidine (Fig. 5).[39] This enzyme recently has been purified and determined to have a molecular mass of 50 kD,[40] in contrast to the enzymes in Table 1 which all have molecular masses of around 100 kD. The sequence of the first seven N-terminal amino acids of the *F. oxysporum* enzyme has been determined,[40] but was insufficient to indicate any relatedness with known saponin detoxifying enzymes or glycosyl hydrolases.

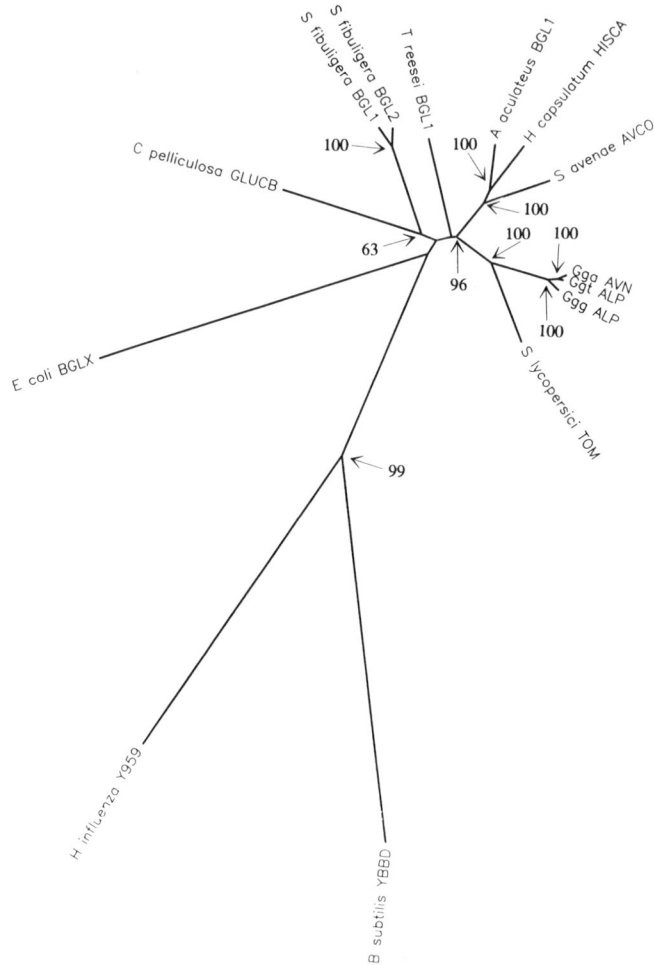

Figure 6. Phenogram indicating relatedness of protein sequences for saponin detoxifying enzymes and other members of the family 3 group of glycosyl hydrolases. The tree was constructed using the PROTDIST, NEIGHBOR-NJ, SEQBOOT, CONSENSE and DRAWTREE programs contained in the Phylogeny Inference Package PHYLIP version 3.5c (J. Felsenstein, University of Washington, Seattle). The numbers at the forks indicate the number of times the group consisting of the species distal to that fork occurred among the tree, out of 100 trees. The abbreviations are as follows, with GenBank accession numbers given in brackets: *Septoria lycopersici* tomatinase, TOM (U35462); *Gaemannomyces graminis* var. *avenae* avenacinase, *Gga* AVN (U35463); *G. graminis* var. *tritici* and *graminis* avenacinase-like proteins, *Ggt* ALP, *Ggg* ALP (Bryan, Daniels and Osbourn, unpublished); *Septoria avenae* avenacosidase, AVCO (Wubben, Daniels and Osbourn, unpublished); *Histoplasma capsulatum* H-antigen, HISCA (U20346); β-glucosyl hydrolases from *Aspergillus aculateus*, BGL1 (D64088), *Trichoderma reesei* BGL1 (U09580), *Saccharomycopsis fibuligera* BGL1 and BGL2 (M22475, M22476), *Candida pelliculosa* (X02903), *Escherichia coli* BGLX (D64088), *Haemophilus influenzae* Y959 (L45597), *Bacillus subtilis* YBBD (L19954).

SUMMARY

A number of lines of evidence are now emerging to indicate that saponins are likely to contribute to disease resistance for at least some plant/fungus interactions. If such compounds prove to have general significance for plant protection, then the manipulation of plant biosynthetic pathways to give enhanced levels of saponins, or to produce novel or altered saponins, may offer a new route towards effective disease control. A complementary strategy may involve identification of inhibitors of saponin detoxifying enzymes produced by pathogens. These enzymes are extracellular, making them ideal targets for fungicides.[6,7]

ACKNOWLEDGMENTS

The Sainsbury Laboratory is supported by the Gatsby Charitable Foundation. JP Wubben and JP Carter are supported by the European grants CHRX-CT93-0244 and BIO2-CT93-3001, respectively. Research in the authors' laboratory has also been supported by the UK Biotechnology and Biological Sciences Research Council.

REFERENCES

1. PRICE, K.R., JOHNSON, I.T., FENWICK, G.R. 1987. The chemistry and biological significance of saponins in food and feedingstuffs. CRC Critical Reviews in Food Science Nutrition 26:27–133.
2. HOSTETTMANN, K.A., MARSTON, A. 1995. Saponins. Chemistry and Pharmacology of Natural Products. Cambridge University Press, Cambridge.
3. FENWICK, G.R., PRICE, K.R., TSUKAMOTA, C., OKUBO, K. 1992. Saponins. In: Toxic Substances In Crop Plants, J.P. D'Mello, C.M. Duffus and J.H Duffus, (eds.), The Royal Society of Chemistry, Cambridge, pp. 285–327.
4. HOSTETTMANN, K., HOSTETTMANN, M., MARSTON, A. 1991. Saponins. Methods in Plant Biochemistry 7:435–471.
5. RODDICK, J.G. 1974. The steroidal glycoalkaloid α-tomatine. Phytochemistry 13:9–25.
6. OSBOURN, A.E. 1995. Saponins and plant defense—A soap story. Trends in Plant Science 1:4–9.
7. OSBOURN, A.E. 1996. Preformed antimicrobial compounds and plant defense against fungal attack. The Plant Cell 8:1821–1831.
8. SCHÖNBECK, F., SCHLÖSSER, E. 1976. Preformed substances as potential protectants. In: Physiological Plant Pathology, R. Heitefuss, P.H. Williams, (eds.), Springer Verlag, Berlin, pp. 653–678.
9. LÜNING, H.U., SCHLÖSSER, E. 1975. Role of saponins in antifungal resistance V. Enzymatic activation of avenacosides, Zeitschrift für Pflanzenkrankheit und Pflanzenschutz 82:699–703.
10. NISIUS, H. 1988. The stromacentre in Avena plastids: An aggregation of β-glucosidase responsible for the activation of oat-leaf saponins. Planta 173:474–481.
11. INOUE, K., EBIZUKA, Y.. 1996. Purification and characterization of furostanol glycoside 26-O-β-glucosidase from *Costus speciosus* rhizomes. FEBS Letters 378:157–160.

12. GUS-MAYER, S., BRUNNER, H., SCHNEIDER-POETSCH, H.A.W., LOTTSPEICH, F., ECKERSKORN, C., GRIMM, R., RÜDIGER, W. 1994. The amino acid sequence previously attributed to a protein kinase or a TCP1-related molecular chaperone and co-purified with phytochrome is a β-glucosidase. FEBS Letters 347:51–54.
13. GUS-MAYER, S., BRUNNER, H., SCHNEIDER-POETSCH, H.A.W., RÜDIGER, W. 1994. Avenacosidase from oat: Purification, sequence analysis and biochemical characterization of a new member of the BGA family of β-glucosidases. Plant Molecular Biology 26:909–921.
14. TSCHESCHE, R., TAUSCHER, M., FEHLHABER, H.W., WULFF, G. 1969. Avenacosid A, ein bisdesmosidisches Steroidsaponin aus *Avena sativa*. Chemische Berichte 102:2072–2082.
15. TSCHESCHE, R., LAUVEN, P. 1971. Avenacosid B, ein zweites bisdesmosidisches Steroidsaponin aus *Avena sativa*. Chemische Berichte 104:3549–3555.
16. GOODWIN, R.H., POLLOCK, B.M. 1954. Studies on roots. I. Properties and distribution of fluorescent constituents in *Avena* roots. Amer. J. Bot. 4:516–520.
17. MAIZEL, J.V., BURKHARDT, H.J., MITCHELL, H.K. 1964. Avenacin, an antimicrobial substance isolated from *Avena sativa*. I. Isolation and antimicrobial activity. Biochemistry 3:424–431.
18. CROMBIE, W.M.L., CROMBIE, L. 1986. Distribution of avenacins A-1, A-2, B-1 and B-2 in oat roots: Their fungicidal activity towards 'take-all' fungus. Phytochemistry 25:2069–2073.
19. TURNER, E.M. 1953. The nature of resistance of oats to the take-all fungus. J. Exp. Bot. 4:264–271.
20. OSBOURN, A.E., CLARKE, B.R., LUNNESS, P., SCOTT, P.R., DANIELS, M.J. 1994. An oat species lacking avenacin is susceptible to infection by *Gaeumannomyces graminis* var. *tritici*. Physiol. Mol. Plant Pathol. 45:457–467.
21. CROMBIE, L., CROMBIE, W.M.L., WHITING, D.A. 1984. Structures of the four avenacins, oat root resistance factors to 'take-all' disease. J. Chem. Soc. Chem. Communications 4:246–248.
22. CROMBIE, L., CROMBIE, W.M.L., WHITING, D.A. 1986. Structures of the oat root resistance factors to take-all disease, avenacins A-1, A-2, B-1 and B-2 and their companion substances. J. Chem. Soc. Perkins I:1917–1922.
23. TURNER, E.M. 1961. An enzymic basis for pathogen specificity in *Ophiobolus graminis*. J. Exp. Bot. 12:169–175.
24. CROMBIE, W.M.L., CROMBIE, L., GREEN, J.B., LUCAS, J.A. 1986. Pathogenicity of take-all fungus to oats: Its relationship to the concentration and detoxification of the four avenacins. Phytochemistry 25:2075–2083.
25. OSBOURN, A.E., CLARKE, B.R., DOW, J.M., DANIELS, M.J. 1991. Partial characterization of avenacinase from *Gaeumannomyces graminis* var. *avenae*. Physiol. Mol. Plant Pathol. 38:301–312.
26. BOWYER, P., CLARKE, B.R., LUNNESS, P., DANIELS, M.J., OSBOURN, A.E. 1995. Host range of a plant pathogenic fungus determined by a saponin detoxifying enzyme. Science 267:371–374.
27. TURNER, E.M. 1940. *Ophiobolus graminis* Sacc. var. *avenae* var. N., as the cause of take all or whiteheads of oats in Wales. Trans. Br. Mycol. Soc. XXIV:269–281.
28. BRYAN, G.T. 1995. The *Gaeumannomyces-Phialophora* complex - Molecular Taxonomy and Characterization of Avenacinase-like Proteins. Ph.D. thesis, University of East Anglia.
29. WUBBEN, J.P., PRICE, K.R., DANIELS, M.J., OSBOURN, A.E. 1996. Detoxification of oat leaf saponins by *Septoria avenae*. Phytopathology 86:986–992.
30. ARNESON, P.A., DURBIN, R.D. 1968. Studies on the mode of action of tomatine as a fungitoxic agent. Plant Physiology 43:683–686.
31. ARNESON, P.A., DURBIN, R.D. 1967. Hydrolysis of tomatine by *Septoria lycopersici*; a detoxification mechanism. Phytopathology 57:1358–1360.
32. OSBOURN, A.E., BOWYER, P., LUNNESS, P.A., CLARKE, B.R., DANIELS, M.J. 1995. Fungal pathogens of oat roots and tomato leaves employ closely related enzymes to detoxify host plant saponins. Molec. Plant-Microbe Interactions 8:971–978.

33. SANDROCK, R.W., DELLAPENNA, D., VANETTEN, H.D. 1995. Purification and characterization of β_2-tomatinase, an enzyme involved in the degradation of α-tomatine and isolation of the gene encoding β_2-tomatinase from *Septoria lycopersici*. Molec. Plant Microbe Interactions 8:960–970.
34. DOW, J.M., CALLOW, J.A. 1978. A possible role for α-tomatine in the varietal-specific resistance of tomato to *Cladosporium fulvum*. Phytopathologische Zeitschrift 92:211–216.
35. HENRISSAT, B. 1991. A classification of glycosyl hydrolases based on amino acid sequence similarities. Biochem. J. 280:309–316.
36. HENRISSAT, B., BAIROCH, A. 1993. New families in the classification of glycosyl hydrolases based on amino acid sequence similarities. Biochem. J. 293:781–788.
37. DEEPE, G.S., DUROSE, G.G. 1995. Immunobiological activity of recombinant H antigen from *Histoplasma capsulatum*. Infection and Immunity 63:3151–3157.
38. CASTLE, L.A., SMITH, K.D., MORRIS, R.O. 1992. Cloning and sequencing of an *Agrobacterium tumefaciens* β-glucosidase gene involved in modifying a *vir*-inducing plant signal molecule. J. Bacter. 174:1478–1486.
39. FORD, J.E., McCANCE, D.J., DRYSDALE, R.B. 1977. The detoxification of α-tomatine by *Fusarium oxysporum* f.sp. *lycopersici*. Phytochemistry 16:545–546.
40. LAIRINI, K., PEREZ-ESPINOSA, A., PINEDA, M., RUIZ-RUBIO, M. 1996. Purification and characterization of tomatinase from *Fusarium oxysporum* f.sp. *lycopersici*. Appl. Environ. Microbiol. 62:1604–1609.

Chapter Two

ROLE OF TOXINS IN PLANT MICROBIAL INTERACTIONS

Susan P. McCormick, Thomas M. Hohn, Anne E. Desjardins,
Robert H. Proctor, and Nancy J. Alexander

Mycotoxin Research Unit
USDA-ARS-NCAUR
Peoria, Illinois, 61604

Introduction 17
Mycotoxins 17
Biosynthesis of Trichothecenes 19
Role in Plant Diseases 23
Resistance 24
Plant Interactions 26
Conclusion 27

INTRODUCTION

Both plants and fungi produce a wide variety of low molecular weight natural products. Many of these compounds were once considered secondary metabolites with no particular biological role in the producing organism. Understanding the chemical interactions between plants and microorganisms can be complex and requires an integrated approach.

MYCOTOXINS

Many genera of fungi produce toxic metabolites. Mycotoxins can be narrowly defined as fungal metabolites that are produced in agricultural products at concentrations sufficient to adversely affect vertebrate health,[1] either by directly causing disease symptoms or by less acute means such as adversely affecting growth rate as a result of feed refusal. Mycotoxins comprise a diverse, heterogeneous group of metabolites and, therefore, may be formed via polyketide, terpenoid, or other biosynthetic routes. Common examples of mycotoxins are aflatoxins, zearalenone, ochratoxins, fumonisins, and, trichothecenes (Fig. 1).

Aflatoxins, xanthones produced by *Aspergillus flavus* or *A. parasiticus*, typically in corn and other oil seed crops such as peanuts, cottonseed, and pistachio nuts, have mutagenic, teratogenic, and carcinogenic activity. Aflatoxin B_1 primarily acts as a liver carcinogen, but exposure to aflatoxins may result in a wide variety of symptoms including hemorrhaging, feed refusal, and abortion.[2] Its mutagenic, teratogenic, and carcinogenic effects are due to the formation of an aflatoxin-DNA adduct.[3]

Zearalenone, an acetate-derived mycotoxin produced by several species of *Fusarium* in contaminated corn, wheat, and barley, acts as an estrogen in swine

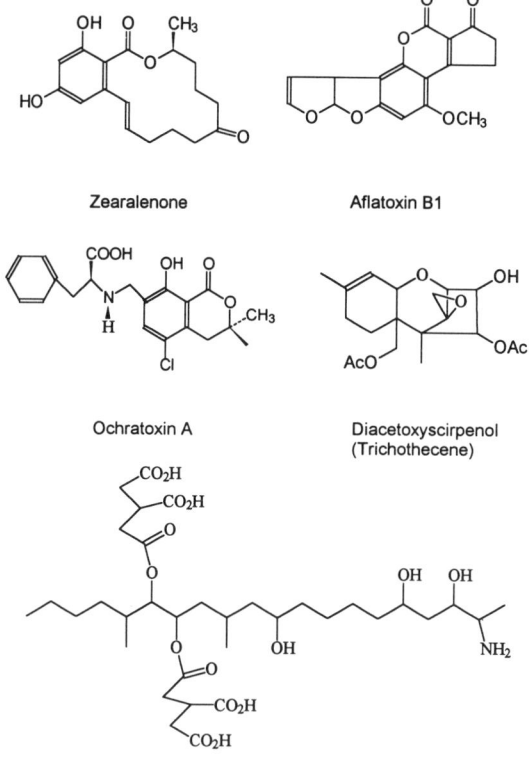

Figure 1. Structures of mycotoxins: aflatoxin B_1, ochratoxin A, zearalenone, diacetoxyscirpenol, and fumonisin B_1.

and can cause severe reproductive problems and a range of symptoms including swollen genitalia, premature mammary gland development, and oestrus and testicular hypoplasia.[4]

Ochratoxins are chlorinated isocoumarins produced by *Aspergillus ochraceus* and *Penicillium viridicatum* on grain, legumes, and other commodities.[5] These compounds are reported to be carcinogenic and teratogenic. Ochratoxins are nephrotoxic and hepatotoxic and may cause symptoms such as retarded growth, gastroenteritis, and polyuria.[6]

Fumonisins are a group of water soluble mycotoxins produced by strains of *Fusarium moniliforme*.[7] They were initially identified as the causative agent of equine leucoencephalomalacia, a disease characterized by liquifaction of brain tissue, that resulted from horses ingesting corn contaminated with *F. moniliforme*.[8] Fumonisins, which are structurally similar to the sphinogolipid sphinganine, act as inhibitors of sphingolipid biosynthesis and cause apoptosis or cell death.[9,10]

Trichothecenes are sesquiterpene epoxides produced by several species of *Fusarium, Myrothecium, Trichothecium, and Stachybotrys,* primarily on grains such as corn, wheat, and barley. Intake of grain contaminated with trichothecenes may result in a variety of symptoms including diarrhea, hemorrhaging, and feed refusal. Trichothecenes also act as acute skin irritants and they act as potent inhibitors of eukaryotic protein synthesis by inhibition of peptidyl transferase.[11,12]

Although trichothecenes share the trichothecane skeleton consisting of three rings with a double bond and an epoxide group, well over 60 of them have been characterized which differ in the number and position of hydroxyl, carbonyl, and ester groups attached to the trichothecane backbone. Both the double bond and the epoxide moieties appear to be required for toxicity, but different substituents can greatly alter the toxicity.[12,13]

BIOSYNTHESIS OF TRICHOTHECENES

The biosynthesis of trichothecenes proceeds from farnesyl pyrophosphate. Farnesyl pyrophosphate is converted to trichodiene by a sesquiterpene cyclase, trichodiene synthase (Fig. 2).[14] Conversion of trichodiene to the more complex trichothecenes, such as T-2 toxin, diacetoxyscirpenol, and deoxynivalenol, requires a series of oxygenations, isomerizations, cyclizations, and esterifications. The sequence of steps of trichothecene biosynthesis shown in Figure 2 was constructed based on feeding experiments with *Fusarium culmorum* that produce 3-acetyldeoxynivalenol[15,16] and with blocked mutant strains derived from a T-2 toxin accumulating stain of *Fusarium sporotrichioides*.[17–19] Three UV mutant strains generated from a T-2 toxin producing strain of *Fusarium sporotrichioides* accumulate trichodiene, decalonectrin, or 4,15-diacetoxyscirpenol and were designated as being blocked at three genes, *Tri4, Tri3,* and *Tri1* respectively.[18]

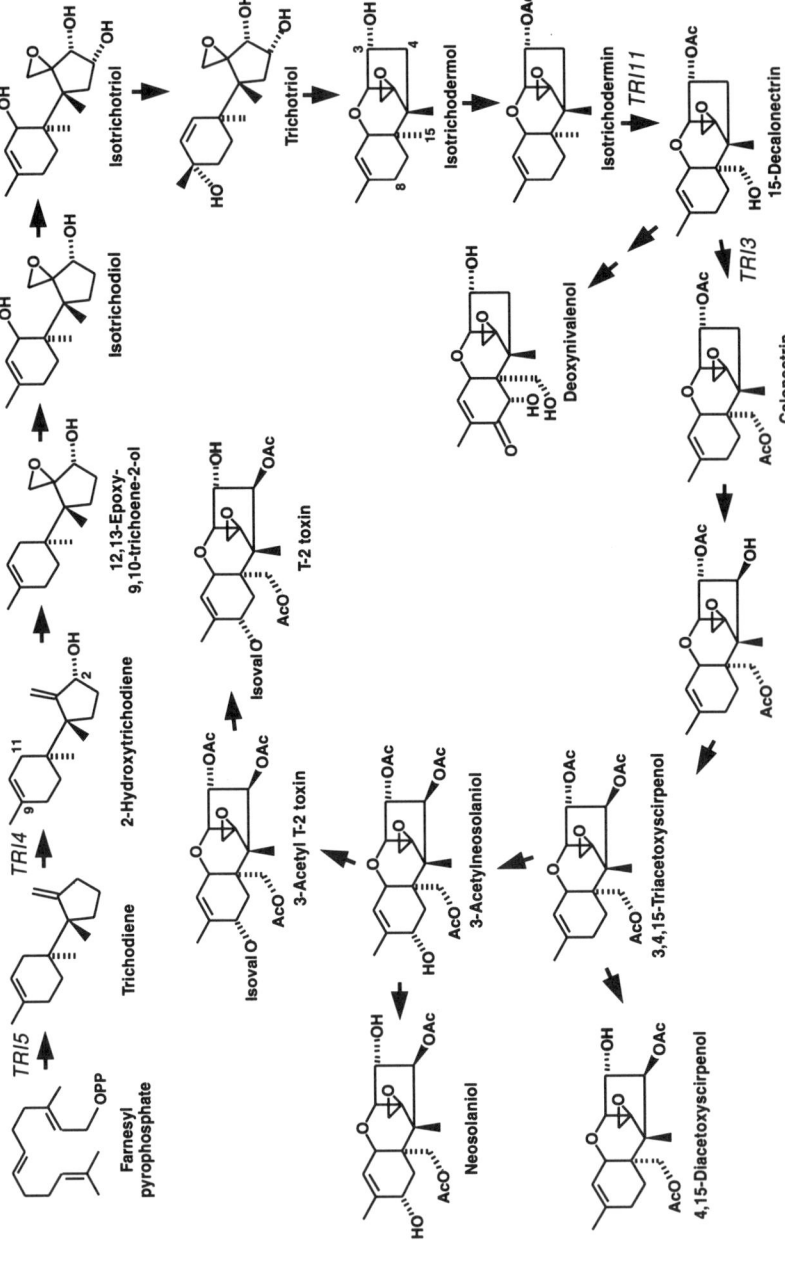

Figure 2. Proposed biosynthetic pathway for trichothecene mycotoxins: T-2 toxin, diacetoxyscirpenol, neosolaniol, and deoxynivalenol.

T-2 toxin, diacetoxyscirpenol, and deoxynivalenol appear to share the early steps in the biosynthetic pathway.

Trichodiene probably undergoes an allylic oxygenation at the C-2 position followed by oxygenation at C-12 and formation of the epoxide.[20] A second allylic oxygenation at C-11 results in the formation of isotrichodiol. This is followed by addition of a hydroxyl group to the C-3 position to form isotrichotriol. Isotrichotriol isomerizes to trichotriol. Interestingly, trichotriol, which lacks the tetrafuran ring but has double bond and epoxide moieties, was found to have embryotoxicity.[21] Trichotriol cyclizes to form the first pathway intermediate with the entire trichothecene skeleton, isotrichodermol. Isotrichodiol can undergo a similar isomerization and cyclization to trichothecene, however, trichothecene is not a biosynthetic intermediate of any *Fusarium* trichothecenes,[18,22] which typically have a substituent at the C-3 position. Trichothecene may serve as a biosynthetic intermediate of other trichothecenes produced by *Trichothecium* and *Myrothecium* which lack hydroxylation at C-3.[17]

Isotrichodermol is acetylated to form isotrichodermin. Isotrichodermin undergoes an oxygenation at C-15 to form decalonectrin. Decalonectrin is the branch point between the pathway for synthesis of deoxynivalenol in *F. culmorum*,[23] and of T-2 toxin, diacetoxyscirpenol, and neosolaniol in *F. sporotrichioides*.[19] Results of pulse labeling experiments of calonectrin to *Fusarium culmorum* suggested that deacetylation at C-15 may be required for conversion to 3-acetyldeoxynivalenol.[23] In *F. sporotrichioides*, decalonectrin is acetylated at C-15 to form calonectrin.[24] Calonectrin undergoes a hydroxylation at C-4 to form 3,15-diacetoxyscirpenol. Acetylation at C-4 precedes oxygenation at C-8. Finally, the C-8 hydroxyl group is esterified with an isovalerate group. The three toxins typically produced by *F. sporotrichioides*, T-2 toxin, neosolaniol, and 4,15-diacetoxyscirpenol, are all produced by deacetylation at the C-3.

Trichodiene synthase, which catalyzes the conversion of farnesyl pyrophosphate to trichodiene, has been purified and characterized.[14] The cloning of the trichodiene synthase gene (*Tri5*) has greatly facilitated understanding of the biosynthesis of these toxins, the organization and function of trichothecene genes, and the role of the trichothecenes in plant disease. The function of the trichothecene genes was determined with transformation mediated gene disruption experiments in which *Fusarium sporotrichioides* was transformed with a vector containing promoter 1 from *Cochliobolus heterostrophus*, a hygromycin selectable marker, and a doubly truncated portion of the gene of interest (Fig. 3). Disruption of *Tri5* results in transformants that produce no trichothecene metabolites.[25]

A cosmid library was generated and screened for cosmid clones carrying the *Tri5* gene which were able to complement the three UV mutant strains. Two overlapping cosmid clones (25kB) were able to complement and restore T-2 toxin production in the UV mutant strain blocked at *Tri3* which normally accumulates decalonectrin and in the UV mutant strain blocked at *Tri4* which normally accumulates trichodiene. This indicated that at least *Tri3, Tri4,* and *Tri5* were organized in a gene cluster.[26] *Fusarium* trichothecene biosynthetic pathway genes are not unique in being clustered. Several biosynthetic genes for aflatoxin also have been shown to be clustered.[27]

Mutants generated by gene disruption of *Tri3* and *Tri4* were identical to the UV mutant strains. Both *Tri4* UV mutants and transformants accumulate the parent hydrocarbon, trichodiene. Sequence alignments suggested that *Tri4* encodes a P450 oxygenase responsible for the first oxygenation step or steps.[28] Cell-free oxygenation of trichodiene and identification of the earliest oxygenated intermediates has been difficult.[29] A dioxygenated intermediate, 12,13-epoxy-9,10-trichoene-2-ol, has been isolated from *Fusarium culmorum*.[30] Heterologous expression of *Tri4* in yeast may permit the identification of either 12,13-epoxy-9,10-trichoene-2-ol or 2-hydroxytrichodiene as the actual product of the *Tri4* oxygenase.

UV mutants blocked at *Tri3* and mutants generated by gene disruption of *Tri3* accumulate decalonectrin, suggesting that acetylation at C-15 is blocked. Whole cell and cell free feeding experiments have clearly demonstrated that *Tri3* encodes an acetyltransferase specific to the C-15 position (Fig. 2).[24]

In order to characterize additional trichothecene biosynthetic genes,[31] the nucleotide sequence of DNA flanking *Tri5* was analyzed for the presence of open reading frames (ORF). The ORF identified were then compared to a protein sequence data base. The suspected coding regions were used to probe Northern blots. Finally, truncated coding regions were generated for gene disruption experiments. To date, approximately 25kB of DNA around *Tri5* have been sequenced and found to contain nine genes related to trichothecene biosynthesis (Fig. 4).

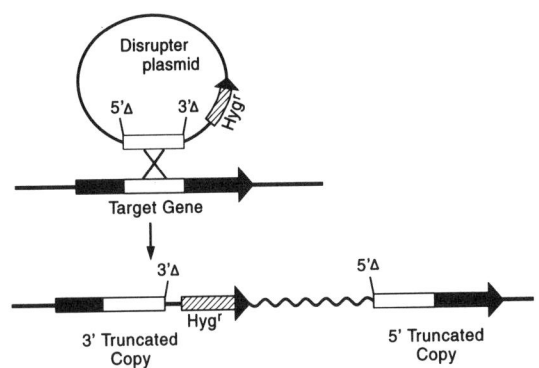

Figure 3. Schematic for gene disruption of fungal mycotoxin genes.

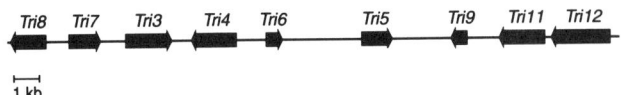

Figure 4. Gene map for trichothecene biosynthetic genes.

Tri6 encodes a protein involved in transcriptional regulation of trichothecene biosynthesis. The protein contains a zinc finger motif, characteristic of DNA binding proteins. Disruption of *Tri6* blocked T-2 toxin accumulation and greatly reduced the expression of other trichothecene biosynthetic genes.[32] *Tri7* apparently encodes a second transacetylase. Transformants produced with a disrupted *Tri7* sequence accumulate HT-2 toxin, which differs from T-2 toxin in the presence of an acetyl group at C-4. Transformants with a disrupted *Tri8* accumulate diacetoxyscirpenol and are chemically indistinguishable from *Tri1* UV blocked mutants. Feeding experiments and sequence homologies suggest that *Tri1* encodes a P450 oxygenase and that *Tri8* encodes the gene responsible for esterification of the C-8 hydroxyl group. *Tri11* encodes an additional P450 oxygenase, this one specific for the C-15 position.[33,34]

ROLE IN PLANT DISEASES

Although mycotoxins are by definition characterized by their ability to cause ill health in vertebrates, many mycotoxin producing organisms are destructive plant pathogens. Several are reported to be phytotoxic,[35] and the biochemical mode of action of mycotoxins, such as trichothecenes which disrupt protein synthesis and fumonisins which disrupt lipid metabolism, suggests that they could also play an important role in plant diseases. The availability of mutants generated by gene disruption has made it possible to carefully assess the role of toxins in plant diseases.

The possible role of trichothecenes in plant disease was initially assessed with UV blocked mutant strains of *Fusarium sporotrichioides*.[36] Inoculation of parsnip root discs with a T-2 toxin producing strain of *Fusarium sporotrichioides* caused extensive root rot. UV mutant strains blocked at *Tri4* that accumulate only the hydrocarbon trichodiene were unable to cause parsnip root rot, and the mutant that produced decalonectrin caused root rot of an intermediate severity.

Mutants generated by disruption of the sesquiterpene cyclase gene (*Tri5*) in a number of *Fusarium* species have made it possible to assess the role of trichothecene toxins, either in causing plant diseases such as potato dry rot and wheat head scab or in contributing to their severity. *Fusarium sambucinum* (*Gibberella pulicaris*) produces the trichothecene, 4,15-diacetoxyscirpenol and is a causative agent of potato dry rot. Using the gene disruption strategy (Fig. 3), transformants blocked at *Tri5* which produce no trichothecene metabolites were generated and tested on potato tubers and parsnip root slices.[19,37] Although presence of toxin increased the severity of disease on parsnip slices, there was essentially no difference in the rate of growth between toxigenic and nontoxigenic strains inoculated on potato tuber slices. This latter observation suggests that trichothecene toxins are not required to produce potato dry rot symptoms and that trichothecenes may have a role in some plant diseases but not in others.

Potato tuber tissue was found to rapidly metabolize diacetoxyscirpenol,[38] which would effectively limit the exposure of the plant tissue to the toxin.

Fusarium graminearum (*Gibberella zeae*), which produces deoxynivalenol and acetylated deoxynivalenol, causes a variety of plant diseases including seedling blights, root rots, and ear and head blight of wheat, barley, rye, maize, rice, and other grains. Wheat head blight, or wheat head scab, results in premature bleaching of the heads, reduced yield, reduced seed weight, and grain that is unsaleable if contaminated with the trichothecene deoxynivalenol. The gene disruption strategy of *Tri5* (Fig. 3) was used to produce non-toxigenic mutants of *G. zeae*.[39] The virulence of these non-producing mutants was significantly reduced in growth chamber studies of wheat seedling blight and head scab. A two-year field experiment with toxigenic and non-toxigenic strains of *G. zeae*, in which wheat heads were assessed for spread of the disease, and seed was assessed for toxin, seed weight, seed yield, and seed viability, demonstrated that trichothecenes are a virulence factor for wheat head scab.[40]

The field experiments were designed to determine if trichothecene production contributes to the virulence of *Fusarium graminearum*. The results confirmed that production of trichothecene toxins significantly increases the ability of *Fusarium graminearum* to cause wheat head scab and also suggested that the toxins may render wheat more susceptible to *Fusarium* wheat head scab. This raises the interesting possibility that if wheat were made resistant to trichothecene toxins, it might also be resistant to wheat head scab.

RESISTANCE

Genes that confer resistance to trichothecenes and that may, therefore, confer resistance to the associated plant diseases are of great interest. Since trichothecenes function not only as vertebrate and plant toxins but also as antibiotics and inhibitors of eukaryotic protein synthesis, the fungi that produce these mycotoxins require a mechanism to protect themselves from the toxins. One possible means of resistance to trichothecenes is modification of the biochemical target. One study in *Trichothecium roseum* has suggested that ribosomes are methylated and thereby prevent or alter binding of the toxin.[41] A second method of resistance is the use of transporter systems or molecular pumps. A model gene for this type of resistance, *PDR5*, has been characterized in *Saccharomyces ceriviseae*.[42] *PDR5* is a member of the ABC super family of transporters that confer multidrug resistance. Yeast containing *PDR5* are able to grow in the presence of high concentrations of antibiotics including trichothecenes. Disruption of the *PDR5* gene renders yeast extremely sensitive to exposure to trichothecenes such as diacetoxyscirpenol and T-2 toxin.

A *Fusarium* (*Tri12*) gene was identified within the trichothecene biosynthetic cluster that encodes a protein with some sequence similarities to other multidrug resistance proteins and to other transporter proteins in yeast and bacteria.[43] Gene disruption of *Tri12* resulted in transformants with greatly reduced production of toxins. These *Tri12⁻* transformants did not exhibit a greater sensitivity to the toxins, however, yeast transformants lacking *PDR5*, and therefore sensitive to trichothecenes, in which *Fusarium Tri12* had been inserted, were able to grow in the presence of trichothecenes. The *PDR5⁻/Tri12⁺* transformants did not have the same level of resistance to trichothecenes as the parent *PDR5* strain. The results suggested that *Tri12* may encode a transporter protein but that this transport system is probably not a major contributor to resistance.

A third method of achieving resistance to trichothecenes is by detoxification of the molecule—either by catabolism or chemical modification to a less toxic compound. Microorganisms or plants might serve as a source of this type of resistance gene. Potato leaf and tuber tissues were found to be able to rapidly detoxify 4,15-diactoxyscirpenol by deacetylation and further metabolism.[38]

In order to identify and isolate native resistance genes that might chemically modify the toxins produced by *Fusarium*, a *F. graminearum* cDNA library was screened in a strain of yeast sensitive to trichothecenes. A single gene, *Tri101* was identified with this method and yeast transformed with *Tri101* were able to grow in the presence of T-2 toxin. Further examination indicated that the yeast containing *Tri101* converted T-2 toxin to a less toxic product, 3-acetyl T-2 toxin, suggesting that the gene encoded a transacetylase.[44] Recent experiments in *F. sporotrichioides* have identified a biosynthetic gene, *Tri^r*, which encodes isotrichodermol 3-O-acetyltransferase. There does not appear to be strong substrate specificity, and yeast strains containing *Tri^r* were able to acetylate, and apparently detoxify, T-2 toxin, deoxynivalenol, and isotrichodermol. The toxin-sensitive strain of yeast (*PDR5⁻*) was significantly less sensitive to trichothecenes acetylated at the C-3 position than the corresponding trichothecenes containing a C-3 hydroxyl group.

O-Acetyltransferases are involved in the biosynthesis of a variety of natural products. The hydroxyl groups of terpenoids are frequently acetylated resulting in decreased polarity and, in many cases, altered bioactivity. Previous studies have shown that, in general, removal of trichothecene acetyl groups results in a less toxic compound. The trichothecenes, T-2 toxin and HT-2 toxin, are identical except for the presence of an acetyl group at the C4 position of T-2 toxin. However, the LD_{50} for intraperitoneal injection of HT-2 toxin in mice is almost double that observed for T-2 toxin.[13] *Tri^r* and *Tri101* encode transacetylases that appear to be involved in detoxification. Examination of the biosynthetic scheme for T-2 toxin (Fig. 2) shows that essentially all of the biosynthetic intermediates of T-2 toxin are acetylated at the C-3 position but that the end products, T-2 toxin, 4,15-diacetoxyscirpenol, neosolaniol, and deoxynivalenol have a free hydroxyl group at C-3. This suggests that acetylation of the C-3 position occurs early in the pathway and may act as a protective resistance mechanism for the producing fungus. Deacetylation only occurs at the time of release of the more toxic product.

PLANT INTERACTIONS

Still another effective resistance mechanism is to block synthesis of the mycotoxin by the fungus. Plants have a variety of chemical defense mechanisms. A survey of a variety of shikimate acid derivatives, including flavonoids and coumarins and furanocoumarins, identified several compounds that effectively block trichothecene biosynthesis, apparently by acting as inhibitors of P450 oxygenases.[45] Addition of small concentrations of the furanocoumarin, xanthotoxin, to trichothecene-producing *Fusarium* cultures results in the accumulation of large quantities of the hydrocarbon, trichodiene.[16] In the *Pastinaca–Fusarium* interaction,[36] *Fusarium pulicaris* manufactures the trichothecene, diacetoxyscirpenol, which acts as a virulence factor for parsnip root rot. The parsnip tissue produces the furanocoumarin phytoalexin, xanthotoxin, in response to the infection. The xanthotoxin blocks the trichothecene biosynthetic pathway and stops toxin production. In the experiments described by Desjardins et al.,[36] the *Fusarium* stays one step ahead of the parsnip defense mechanism.

A complex interaction also occurs with *Fusarium* dry rot of potatoes. Several sesquiterpene phytoalexins have been identified and characterized in potato and related food crops. Potato tubers typically produce varying amounts of rishitin, lubimin, and solavetivone when challenged with an infection or artificially with arachidonic acid. Many wild potato species are highly resistant to *Fusarium* and other important potato pathogens, and this resistance may be due to either production of high levels of sesquiterpene metabolites or efficient metabolism of fungal toxins.[38] The sesquiterpene phytoalexins do not appear to block toxin synthesis but they are able to inhibit growth of some sensitive

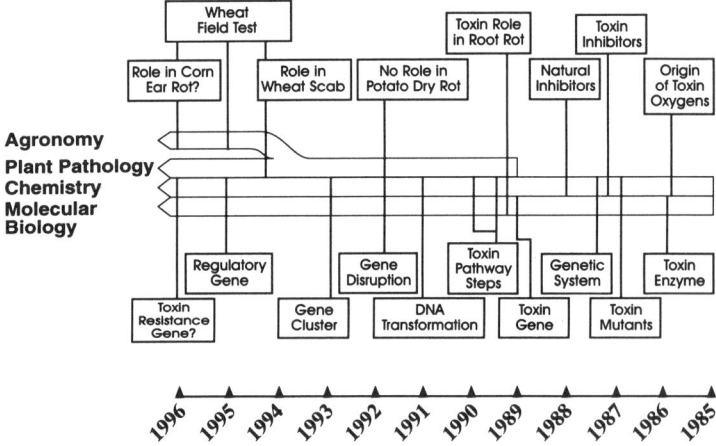

Figure 5. Strategy for addressing biosynthesis, regulation of trichothecene mycotoxins, and their role in plant disease.

Fusarium strains and secondarily lessen toxin production. Virulent strains of *Fusarium* are able to metabolize the potato phytoalexins.[46,47] A survey of 46 potato cultivars and breeding selections including wild potato germplasm showed that there was no obvious correlation between quantity of total or individual phytoalexins and resistance to *Fusarium* dry rot.[48]

The sesquiterpene phytoalexins produced in response to golden nematode infection, however, may be an important factor in the tuber resistance to that pathogen. Wild potato varieties or clones derived from *Andigena* produce a high relative ratio of the phytoalexin, solavetivone, and have marked resistance to golden nematode.[48] The genes responsible for resistance to golden nematode, as well as other resistance genes are located on chromosome V, which also holds genes for sesquiterpene biosynthesis. Bioassays of nematodes with purified phytoalexins have been inconclusive. It is possible that nematode resistance is genetically linked to nematode resistance rather than being a result of phytoalexin production.

CONCLUSION

A coordinated effort using the tools of chemistry, molecular biology, biochemistry, and plant pathology and agronomy has facilitated the study of the role of *Fusarium* mycotoxins in plant diseases. This approach (Fig. 5) has been used over the last ten years to understand better the role of trichothecene toxins in several plant diseases and to devise improved strategies for their elimination from animal and human food sources.

REFERENCES

1. HAYES, A.W. 1981. Mycotoxin Teratogenicity and Mutagenicity. CRC Press, Boca Raton, 121 p.
2. BILGRAMI, K.S., SINHA, K.K. 1992. Aflatoxins: Their biological effects and ecological significance. In: Handbook of Applied Mycology Vol. 5 Mycotoxins in Ecological systems, (D. Bhatnagar, E.B. Lillihoj and D.K. Arora, eds.), Marcel Dekker, New York, pp.59–86.
3. BAILEY, E.A., IYER, R.S., STONE, M.P., HARRIS, T.M., ESSIGMANN, J.M. 1996. Mutational properties of the primary aflatoxin B1-DNA adduct. Proc. Natl. Acad. Sci. 93:1535–1539.
4. SHARMA, R.P., SALUNKHE, D.K. 1991. Introduction to Mycotoxins. In: Mycotoxins and Phytoalexins, (R.P. Sharma, and D.K. Salunkhe, eds.), CRC Press, Boca Raton, pp. 339–359.
5. XIAO, H., MADHYAS, S., MARQUARDT, R.R., LI, S., VODELA, J.K., FROHLICH, A.A., KEMPPAINEN, B.W. 1995. Toxicity of ochratoxin A, its opened lactone and several of its analogs: Structure activity relationships. Toxicology 137:182–192.
6. MILLER, J.D. 1995. Fungi and mycotoxins in grain: Implications for stored product research. J. Stored Prod. Res. 31:1–16.
7. BEZUIDENHOUT, S.C., GELDERBLOM, W.C.A., GORST-ALLMAN, C.P., HORAK, R.M., MARASAS, W.F.O., SPITELLER, G., VLEGGAAR, R. 1988. Structure elucidation of the fumonisins, mycotoxins from *Fusarium moniliforme*. J. Chem. Soc. Chem. Comm. 743–745.

8. WILSON, T.M., ROSS, P.F., NELSON, P.E. 1991. Fumonisin mycotoxins and equine leukoencephalomalacia. J. Am. Vet. Med. Assoc. 198:1104–1105.
9. WANG, E., ROSS, P.F., WILSON, T.M., RILEY, R.T., MERRILL, A.H. 1991. Inhibition of sphingolipid biosynthesis by fumonisins: Implications for diseases associated with *Fusarium moniliforme*. J. Biol. Chem. 26:14486–14490.
10. WANG, H., JONES, C., CIACCI-ZANELLA, J., HOLT, T., GILCHRIST, D.G., DICKMAN, M.B. 1996. Fumonisins and *Alternaria alternata* lycopersici toxins: Sphinganine analog mycotoxins induce apoptosis in monkey kidney cells. Proc. Natl. Acad. Sci. 93:3461–3463
11. McLAUGHLIN, C.S., VAUGHN, M.H., CAMPBELL, J.M., WEI, C.M., STAFFORD, M.E., 1977. Inhibition of protein synthesis by trichothecenes. In: Mycotoxins in Human and Animal Health. (J.V. Rodricks, C.W. Hesseltine and M.A. Mehlman, eds.), Pathotoxin Publishers, Park Forest, IL, pp. 263–273.
12. UENO, Y. 1977. Mode of action of trichothecenes. Pure Appl. Chem. 49:1737–1745.
13. UENO, Y. 1980. Toxicological evaluation of trichothecene mycotoxins. In: Natural Toxins, (D. Eaker, and T. Wadstrom, eds.), Pergamon Press, New York, pp. 663–671.
14. HOHN, T.M., VANMIDDLESWORTH, F. 1986. Purification and characterization of the sesquiterpene cyclase trichodiene synthase from *Fusarium sporotrichioides*. Arch. Biochem. Biophys. 251:756–761.
15. ZAMIR, L.O., DEVOR, K.A., SAURIOL, F. 1991. Biosynthesis of trichothecene: oxygenation steps post-trichodiene. J. Chem. Soc., Chem. Commun. 1033–1034.
16. HESKETH, A.R., GLEDHILL, L., MARSH, D.C., BYCROFT, B.W., DEWICK, P.M., GILBERT, J. 1991. Biosynthesis of trichothecene mycotoxins: Identification of isotrichodiol as a post-trichodiene intermediate. Phytochemistry 30:2237–2243.
17. BEREMAND, M.N., McCORMICK, S.P. 1992. Biosynthesis and regulation of trichothecene production by *Fusarium* species. In: Handbook of Applied Mycology Vol. 5 Mycotoxins in Ecological systems,. (D. Bhatnagar, E.B. Lillihoj and D.K. Arora, eds.), Marcel Dekker, New York, pp. 359–384.
18. McCORMICK, S.P., TAYLOR, S.L, PLATTNER, R.D., BEREMAND, M.N. 1990. Bioconversion of possible T-2 toxin precursors by a mutant strain of *Fusarium sporotrichiodes* NRRL 3299. Appl. Environ. Microbiol. 56:702–706.
19. DESJARDINS, A.E., HOHN, T.M., McCORMICK, S.P. 1993. Trichothecene biosynthesis in *Fusarium* species: Chemistry, Genetics, and Significance. Microbiol Rev. 57:595–604.
20. ZAMIR, L.O., DEVOR, K.A., NIKOLAKAKIS, A., NADEAU, Y., SAURIOL, F. 1992. Structures of new metabolites from *Fusarium* species: An apotrichothecene and oxygenated trichodienes. Tetrahedron Lett. 33:5181–5184.
21. CORLEY, D.G., ROTTINGHAUS, G.E., TEMPESTA, M.S., 1987. Toxic trichothecenes from *Fusarium sporotrichioides* (MC-72083). J. Org. Chem. 52:4405–4408.
22. ZAMIR, L.O., DEVOR, K.A., NIKOLAKAKIS, A., SAURIOL, F. 1990. Biosynthesis of *Fusarium culmorum* trichothecenes, the role of isotrichodermin and 12,13-epoxytrichothec-9-ene. J. Biol. Chem. 265:6713–6725.
23. ZAMIR, L.O., NIKOLAKAKIS, A., DEVOR, K.A., SAURIOL, F. 1996. Biosynthesis of the trichothecene 3-acetyldexoynivalenol. Is isotrichodermin a biosynthetic precursor? J. Biol. Chem. 271:27353–27359.
24. McCORMICK, S.P., HOHN, T.M., DESJARDINS, A.E. 1996. Isolation and Characterization of *Tri3*, a gene encoding 15-O-Acetyltransferase from *Fusarium sporotrichioides*. Appl. Environ. Microbiol. 62:353–359.
25. HOHN, T.M., DESJARDINS, A.E. 1992. Isolation and gene disruption of the *Tox5* gene encoding trichodiene synthase in *Gibberella pulicaris*. Mol. Plant-Microbe Interact, 5:249–256.
26. HOHN, T.M., McCORMICK, S.P., DESJARDINS, A.E. 1993. Evidence for a gene cluster involving trichothecene pathway biosynthetic genes in *Fusarium sporotrichioides*. Curr. Genet. 24:291–295.

27. SKORY, C.D., CHANG, P.K., CARY, J., LINZ, J.E. 1992. Isolation and characterization of a gene from *Aspergillus parasiticus* associated with the conversion of versicolorin A to sterigmatocystin in aflatoxin biosynthesis Appl. Environ. Microbiol. 58:3527–3537.
28. HOHN, T.M., DESJARDINS, A.E., McCORMICK, S.P. 1995. The *Tri4* gene of *Fusarium sporotrichioides* encodes a cytochrome P450 monooxygenase in trichothecene biosynthesis. Molecular and General Genetics 248:95–102.
29. GLEDHILL, L., HESKETH, A.R., BYCROFT, B.W., DEWICK, P.M., GILBERT, J., 1991. Biosynthesis of trichothecene mycotoxins: Cell-free epoxidation of a trichodiene derivative. FEMS Microbiology Letters 81:241–246.
30. ZAMIR, L.O., DEVOR, K.A., MORIN, N., SAURIOL, F. 1991. Biosynthesis of trichothecenes: Oxygenation steps post-trichodiene. J. Chem. Soc., Chem. Comm. 1033–1034.
31. HOHN, T.M., DESJARDINS, A.E., McCORMICK, S.P., PROCTOR, R.H. 1995. Biosynthesis of trichothecenes, genetic and molecular aspects: Issues involving toxic microorganisms. In: Molecular Approaches to Food Safety. (M. Eklund, J.L. Richard and M. Katsutoshi, eds.). Alaken, Inc., Fort Collins, pp. 239–247.
32. PROCTOR, R.H., HOHN, T.M., McCORMICK, S.P., DESJARDINS, A.E. 1995. *Tri6* encodes an unusual zinc finger protein involved in the regulation of trichothecene biosynthesis in *Fusarium sporotrichioides*. Appl. Environ. Microbiol. 61:1923–1930.
33. ALEXANDER, N.J., HOHN, T.M., McCORMICK, S.P. 1998. The *Tri11* gene *of Fusarium sporotrichioides* encodes a cytochrome P450 monooxygenase responsible for the hydroxylation of C-15 in the trichothecene biosynthetic pathway. Appl. Environ. Microbiol. 64:221–225
34. McCORMICK, S.P., HOHN, T.M. 1997. Accumulation of trichothecenes in liquid cultures of a *Fusarium sporotrichioides* mutant lacking a functional trichothecene C-15 hydroxylase. Applied Environ. Microbiol. 63:1685–1688.
35. CUTLER, H. 1988. Trichothecenes and their role in the expression of plant disease. In: Biotechnology of Crop Protection ACS Symp. 379, (P. Hedin, J. Menn and R. Hollingsworth, eds.) Amer. Chem. Soc., Washington, DC, pp. 50–72.
36. DESJARDINS, A.E., SPENCER, G.F., PLATTNER, R.D., BEREMAND, M.N. 1989. Furanocoumarin phytoalexins, trichothecene toxins and infection of *Pastinaca sativa* by *Fusarium sporotrichioides* Phytopathology 79:170–175.
37. DESJARDINS, A.E., HOHN, T.M., McCORMICK, S.P. 1992. Effect of gene disruption of trichodiene synthase on the virulence of *Gibberella pulicaris*. Mol. Plant-Microbe Interact. 5:214–222.
38. DESJARDINS, A.E., CHRIST-HARNED, E.A., McCORMICK, S.P., SECOR, G.A.. 1993. Population structure and genetic analysis of field resistance to thiabendazole in *Gibberella pulicaris* from potato tubers. Phytopathology 83: 164–170.
39. PROCTOR, R.H., HOHN, T.M., McCORMICK, S.P. 1995. Reduced virulence *of Gibberella zeae* caused by disruption of a trichothecene toxin biosynthetic gene. Mol. Plant-Microbe Interact. 8:593–601.
40. DESJARDINS, A.E., PROCTOR, R.H., BAI, G.H., McCORMICK, S.P., SHANER, G., BUECHLEY, G., HOHN, T.M. 1996. Reduced virulence of trichothecene-non-producing mutants of *Gibberella zeae* in wheat field tests. Mol. Plant-Microbe Interact. 9:775–781.
41. IGLESIAS, M., BALLESTA, J.P.G. 1994. Mechanism of resistance to the antibiotic trichothecin in the producing fungi. Eur. J. Biochem. 223: 447–453.
42. BALZI, E., WANG, M., LETERME, S., VAN DYCK, L., GOFFEAU, A. 1994. PDR5, a novel yeast multidrug resistance conferring transporter controlled by the transcription regulator PDR1. J. Biol. Chem. 269: 2206–2214.
43. ALEXANDER, N.J., PROCTOR, R.H., McCORMICK, S.P., PLATTNER, R.D. 1997. Genetic and molecular aspects of the biosynthesis of trichothecenes by *Fusarium*. Cereal Research Communications. 25: 315–320.
44. KIMURA, M., KANEKO, I., KOMIYAMA, M., TAKATSUKI, A., KOSHINO, H., YONEYAMA, K., YAMAGUCHI, I. 1998. Trichothecene 3-O-acetyltransferase protects both the

producing organism and transformed yeast from related mycotoxins. J. Biol. Chem. 273:1654–1661.
45. DESJARDINS, A.E., PLATTNER, R.D., SPENCER, G.F. 1988. Inhibition of trichothecene toxin biosynthesis by naturally occurring shikimate aromatics. Phytochemistry 27:767–771.
46. GARDNER, H.W., DESJARDINS, A.E., McCORMICK, S.P. 1994. Detoxification of the potato phytoalexin rishitin by *Gibberella pulicaris*. Phytochemistry 37:1001–1005.
47. DESJARDINS, A.E., GARDNER, H.W., PLATTNER, R.D. 1989. Detoxification of the potato phytoalexin lubimin by *Gibberella pulicaris*. Phytochemistry 28:431–437.
48. DESJARDINS, A.E., McCORMICK, S.P., CORSINI, D.L. 1995. Diversity of sesquiterpenes in 46 potato cultivars and breeding selections. J. Agric. Food Chem. 43:2267–2272.

Chapter Three

ACTIVE OXYGEN IN FUNGAL PATHOGENESIS OF PLANTS

The Role of Cercosporin in *Cercospora* Diseases

Margaret E. Daub, Marilyn Ehrenshaft, Anne E. Jenns, and Kuang-Ren Chung

Department of Plant Pathology
North Carolina State University
Raleigh, North Carolina 27695-7616

Introduction 31
Photosensitization 34
Cercosporin Production by *Cercospora* Species 35
Cercosporin as a Photosensitizer 36
Cercosporin Toxicity to Plants 37
Roie of Cercosporin in Disease 38
Cellular Resistance to Photosensitizers 39
Cercospora Resistance to Cercosporin 40
Cercosporin-Sensitive Mutants 44
Isolation and Characterization of Cercosporin-Resistance Genes 45
Summary 51

INTRODUCTION

The term "active oxygen species" generally refers to both radical and non-radical derivatives of oxygen which are highly reactive in biological systems. Ground state molecular oxygen is relatively unreactive in cells due to its "triplet" conformation (the presence of two unpaired electrons of parallel spin). Since most organic molecules in biological systems have paired electrons with opposite spins ("singlet" conformation), reactions of ground state oxygen with

Phytochemical Signals and Plant–Microbe Interactions, edited by Romeo *et al.*
Plenum Press, New York, 1998.

other compounds are generally spin-restricted, and often occur one electron at a time. Thus, metabolic reactions involving oxygen often result in the production of radical or reduced intermediates,[1] including superoxide ($O_2^{\cdot -}$), hydrogen peroxide (H_2O_2), and the hydroxyl radical (OH·). Another activated oxygen species, singlet oxygen ($^1O_2$), is not a free radical, but the lack of any spin restriction makes it highly reactive with biological molecules. Although toxic to cells, all of these forms of oxygen occur in biological systems and have increasingly been shown to play important roles in normal cellular processes, including normal metabolic and biosynthetic reactions, signaling, and cell defense.

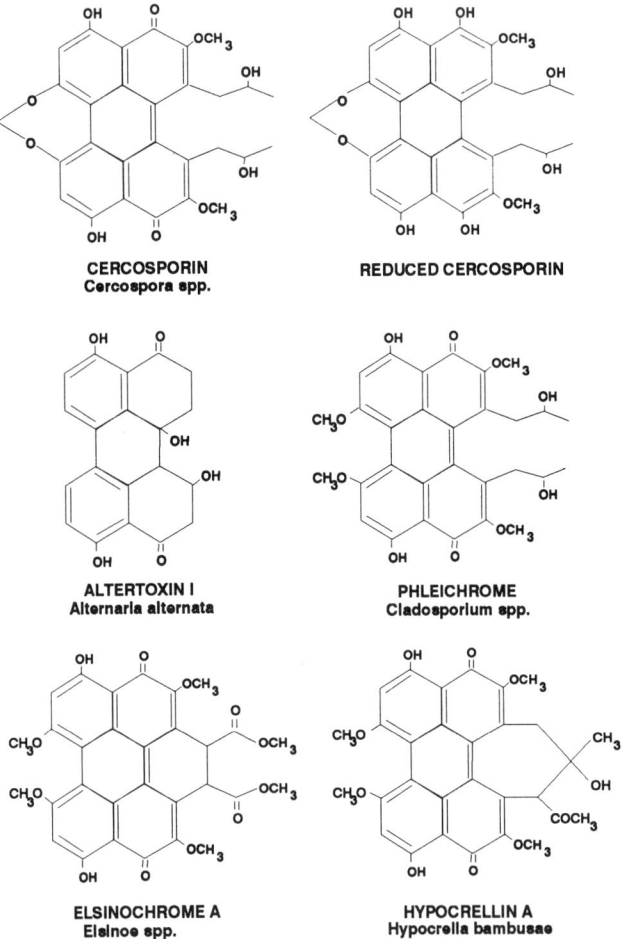

Figure 1. Structures of cercosporin, reduced cercosporin, and perylenequinone toxins produced by plant pathogenic fungi.

In recent years, numerous studies have demonstrated that plants use activated forms of oxygen as defense mechanisms against pathogen attack.[2-4] These studies have documented a strong correlation between production of activated oxygen species and the occurrence of the hypersensitive reaction (HR), the most common and dramatic defense reaction in plants, as well as in systemic acquired resistance (SAR), a non-specific induced resistance response. Multiple roles for active oxygen in plant defense have been suggested, including acting as second messengers, as substrates in synthesis of structural defense components, and as the toxic agents that cause plant cell and pathogen death. The production of active oxygen species by plants is particularly intriguing as it provided the first major link between plant and animal defense responses since animal cells also utilize active oxygen in defense, both as second messengers and as cytotoxic agents.[5,6]

Less known but equally intriguing is the fact that some fungal pathogens of plants have evolved to utilize active oxygen in pathogenesis of their hosts. These pathogens produce photoactivated perylenequinone toxins (Fig. 1) which, upon illumination, generate activated forms of oxygen which serve as cytotoxic agents during pathogenesis. To date, such toxins have been isolated from plant pathogenic fungi in at least eight genera (Table 1). Many of these compounds have been studied, not for their involvement in plant disease, but because of concerns about potential human toxicity in contaminated foods[7,8] or because of their potential as pharmaceuticals. Although evidence for a role of these toxins in disease is so far limited to studies on *Cercospora* and *Alternaria* species, production of photoactivated perylenequinones by so many fungal plant pathogens suggests that photosensitization is an important pathogenesis factor, particularly for fungi which parasitize aerial surfaces of plants.

Table 1. Production of photoactivated perylenequinones by plant pathogenic fungi

Fungus	Phototoxin	Reference
Cercospora species	cercosporin	9
Alternaria species	alteichin	10
	altertoxins I, II, III	11
	alterlosin I, II	12
	stemphyltoxin III	7
Stemphylium botryosum	stemphyltoxin III	13
Cladosporium species	phleichrome	14
	entisophleichrome	15
	cladochrome	13
Elsinoe species	elsinochromes	17
Shiraia bambusicola	shiraiachromes	18
Hypocrella bambusae	hypocrellins	19
Mycosphaerella fijiensis	cercosporin	Smith and Daub, unpublished

This review will focus on what is known about the role of active oxygen in pathogenesis in one group of plant pathogenic fungi, members of the genus *Cercospora*. These fungi produce a perylenequinone toxin, cercosporin, which kills plant cells via photoactivation and production of active oxygen species. Studies in this lab and others have defined the mode of toxicity of cercosporin on plant cells, and the critical role played by this toxin in plant pathogenesis. Most recently, fungal genes which encode resistance to cercosporin have been isolated and characterized. Understanding of the function of these genes is leading to a fundamental understanding of the basis of cellular resistance to active oxygen. These genes also have the potential to provide novel control strategies for important pathogens of plants.

PHOTOSENSITIZATION

Cercosporin and other perylenequinone toxins are photosensitizing compounds. Photosensitizers are a group of structurally-diverse chemicals which are classified together based on their common ability to be activated by light and to produce activated oxygen species. During light exposure, a photosensitizer is converted to an electronically-activated triplet state. The triplet photosensitizer then reacts in one of two ways. It may react by electron-transfer (radical) reactions via a reducing substrate (Type I reaction), leading to the production of a variety of free radicals including lipid free radicals and a diversity of active oxygen species such as $O_2^{\cdot -}$, H_2O_2, and $OH\cdot$.[21] The triplet sensitizer may also react directly with oxygen by an energy-transfer mechanism (Type II reaction),[22,23] leading to the production of $^1O_2$. In cells, photosensitizers cause damage to membranes, proteins, and DNA, with the type of damage being determined by where the photosensitizer molecule localizes in the cell, such as in membranes, the cytoplasm, or nucleus.[24]

Photosensitizers are common in nature. Plants are major producers of them, including compounds such as chlorophyll, coumarins, thiophenes, and acetylenes.[25] Medically-important compounds such as riboflavin and porphyrins, and dyes such as methylene blue, toluidine blue, eosin Y, rose bengal, and acridine orange are also photosensitizers.[23] Hypericin, an extended quinone photosensitizer produced by plants in the genus *Hypericum* (St. John's wort), causes a disease called hypericism in animals that feed on the plants and are exposed to light.[26] Ingestion of sufficient doses of hypericin results in skin irritations, changes in behavior, convulsions, and can lead to death in severe cases. Furanocoumarins such as psoralens also cause toxicity in humans and animals which come in contact with psoralen-containing plants.[25] Other photosensitizers can be found in protozoans where they function in photomovement.[27,28] Photosensitizers are also being investigated for their use as pesticides and pharmaceuticals. Photosensitizers such as the halogenated xanthine dyes, terthiophenes, and several porphyrin derivatives are actively being investigated as herbicides and insecticides.[25] The most intensive investigations of photosensitizers for medical uses is as antiviral agents[29,30] and as anti-cancer agents. For example, hypocrellins, cercosporin analogues which are produced by the bamboo pathogen *Hypocrella bambusae*, are being studied for use in photodynamic tumor therapy, a process which utilizes photosensitizers and targeted laser irradiation to kill tumor cells.[19] Thus, photosensitizers have wide distribution and function in nature.

CERCOSPORIN PRODUCTION BY *CERCOSPORA* SPECIES

Cercosporin is produced by members of the fungal genus *Cercospora*, a widespread and highly successful group of pathogens which cause leaf spot and blight diseases on many important crop species including corn, sugar beet, banana, tobacco, coffee, soybean, rice, and several ornamental, vegetable, and weed species. *Cercospora* species cause major economic problems because they have a world-wide distribution and broad host range. Also, it has been difficult to identify adequate levels of resistance to *Cercospora* pathogens in many host plant species, thus, chemical pesticides are often needed to control these diseases.

The success of this group of pathogens appears to be due to their production of the photosensitizing toxin, cercosporin. Cercosporin was first isolated in the 1950s from the mycelium of the soybean pathogen *Cercospora kikuchii*.[31-33] It has since been isolated from many species in the genus and from plants infected with *Cercospora* pathogens.[34] Cercosporin's structure (Fig. 1) was determined independently by Lousberg et al. and Yamazaki and Ogawa.[35,36] Its stereochemistry and spectral properties have also been studied.[36-38] Cercosporin is red in color, is only sparingly soluble in water, but is readily soluble in base and in several organic solvents. Because of its limited water-solubility, when secreted by the fungus during growth in culture, red crystals of cercosporin accumulate in the culture medium and are readily visible.

The biosynthesis of cercosporin by the fungus has been studied. Okubo et al.[39] determined that cercosporin synthesis begins with the condensation of acetate and malonate, and proposed that it was synthesized via the polyketide pathway, an active pathway of secondary metabolism in fungi. Efforts to isolate genes involved in cercosporin biosynthesis took advantage of the fact that (in addition to the light requirement for cercosporin activity) cercosporin biosynthesis also requires light.[40-43] Subtractive hybridization techniques were used to isolate six cDNA clones from *C. kikuchii* whose expression was up-regulated in light-grown culture.[41] Transcripts from the genes corresponding to these six cDNA clones accumulated from approximately 2-fold to 20-fold more in the light than in the dark. Expression of the most highly light-enhanced of the cDNA clones, called cLE6 (*l*ight *e*nhanced clone #6), correlated well with cercosporin production. A kinetic analysis showed that accumulation of cLE6-specific RNA closely parallels cercosporin accumulation in light-grown culture of wild-type *C. kikuchii*. Furthermore, in mutants which are deficient or altered in cercosporin biosynthesis, cLE6 RNA levels fluctuate in close harmony with cercosporin concentrations.

Because of the strongly suggestive correlations between cLE6 expresssion and cercosporin accumulation, a genomic clone corresponding to cLE6 was isolated, and both the cDNA and genomic clones were sequenced.[44] Database analysis indicated that the LE6 gene encodes a protein containing domains with high homology to antibiotic/toxin pumps. These are proteins which are involved in transport of antibiotics and toxins, and are also believed to contribute to autoresistance by exporting the toxic product away from the producing organism. To correlate a phenotype with the LE6 gene, targeted gene disruption experiments were performed. Targeted gene disruption involves isolation of a gene, construction of a mutated copy of that gene, and transformation of the mutant construct back into the fungus. As transformation in filamentous fungi occurs by integration into chromosomes, frequently at homologous sequences, transformants can be recovered where the mutant copy of the gene replaces the endogenous copy. Disruption of the LE6 gene resulted in a 95% reduction in cercosporin accumulation in the light, and also reduced (by *ca*. 50%) the ability of the disruption mutants to grow in the presence of exogenous cercosporin. Confirmation of the role of this protein in exporting cercosporin by direct measurement of export efficiency in wild-type and disrupted strains has not yet been done.

Mutants also have been isolated which will have utility in the study of cercosporin biosynthesis. UV mutagenesis was used to produce three strains of *C. kikuchii* which accumulated approximately 100-fold lower concentrations of cercosporin than the wild-type strain from which they were derived.[45] Two spontaneous mutants whose production of cercosporin is regulated by medium composition and by host plant extracts have also been isolated and characterized.[46] These mutants have been used to test the requirement for cercosporin in the development of *C. kikuchii* infection of soybean (see section on role of cercosporin in disease). These strains also may prove useful in the future in defining cercosporin biosynthesis intermediates, and as recipient strains for transformation of a genomic library leading to the identification of genes required for cercosporin biosynthesis.

CERCOSPORIN AS A PHOTOSENSITIZER

Cercosporin's photosensitizing activity was first reported by Yamazaki et al.[47] who demonstrated that cercosporin toxicity to mice and bacteria was light dependent. They also demonstrated a requirement for oxygen by assaying photooxygenation of dimethylfuran. Since that report, there have been many studies that have documented the production of active oxygen species by illuminated cercosporin. Production of $^1O_2$ by cercosporin has been measured through analysis of oxidation products of $^1O_2$-trapping compounds (cholesterol and 2,5-bis hydroxymethyl furan),[48,49] as well as directly by measurement of luminescence at 1280 nm (diagnostic for $^1O_2$) following laser excitation of cercosporin.[50,51] Cercosporin is also capable of producing $O_2^{\cdot-}$ (assayed by the superoxide dismutase-inhibitable reduction of p-nitro blue tetrazolium), but only if incubated in the light in the presence of reducing agents such as methionine, ergothioneine, and urate.[48,49]

Although *in vitro* assays have demonstrated the production of both $^1O_2$ and $O_2^{\cdot-}$ by cercosporin, the primary active oxygen species involved in *in vivo* toxicity is not yet known with certainty. Evidence to date, however, suggests that $^1O_2$ is the major agent responsible for toxicity. Cercosporin is a highly efficient producer of $^1O_2$ (quantum yield of $^1O_2$ formation = 0.81 ± 0.07).[50] In contrast, production of $O_2^{\cdot-}$ by cercosporin in the presence of methionine was approximately one third of that produced using riboflavin (a good type I photosensitizer), suggesting that cercosporin is not a potent producer of $O_2^{\cdot-}$.[48] Cercosporin killing of suspension-cultured plant cells was significantly inhibited when either of two different $^1O_2$ quenchers (1,4-diazabicyclo octane and the carotenoid carboxylic acid bixin) were added to the medium, suggesting the importance of $^1O_2$ production in *in vivo* toxicity.[9] There is so far little if any evidence for a role of cercosporin-produced $O_2^{\cdot-}$ in cellular toxicity. Tobacco cell culture mutants that were selected for resistance to paraquat and had elevated levels of catalase and superoxide dismutase activity showed no or only slight increases in resistance to cercosporin,[52-54] suggesting that $O_2^{\cdot-}$ may not be important in cercosporin toxicity *in vivo*. Thus, to date, evidence suggests that $^1O_2$ formation may be the major mechanism by which cercosporin exerts its toxicity. Definitive support for this hypothesis awaits direct measurements of $^1O_2$ and $O_2^{\cdot-}$ production in cells.

CERCOSPORIN TOXICITY TO PLANTS

The toxic effects of cercosporin on plant cells have been well documented and demonstrate that cercosporin kills plant cells by peroxidation of the membrane lipids leading to membrane breakdown. Cercosporin is a small (534 mw), lipid-soluble molecule which readily penetrates into plant cells. Alterations in rotation of membrane electron spin resonance (ESR) spectroscopy probes were seen in cells that were treated with cercosporin, but protected against toxicity by incubation in the dark.[55] These results suggested that cercosporin localizes in membranes. Cercosporin treatment of plant tissues and cells in the light results in rapid ion leakage and bursting of protoplasts, both of which can be inhibited by antioxidants and $^1O_2$ quenchers.[56,57] These effects are rapid. For example, statistically significant increases in the rate of electrolyte leakage from leaf disks of tobacco occur within 1–2 minutes of exposure to cercosporin in the light.

Membrane damage is due to peroxidation of the membrane lipids. Cells damaged by cercosporin show an accumulation of lipid peroxidation products such as malondialdehyde and ethane,[56,58] and a marked increase in the ratio of saturated to unsaturated fatty acids, indicating a selective breakdown of the unsaturated fatty acids, consistent with lipid peroxidation.[55] Increases in O_2 consumption by plant and rat tissues and cells have also been measured.[58] Cercosporin treatment causes a decrease in plant protoplast membrane fluidity and an apparent increase in the membrane phase transformation temperature, both measured by ESR spectroscopy with stearic acid spin labels.[55] Fluidity changes were detected with probes which localize at different regions of the membrane, a result consistent with peroxidative damage to the fatty acid chains. All of the types of changes documented with cercosporin are known to occur in membranes damaged by lipid peroxidation.[59,60]

The membrane damaging activity of cercosporin is consistent with ultrastructural studies of diseased tissue and with the known mode of parasitism by this fungus. Ultrastructural studies on sugar beet tissue either treated with cercosporin or infected with *Cercospora beticola* demonstrated that the primary ultrastructural symptom was a breakdown of the plasmalemma, tonoplast, and organelle-bounding membranes.[61,62] *Cercospora* pathogens penetrate host leaves through the stomata, colonize the intercellular spaces of the leaf, and remain external to the cells. Thus, they need a mechanism for obtaining nutrients from their host cells. Cercosporin's membrane damaging activity allows for leakage of nutrients from the cells and into the intercellular spaces, providing the fungus with the substrates necessary for growth and reproduction.

ROLE OF CERCOSPORIN IN DISEASE

There have been many investigations of cercosporin's possible role in disease development, and all indicate that cercosporin is required for successful pathogenesis of plants. As described above, *Cercospora*-infected leaf tissues show membrane damage as a major ultrastructural symptom,[61] consistent with cercosporin's mode of action. Also, cercosporin is readily recovered from infected plant material, demonstrating its production during host colonization.[34] The importance of light in symptom development, a phenomenon widely noted for diseases on several hosts, remained a mystery until the discovery of cercosporin. On coffee, shading reduced the penetration of *Cercospora coffeicola* through the stomata as well as the number of lesions which eventually developed on infected leaves.[63] Symptoms caused by *C. beticola* on sugar beet leaves were both delayed and less severe on plants incubated under low light conditions.[64] Unlike the results with coffee, low light did not affect the number of stomates penetrated, and, thus, appeared to exert an effect after the fungus penetrated into leaves. Symptoms caused by *Cercospora musae* (*Mycosphaerella musicola*), the causal agent of the yellow Sigatoka disease of banana, only developed on banana leaves exposed to high light intensities.[65] As with sugar beet, the light requirement did not affect stomatal penetration, but was required for normal lesion development after the fungus penetrated into the leaf. This latter work confirmed observations that symptoms of the disease did not develop on banana leaves that were shaded, an effect so striking that growth of bananas under the shade of coconut palms was suggested in the 1940s as a means of disease control.[66] Symptom expression (lesion number and the proportion of the leaf surface with lesions) on cotton infected by *Alternaria alternata*, another perylenequinone toxin-producing fungus, also was shown to be greatly enhanced by exposure to sunlight.[67] The investigators also noted that symptomless infection was common in leaves not exposed to sunlight, and proposed that sunlight caused symptomless infections to develop into visible lesions. The results of these studies indicate that in at least some hosts, the fungi are capable of germinating and penetrating into their host leaves, but that colonization, cell death, and subsequent necrotic lesion formation and blighting of leaf tissue requires the activity of a metabolite dependent on light for toxicity.

Inoculation experiments with the cercosporin-deficient mutants described earlier (see section on cercosporin production by *Cercospora* species) also support a role for cercosporin in disease. Mutants were tested for pathogenicity using fungal plug and mycelium inoculations on leaves of two cultivars of soybean. The three UV-induced mutants which accumulated 100-fold lower concentrations of cercosporin in culture were severly reduced in pathogenicity. Disease incidence (% of infected leaves/inoculated leaves) was between 0–3% for the mutants compared to 88% for the parental, wild-type strain.[45] The spontaneous, medium-regulated mutants, in contrast, were as pathogenic as the wild-type strain. However, cercosporin was isolated from leaves inoculated with the two medium-regulated mutants, indicating normal cercosporin production by these mutants during colonization of soybean leaves. Cercosporin could not be isolated from leaves inoculated with the non-pathogenic UV mutants. Thus, production of cercosporin appears to be required for the normal development of lesions on infected plants.

CELLULAR RESISTANCE TO PHOTOSENSITIZERS

As indicated above, photosensitizers are commonly-occurring compounds. They are natural products produced by plants, microorganisms, and protozoans, are used as dyes and reagents in laboratories and are finding increasing use as agricultural pesticides and pharmaceuticals.[25,29] Despite their common occurrence, naturally occurring resistance to photosensitizers is rare in nature, and little is known about cellular defenses against them. Plants are a major producer of photosensitizers,[68] but little is known about how they defend themselves against autotoxicity. Behavioral and light-avoidance responses are a common method of protection for protozoans and insects.[27,69] A number of species of insects feed on photosensitizer-containing plants. These insects exhibit protective responses such as feeding within rolled leaves or secreting protective webs.[69] Some insects contain pigments that block transmission of activating wavelengths of light whereas others are able to metabolize photosensitizing compounds produced by the host plant. The protozoan *Blepharisma* produces a photosensitizing pigment blepharismin which serves as a receptor pigment and is involved in photomovement. *Blepharisma* are normally bottom dwellers and are light-sensitive. When exposed to high light intensities, these organisms normally protect themselves by photophobic responses and shading under debris. They are also known to irreversibly oxidize blepharismin to an inactive form upon exposure to high light intensities.[27]

Studies have also identified compounds which quench $^1O_2$ and triplet state photosensitizers, or block the formation of active oxygen species by the photosensitizers. A diversity of compounds have been identified which quench $^1O_2$.[70] However, only a few of these are found in biological systems, and for those that are, few have been documented to play a role in cell defense. The best characterized compounds responsible for photosensitizer protection are carotenoids.[71,72] Carotenoids quench both $^1O_2$ and the triplet state of photosensitizers. Further, they are the most efficient quenchers identified of those that exist in biological systems.[70,71,73] For example, carotenoid pigments are the major means used by plants to protect themselves against $^1O_2$ that is produced by chlorophyll as an unavoidable by-product of photosynthesis.[74] Other cellular $^1O_2$ quenchers include thiols, some amino acids (histidine, methionine, and tryptophan), and other compounds including some peptides, amines, and phenols.[70,75–77] Except for carotenoids, the effectiveness of these quenchers is usually tested *in vitro*, thus, their precise role in photosensitizer resistance of living organisms is not well understood. Antioxidant enzymes such as superoxide dismutase, catalase, and peroxidases are known as important defenses against elevated oxygen levels and oxidants such as ozone and paraquat.[78–80] To date, however, there is little evidence that these enzymes protect cells against photosensitizers, perhaps due to the production of $^1O_2$ which is not quenched by such enzymes.

CERCOSPORA RESISTANCE TO CERCOSPORIN

Cercosporin has almost universal toxicity, with activity documented against mice, bacteria, fungi, and all plants thus far tested with the exception of a single wild rice species.[40,42,47,81,82] *In vitro* studies of cercosporin and other perylenequinones have documented toxicity to human tumor cells and to viruses,[19,30] and also the ability to inactivate protein kinase C.[83] In this lab, we have attempted to obtain cercosporin-resistant plants and fungi through mutagenesis and selection of cells in culture without success.[34] Cercosporin resistance appears to be limited to *Cercospora* fungi and other perylenequinone-producing fungi. These observations suggest that cercosporin and other perylenequinone toxins are almost universally cytotoxic, and that resistance is an active function specific to the few resistant organisms.

Cercospora species can produce up to mM concentrations of cercosporin in light-grown cultures without any measurable decreases in fungal growth.[84] *Cercospora* species also grow normally on medium supplemented with high concentrations (10–100 μM) of other $^1O_2$-generating photosensitizers such as hematoporphyrin, methylene blue, and eosin Y.[34,85,86] This resistance is surprising given the toxicity of $^1O_2$. $O_2^{\cdot-}$ and other radical and reduced forms of active oxygen are normal biproducts of cell metabolic reactions, and cells have potent defenses against them. With the exception of photosynthesis, however, the normal and routine production of $^1O_2$ in cells has not been extensively documented, thus, $^1O_2$ and $^1O_2$-generating compounds remain problematic for most cells.

The resistance shown by *Cercospora* species to cercosporin and other photosensitizers provides a unique system for understanding the cellular basis of $^1O_2$ resistance. Early work on this question identified many possible mechanisms that proved unimportant in cercosporin resistance. Membrane fatty acids are the target of cercosporin. Although fungi in general have more saturated membranes than do plants, membrane preparations isolated from *Cercospora* mycelium contained a high proportion of linoleic acid (18:2), a fatty acid that was readily susceptible to cercosporin-induced peroxidation.[55] Also, there was no evidence that the fatty acid composition of cultures changed when cultures were induced to produce cercosporin. Assays for activity of superoxide dismutase, peroxidase, or catalase did not detect differences between *Cercospora* species and several cercosporin-sensitive fungi, nor was there any evidence for a difference in general antioxidant activity.[82] Reducing agents and thiols such as ascorbate, cysteine, and reduced glutathione are highly effective in protecting sensitive fungi against cercosporin. However, assays of endogenous levels of these agents and of total soluble and protein thiols demonstrated no differences between *Cercospora* species and cercosporin-sensitive fungi.[86–88]

Carotenoids, though potent quenchers of $^1O_2$ and of triplet states of photosensitizers and the most commonly identified means of defense of many organisms against photosensitizers, also do not appear to function in cercosporin defense in *Cercospora* species. *Cercospora* species produce high levels of β-carotene, and carotenoid-deficient mutants of two fungi, *Neurospora crassa* and *Phycomyces blakesleeanus*, were significantly more sensitive to cercosporin than were wild type strains.[89] However, when targeted gene disruption was used to create *Cercospora nicotianae* mutants deficient in carotenoid biosynthesis, no evidence for a role of carotenoids in *Cercospora* resistance could be found. The gene for the enzyme phytoene dehydrogenase (which catalyzes the final step in production of the colored carotenoid pigments from colorless precursors) was isolated from *C. nicotianae*, and a disrupted version was transformed into the wild type strain.[90,91] Three mutants out of approximately 30 screened were identified which were deficient in β-carotene production. These mutants were screened for resistance to cercosporin, both by inducing endogenous cercosporin synthesis and by growth on medium amended with cercosporin. The disruption mutants were no less resistant to cercosporin or to five other $^1O_2$-generating photosensitizers than was the β-carotene-producing wild type strain. This result was surprising as addition of the carotenoid bixin protects cultured tobacco cells against cercosporin.[9] Also, resistance of a wild rice species to cercosporin and *Cercospora oryzae* infection was correlated with carotenoid content.[81] The lack of protection in *Cercospora* against cercosporin by endogenous β-carotene production may be due to localization. Fungal carotenoids are generally localized in lipid bodies in the cytoplasm,[92] and would be unable to protect against cercosporin in membranes. The inability of plant carotenoids to protect against cercosporin similarly may be due to localization of carotenoids in chloroplasts and other plastids.

To date, the only cellular mechanism which has been correlated with cercosporin resistance in *Cercospora* species is the ability of the fungus to transiently reduce and detoxify cercosporin.[51,85,88] The hypothesis that cercosporin resistance was due to reduction was initially investigated because fluorescence microscopy studies revealed that *Cercospora* hyphae emitted a green fluorescence in the presence of cercosporin. Cercosporin is red in color and has a red fluorescence. This observation suggested that cercosporin may be modified when in contact with hyphae. In addition, some photosensitizers are converted to colorless "leuco" forms when reduced. These reduced forms do not absorb light and are not photoactive. Photosensitizer reduction, therefore, could be a mechanism of biological detoxification and resistance.

To test this hypothesis, the cell surface reducing ability of cercosporin-resistant and sensitive fungi was estimated by assaying the ability of these fungi to reduce a series of 20 tetrazolium dyes spanning a wide range of redox potentials.[88] Fungal species were cultured on medium containing the different dyes, and the ability of a fungus to reduce a dye was assayed microscopically by the presence of colored crystals of the dye formazans. Species resistant to cercosporin (*Cercospora* species and *A. alternata*) were capable of reducing a significantly greater range of dyes than were sensitive species (*Neurospora crassa*, *Aspergillus flavus*, and *Penicillium* species).

Next, the properties of reduced cercosporin (Fig. 1) were investigated.[51] Cercosporin can be reduced by the addition of a strong reducing agent such as dithionite or zinc dust, however, reduced cercosporin is highly unstable and instantly reoxidizes upon aeration or extraction away from the reducing agent. In order to test the toxicity of reduced cercosporin, two stably-reduced derivatives (a methylated derivative and an acetylated derivative) were synthesized. Reduced cercosporin and the reduced derivatives are green in color (as compared to the red color of cercosporin), absorb approximately half the amount of light as cercosporin, and are highly fluorescent. Dissipation of absorbed light energy by fluorescence would compete with transfer of energy to oxygen to produce $^1O_2$, a hypothesis that was confirmed by direct measurement of $^1O_2$ generation from the reduced derivatives. When relative $^1O_2$ yields were measured via assay of $^1O_2$ luminescence at 1280 nm, the reduced derivatives produced less than 20% (per quantum of light absorbed) of the $^1O_2$ produced by cercosporin. Coupled with the decreased absorbance of light, reduced derivatives are, therefore, inefficient in $^1O_2$ production. The reduced derivatives were also less toxic to growth of sensitive fungi and had less activity in a lipid peroxidation assay.

Confirmation of the ability of *Cercospora* species to reduce cercosporin was obtained using fluorescence microscopy and bandpass filters which allowed differentiation between fluorescence emission from cercosporin and reduced cercosporin.[85] Cercosporin in contact with hyphae was shown to emit a green fluorescence, detectable with a 515–545 nm bandpass filter, characteristic of reduced cercosporin. Red crystals emitting a red fluorescence, detectable with a 575–635 nm bandpass filter (indicative of non-reduced cercosporin), were clustered near the hyphal strands, thus, cercosporin excreted into the culture is not reduced. When hyphae were killed by heat, UV light, or chloroform vapor, only red fluorescence was present, indicating that maintenance of reduced cercosporin is an active function of hyphae. The same assay demonstrated that cercosporin-resistant *A. alternata* was also able to reduce cercosporin, but the cercosporin-sensitive fungi *N. crassa* and *Asp. flavus* were not.

Information on the localization of the cellular proteins or machinery responsible for resistance was obtained by assaying cercosporin resistance of fungal protoplasts. Since cercosporin is lipophilic, localizes in cell membranes, and damages plant cells by breakdown of the plasma membrane, resistance was hypothesized to function at the fungal cell surface, either within the cell wall or membrane or localized between. Protoplasts of *C. nicotianae* were obtained by digesting mycelium with cell wall degrading enzymes.[93] Freshly isolated protoplasts were killed when exposed to cercosporin, but protoplasts began to regain resistance soon after isolation. By 8 hours, 100% of protoplasts were capable of normal growth in the presence of cercosporin. However, resistance was restored prior to regeneration of an intact cell wall. Although assays with fluorescein-labeled lectins demonstrated active synthesis of cell wall carbohydrates within 1 hr. of protoplast isolation, protoplasts remained sensitive to osmotic shock until 12 hrs. after isolation. Electron microscopy of protoplasts at 12 hrs. demonstrated that most lacked a defined cell wall at this time; some had no detectable cell wall and many others were surrounded only by a loose mat of fibrils which did not adhere to the protoplast membrane.[94] It is unlikely that the cell wall carbohydrate polymers that are being synthesized are responsible for resistance. Carbohydrates do not have $^1O_2$ quenching ability,[70] and experiments attempting to protect plant cells against cercosporin by the addition of fungal wall polymers were not successful (unpublished results). The results of this work are consistent with a hypothesis that resistance may be due to a protein that is capable of reducing cercosporin, and that this protein is localized on the outer surface of the fungal membrane (or the inner surface of the wall). The cell wall degrading enzymes used to produce hyphal protoplasts are crude preparations which have numerous side activities including protease activity. The protoplasting procedure could have inactivated such a protein by partial digestion or by removal of carbohydrate moieties critical for activity.

Based on all of these results, a model was constructed[88] which proposes that *Cercospora* species (and other cercosporin-resistant fungi) protect themselves against cercosporin by maintaining cercosporin in a reduced and non-photoactive form. Cercosporin molecules which are excreted into the medium revert to the oxidized, photoactive form involved in infection of host plants. This mode of resistance would be highly effective for a fungus which relies on a potent photosensitizer for survival in nature.

CERCOSPORIN-SENSITIVE MUTANTS

In order to test the hypothesis correlating cercosporin resistance and reduction, and as a first step toward the isolation of genes involved in resistance, mutants of *C. nicotianae* were isolated which are sensitive to cercosporin.[86,87] The mutants were isolated from mycelial protoplasts which were mutagenized with UV light, allowed to regenerate on medium in the dark, and then screened for cercosporin sensitivity by replica-plating colonies to cercosporin-containing medium and incubating in the light. All isolation and screening was done under conditions that suppress endogenous cercosporin synthesis.[43] Out of almost 12,000 protoplast-derived colonies screened, six cercosporin-sensitive (CS) mutants were isolated, and these fell into two phenotypic classes. Five of the mutants (CS2, CS6, CS7, CS8, and CS9, designated class 1) were totally inhibited when grown on medium containing 1–10 µM cercosporin. Growth of the sixth mutant (CS10, designated class 2), was partially inhibited by 10 µM cercosporin, but not by lower concentrations. The class 1 mutants were incapable of reducing cercosporin, an observation which supports the reduction hypothesis. However, the partially-sensitive CS10 was normal in cercosporin-reducing ability.

Further phenotypic characterization was done.[86] All of the mutants were capable of synthesizing cercosporin when grown under conditions that induce cercosporin synthesis. The class 1 mutants stopped growing when cercosporin was produced in the light, but, surprisingly, endogenous cercosporin production appeared to have little effect on growth of CS10. Cercosporin sensitivity was not due to a general sensitivity of any of the mutants to light. As with other sensitive fungi, the mutants could be protected against cercosporin toxicity by the addition of reducing agents such as ascorbate, cysteine, and reduced glutathione. However, none of the mutants was altered in production of these compounds or in levels of total soluble or protein thiols, indicating that resistance is not due to endogenous production of these agents. Also, none of the reducing agents which were capable of protecting against cercosporin toxicity were actually capable of reducing cercosporin due to cercosporin's strong negative redox potential.[95]

The class 1 mutants were significantly altered in pathogenicity on tobacco.[86] CS2 produced an average of 12 lesions per leaf as compared to an average of 144 lesions per leaf for wild type. Mean lesion size was also smaller, with mean lesion diameters of 1.7 and 2.6 mm for CS2 and wild type, respectively. CS10 produced an average of 82 lesions per leaf, and lesion size was comparable to that of wild type (mean 2.5 mm diameter).

The mutants were also tested for resistance to five other $^1O_2$-generating photosensitizers: methylene blue, toluidine blue, rose bengal, eosin Y, and hematoporphyrin.[87] Wild type *C. nicotianae* grows normally on medium supplemented with high concentrations (10–100 µM) of all these photosensitizers, with the exception of rose bengal which shows some toxicity. Sensitivity of mutant CS10 was found to be specific to cercosporin. Growth of this mutant on these photosensitizers was identical to that of wild type. However, the class 1 CS mutants were unable to grow on medium supplemented with any of the other photosensitizers. This level of sensitivity was unexpected, as even the most cercosporin-sensitive fungal species tested to date are capable of at least some growth on these compounds.[82,85] The results suggest that the class 1 mutants are deficient in a gene which mediates resistance to a range of $^1O_2$-generating photosensitizers. This deficiency results in levels of photosensitizer sensitivity far greater than that which occurs in fungi in nature.

The characterization of the class 1 mutants was the first indication that the hypothesis of resistance resulting from transient reduction and detoxification of cercosporin may not be correct. The observation that these mutants are unable to reduce cercosporin supports the hypothesis. However, the complete sensitivity of the mutants to the five other photosensitizers argues against the presence of a specific cercosporin reductase, as a single enzyme would not be expected to recognize the other photosensitizers which differ widely from cercosporin in both structure and redox potential.

ISOLATION AND CHARACTERIZATION OF CERCOSPORIN-RESISTANCE GENES

To isolate the genes involved in cercosporin resistance in *C. nicotianae*, the two classes of cercosporin-sensitive mutants were complemented with a genomic library from the *C. nicotianae* wild type strain.[96] The library was constructed in a bialaphos-resistance-conferring plasmid, pBAR3,[97] modified by addition of a *cos* site. Approximately 4,000 cosmid clones were isolated which provided a 99% probability of representing the entire genome as intact fragments. DNA from pooled clones (96 clones per pool) was transformed into one mutant from each class (CS10 and CS8). Transformants were selected for resistance to bialaphos, and then 300–400 transformants resulting from transformation with each pool of DNA were screened for resistance to cercosporin by growing them in the light on medium containing 10 µM cercosporin. Over 19,000 CS8 and CS10 transformants were screened, and two cosmid clones were identified which complemented the mutants. Cosmid 18E1 restored both cercosporin and photosensitizer resistance to all five of the class 1 mutants, but not to CS10. Cosmid 30H2 restored cercosporin resistance to CS10, but not to the class 1 mutants.

A single gene on the 18E1 cosmid was found to be sufficient for complementing all five class 1 mutant strains (Fig. 2). It was designated *SOR1* (*s*inglet *o*xygen *r*esistance) because it was capable of restoring resistance to both cercosporin and to the other $^1O_2$-generating photosensitizers.[96] The gene was localized to a 2.1 kb subclone of the 18E1 cosmid. Sequence analysis of the gene and

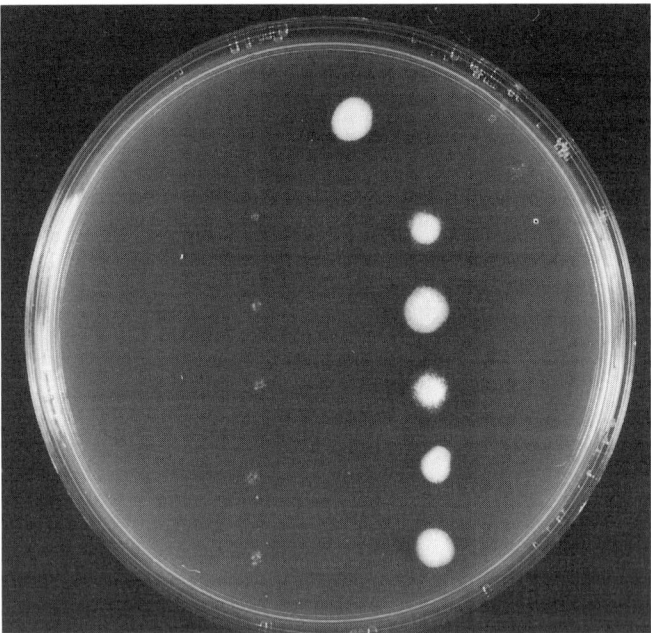

Figure 2. Colony growth of wild type *Cercospora nicotianae*, cercosporin-sensitive CS mutants, and *SOR1*-transformed CS mutants on medium containing 10 µM cercosporin following incubation in the light for 4 days. Center, top = wild type. CS mutants (left column) and *SOR1*-transformed mutants (right column) are from top to bottom: CS2, CS6, CS7, CS8, and CS9.

Table 3. Percent identity between predicted amino acid sequences of SOR1 homologues from different organisms[a]

	Cn	Sp	Hb	Ml	Mt	Hi	Bs	Mj	Y13	Y6	Y14	Mv	Sl
Cercospora nicotianae		74	70	59	60	65	65	57	62	65	64	67	55
Schizosaccharomyces pombe	—		73	68	69	68	65	66	64	66	65	64	59
Hevea brassiliensi	—	—		63	64	65	65	62	59	60	60	61	71
Mycobacterium leprae	—	—	—		80	66	62	54	62	62	61	61	51
Mycobacterium tuberculosis	—	—	—	—		68	63	54	62	63	62	63	51
Haemophilus influenzae	—	—	—	—	—		65	64	63	65	65	60	53
Bacillus subtilis	—	—	—	—	—	—		66	59	57	58	63	53
Methanococcus jannaschii	—	—	—	—	—	—	—		59	56	56	69	51
Yeast (chromosome 13)	—	—	—	—	—	—	—	—		81	81	59	47
Yeast (chromosome 6)	—	—	—	—	—	—	—	—	—		99.7	56	50
Yeast (chromosome 14)	—	—	—	—	—	—	—	—	—	—		56	50
Methanococcus vannillii	—	—	—	—	—	—	—	—	—	—	—		51
Stellaria longipes	—	—	—	—	—	—	—	—	—	—	—	—	

[a]Cn = Cercospora nicotianae, Sp = Schizosaccharomyces pombe, Hb = Hevea brasiliensis, Ml = Mycobacterium leprae, Mt = Mycobacterium tuberculosis, Hi = Haemophilus influenzae, Bs = Bacillus subtilis, Mj = Methanococcus jannaschii, Y13 = Yeast (chromosome 13), Y6 = Yeast (chromosome 6), Y14 = Yeast (chromosome 14), Mv = Methanococcus vannillii, Sl = Stellaria longipes.

also showed that *SOR1* RNA, unlike its yeast homologue, is readily detectable early in the logarithmic phase of the growth cycle and remains at approximately equal amounts as the culture proceeds into stationary phase,[96] consistent with a gene which is needed for resistance to a toxin which is produced early and throughout the growth cycle.

Sequence analysis of *SOR1* ORF-containing fragments amplified by PCR from the class 1 CS mutants has identified the site of the mutation in three mutants. There are single amino acid substitutions in the genes from mutants CS8 and CS9, two substitutions in the CS6 gene, and no mutations in the ORFs of either the CS2 or CS7 gene (Ehrenshaft et al., unpublished).

A homologue to *SOR1* was recently isolated from *A. alternata* (Jenns et al., unpublished). *A. alternata* produces perylenequinone toxins similar in structure to cercosporin, and also shows resistance to cercosporin and other singlet oxygen-generating photosensitizers. A clone containing the *A. alternata SOR1* homologue was isolated from a genomic library[100] via low stringency hybridization. A 6.2 kb subclone was transformed into the *C. nicotianae* class 1 mutant strains. The *A. alternata* gene complemented all five *C. nicotianae* strains at approximately the same frequency as the wild-type *C. nicotianae* gene, not only confirming its identify but also the ability of both the *A. alternata* promoter and ORF to function in *C. nicotianae*. Sequence analysis indicated that there is a 78% nucleic acid identity between the *C. nicotianae* and *A. alternata* ORF's, and that the predicted proteins are highly conserved. Comparison of the homologues and their expression from *Cercospora* and *Alternaria* with those from photosensitizer-sensitive organisms may provide clues to the domains and activity required for photosensitizer resistance.

In addition to *SOR1*, the gene which complements the sole class 2 mutant strain, CS10, also has been partially characterized. CS10 was rescued from partial cercosporin sensitivity to wild type levels of resistance by a 3.4 kb subclone of cosmid 30H2 containing a 1653 bp ORF (Chung et al., unpublished). This gene, designated *CRG1* (cercosporin *r*esistance *g*ene), encodes a protein of 550 amino acid residues with a predicted molecular mass of 60.9 kD. Comparison of the predicted CRG1 protein with proteins and translated sequences in the data base failed to uncover any strong homologies, thus, *CRG1* appears to be unique. Interestingly, like SOR1, CRG1 is predicted to contain one putative transmembrane domain and two putative N-linked glycosylation sites. Southern analysis revealed that *C. nicotianae* contains a single copy of *CRG1*, and that that two other *Cercospora* species, *C. kikuchii* and *C. beticola*, contain homologues.

Sequence analysis of the mutated *CRG1* gene from CS10 did not uncover any change in the ORF, but uncovered a single base change in a putative CAP site in the promoter region. This type of lesion could result in lowered stability of the RNA and/or lowered translation of the gene product, and may account for the partially sensitive phenotype of CS10. Targeted gene disruption experiments are being used to define the phenotype of a *CRG1* null strain.

SUMMARY

Activated forms of oxygen are increasingly being shown to play numerous roles in normal cellular processes in nature. Among these processes are cellular defense reactions where active oxygen is required both for regulation of defense genes and as a participant in the biochemical reactions which occur in defense reactions. The effectiveness of active oxygen in defense of plants is dramatic. Although many pathogens have learned to successfully parasitize plants, successful pathogenesis generally requires that pathogens evade or suppress induction of defense responses.[101] Once a hypersensitive reaction and the accompanying active oxygen production is induced, pathogenesis is rarely successful, a testament to the effectiveness of active oxygen. It is an indication of pathogen plasticity that *Cercospora* pathogens have evolved to take advantage of active oxygen to allow parasitism of plants. The same active oxygen species which stop fungal parasitism by killing both the pathogen and the invaded cell are utilized to induce plant cell death favorable to the development of the fungus.

Understanding the mechanisms by which *SOR1* and *CRG1* function to provide resistance will add considerably to our fundamental understanding of active oxygen and cellular defense. In addition, these genes provide a test system for investigating the utility of pathogen genes in engineering resistance to plant diseases via an approach termed "pathogen-derived resistance". Pathogen-derived resistance is a term coined by Sanford and Johnston[102] to describe a strategy of using genes from a pathogen which negatively regulate critical pathogenesis processes to engineer disease-resistant hosts. The most widespread use of this approach in plant pathology is genetically engineered cross protection for control of virus diseases.[103] To date, there has been little use of genes from plant pathogenic fungi in plant genetic engineering, although a number of possible strategies have been proposed. Antifungal proteins from *Aspergillus*[104] have been suggested as a possible strategy for fungal disease control.[105] The use of toxin detoxification genes from pathogens also has been proposed,[106] and genes for resistance to the toxin tabtoxin (isolated from the bacterial pathogen *Pseudomonas syringae* pv. *tabaci*) have been used to engineer resistance to the wildfire disease in tobacco.[107] Our current goal is to test the utility of *SOR1* and *CRG1* in engineering plants with resistance to *Cercospora* pathogens.

ACKNOWLEDGMENTS

This work has been supported by grants from the National Science Foundation and the United States Department of Agriculture National Research Initiative Competitive Grants Program.

REFERENCES

1. GREEN, M.J., HILL, A.O. 1984. Chemistry of dioxygen. Methods in Enzymology 105:3–22.
2. BAKER, C.J., ORLANDI, E.W. 1995. Active oxygen in plant pathogenesis. Annu. Rev. Phytopathol. 33:299–321.
3. DOKE, N., MIURA, Y., SANCHEZ, L.M., KAWAKITA, K. 1994. Involvement of superoxide in signal transduction: Responses to attack by pathogens, physical and chemical shocks, and UV irradiation. In: Causes of Photooxidative Stress and Amelioration of Defense Systems in Plants, (C.H. Foyer, P.M. Mullineaux, eds.), CRC Press, Inc, Boca Raton, FL, pp. 177–197.
4. MEHDY, M.C., SHARMA, Y.K, KANAGASABAPATHI, S., BAYS, N.W. 1996. The role of activated oxygen species in plant disease resistance. Physiol. Plant. 98:365–374.
5. SUTHERLAND, M.W. 1991. The generation of oxygen radicals during host plant responses to infection. Physiol. Molec. Plant Pathol. 39:79–93.
6. SCHRECK, R., BAERUERLE, P. 1991. A role for oxygen radicals as second messengers. Trends Cell Biol. 1:39–42.
7. DAVIS, V.M., STACK, M.E. 1991. Mutagenicity of stemphyltoxin III, a metabolite of *Alternaria alternata*. Appl. Environ. Microbiol. 57:180–182.
8. STACK, M.E., MAZZOLA, E.P., PAGE, S.W., POHLAND, A.E., HIGHET, R.S., TEMPESTA, M.S., CORELY, D.G. 1986. Mutagenic perylenequinone metabolites of *Alternaria alternata*: Altertoxins I, II, and III. J. Nat. Products 49:866–871.
9. DAUB, M.E. 1982. Cercosporin, a photosensitizing toxin from *Cercospora* species. Phytopathology 72:370–374.
10. ROBESON, D., STROBEL, G., MATUSUMOTO, G.K., FISHER, E.L., CHEN, M.H., CLARDY, J. 1984. Alteichin: An unusual phytotoxin from *Alternaria eichorniae*, a fungal pathogen of water hyacinth. Experientia, 40, 1248–1250.
11. HARTMAN, P.E., SUZUKI, C.K., STACK, M.E. 1989. Photodynamic production of superoxide *in vitro* by altertoxins in the presence of reducing agents. Appl. Environ. Microbiol. 55:7–14.
12. STIERLE, A.C., CARDELLINA II, J.H. 1989. Phytotoxins from *Alternaria alternata*, a pathogen of spotted knapweed. J. Nat. Products 52:42–47.
13. ARNONE, A., ASSANTE, G., DI MODUGNO, V., MERLINI, L, NASINI, G. 1988. Perylenequinones from cucumber seedlings infected with *Cladosporium cucumerinum*. Phytochemistry 6:1675–1678.
14. YOSHIHARA, T., SHIMANUKI, T., ARAKI, T., SAKAMURA, S. 1975. Phleichrome, A new phytotoxic compound produced by *Cladosporium phlei*. Agric. Biol. Chem. 39:1683–1684.
15. ROBESON, D.J., JALAL, M.A.F. 1992. Formation of entisophleichrome by *Cladosporium herbarum* isolated from sugar beet. Biosci. Biotech. Biochem. 56:949–952.
16. OVEREEM, J.C., SIJPESTEIJN, A.K. 1967. The formation of perylenequinones in etiolated cucumber seedlings infected with *Cladosporium cucumerinum*. Phytochemistry 6:99–105.
17. WEISS, U., MERLINI, L., NASINI, G. 1987. Naturally occurring perylenequinones. In: Progress in the Chemistry of Organic Natural Products, (W. Herz, H. Grisebach, G.W. Kirby, C.H. Tamm, eds.), Springer-Verlag, Vienna, Vol. 52, pp. 1–71.
18. WU, H., LAO, X.F., WANG, Q.W., LU, R.R. 1989. The shiraiachromes: Novel fungal perylenequinone pigments from *Shiraia bambusicola*. J. Nat. Prod. 5:948–5951.
19. DIWU, Z. 1995. Novel theraputic and diagnostic applications of hypocrellins and hypericins. Photochem. Photobiol. 61:529–539.
20. DIWU, Z., LOWN, J.W. 1990. Hypocrellins and their use in photosensitization. Photochem. Photobiol. 52:609–616.
21. GIROTTI, A.W. 1990. Photodynamic lipid peroxidation in biological systems. Photochem. Photobiol. 51:497–509.

22. FOOTE, C. S. 1976. Photosensitized oxidation and singlet oxygen: Consequences in biological systems In: Free Radicals in Biology, (W.A. Pryor, ed.), Academic Press, New York, Vol. II, pp. 85–133.
23. SPIKES, J.D. 1989. Photosensitization In: The Science of Photobiology, 2nd Ed., (K.C. Smith, ed.), Plenum Press, New York, pp. 79–110.
24. ITO, T. 1981. Dye binding and photodynamic action. Photochem. Photobiol. 33: 947–955.
25. HEITZ, J.R., DOWNUM, K.R. 1995. Light-Activated Pest Control. American Chemical Society, Washington DC. 279 p.
26. GEISE, A.C. 1980. Hypericism. Photochem. Photobiol. Rev. 5:229–255.
27. GIESE, A.C. 1981. The photobiology of *Blepharisma*. Photochem. Photobiol. Rev. 6:139–80.
28. SONG, P.S., POFF, K.L. 1989. Photomovement In: The Science of Photobiology, (K.C. Smith, ed.), Plenum Press, New York, pp. 305–346.
29. HUDSON, J.B., TOWERS, G.H.N. 1991. Therapeutic potential of plant photosensitizers. Pharmac. Ther. 49:181–222.
30. HUDSON, J.B., ZHOU, J., CHEN, J., HARRIS, L., YIP, L., TOWERS, G.H.N. 1994. Hypocrellin, from *Hypocrella bambusae*, is phototoxic to human immunodeficiency virus. Photochem. Photobiol. 60:253–255.
31. DEUTSCHMANN, F. 1953. Uber die "purple stain" krankheit der sojabohne und die farbstoffbildung ihres erregers (*Cercospora kiluchii* Mats et Tom). Phytopathol. Z. 20:297–310.
32. KUYAMA, S., TAMURA, T. 1957. Cercosporin. A pigment of *Cercospora kikuchii* Matsumoto et Tomoyasu. I. Cultivation of fungus, isolation and purification of pigment. J. Am. Chem. Soc. 79:5725–26.
33. KUYAMA, S., TAMURA, T. 1957. Cercosporin. A pigment of *Cercospora kikuchii* Matsumoto et Tomoyasu. II. Physical and chemical properties of cercosporin and its derivatives. J. Am. Chem. Soc. 79:5726–29.
34. DAUB, M.E. 1987. The fungal photosensitizer cerosporin and its role in plant disease. In: Light-Activated Pesticides, (J.R.Heitz, K.R. Downum, eds.), American Chemical Society, Washington, DC, pp. 271–80.
35. LOUSBERG, R.J.J.CH., WEISS, U., SALMINK, C.A., ARNONE, A., MERLINI, L., NASINI, G. 1971. The structures of cercosporin, a naturally occurring quinone. Chem. Commun. 71:1463–64.
36. YAMAZAKI, S., OGAWA, T. 1972. The chemistry and stereochemistry of cercosporin. Agric. Biol. Chem. 36:1707–18.
37. NASINI, G.L., MERLINI, L., ANDRETTI, G.D., BOCELLI, G., SGARABOTTO, P. 1982. Stereochemistry of cercosporin. Tetrahedron 38:2787–2796.
38. WOLFBEIS, O.S., FÜRLINGER, E. 1983. Absorption, fluroescence and fluorimetric detection limits of naturally occurring quinoid antibiotics and dyes. Mikrochim. Acta 3:385–98.
39. OKUBO, A., YAMAZAKI, S., FUWA, K. 1975. Biosynthesis of cercosporin. Agric. Biol. Chem. 39:1173–1175.
40. BALIS, C., PAYNE, M.G. 1971. Triglycerides and cercosporin from *Cercospora beticola*: Fungal growth and cercosporin production. Phytopathology 61:1477–1484.
41. EHRENSHAFT, M., UPCHURCH, R.G. 1991. Isolation of light-enhanced cDNA clones of *Cercospora kikuchii*. Appl. Environ. Microbiol. 57:2671–2676.
42. FAJOLA, A.O. 1978. Cercosporin, a phytotoxin from *Cercospora* species. Physiol. Plant Pathol. 13:157–164.
43. JENNS, A.E., DAUB, M.E., UPCHURCH, R.G. 1989. Regulation of cercosporin accumulation in culture by medium and temperature manipulation. Phytopathology 79:213–219.
44. CALLAHAN, T.M., EHRENSHAFT, M., UPCHURCH, R.G. 1993. Sequence and functional analysis of light enhanced clone cLE6 of *Cercospora kikuchii*. Phytopathology 83:1422 (Abstract).

45. UPCHURCH, R.G., WALKER, D.C., ROLLINS, J.A., EHRENSHAFT, M. ,DAUB, M.E. 1991. Mutants of *Cercospora kikuchii* altered in cercosporin synthesis and pathogenicity. Appl. Environ. Microbiol. 57:2940–2945.
46. EHRENSHAFT, M., UPCHURCH, R.G. 1993. Host protein(s) induces accumulation of the toxin cercosporin and mRNA in a phytopathogenic strain of *Cercospora kikuchii*. Physiol. Mol. Plant Path. 43:95–107.
47. YAMAZAKI, S., OKUBE, A., AKIYAMA, Y., FUWA, K. 1975. Cercosporin, a novel photodynamic pigment isolated from *Cercospora kikuchii*. Agric. Biol. Chem. 39:287–288.
48. DAUB, M.E., HANGARTER, R.P. 1983. Production of singlet oxygen and superoxide by the fungal toxin, cercosporin. Plant Physiol. 73:855–857.
49. HARTMAN, P.E., DIXON, W.J., DAHL, T.A., DAUB, M.E. 1988. Multiple modes of photodynamic action by cercosporin. Photochem. Photobiol. 47:699–703.
50. DOBROWOLSKI, D.C., FOOTE, C.S. 1983. Chemistry of singlet oxygen. 46. Quantum yield of cercosporin-sensitized singlet oxygen formation. Angewante Chemie 95:729–30.
51. LEISMAN, G.B., DAUB, M.E. 1992. Singlet oxygen yields, optical properties, and phototoxicity of reduced derivatives of the photosensitizer cercosporin. Photochem. Photobiol. 55:373–379.
52. HUGHES, K., NEGROTTO, D., DAUB, M., MEEUSEN, R. 1984. Free radical stress response in paraquat-sensitive and resistant tobacco plants. Environ. Exp. Bot. 24:151–157.
53. HUGHES, K.W., HOLTON, R.W. 1981. Levels of superoxide dismutase, peroxidase, and catalase in a tobacco cell line selected for herbicide resistance. In Vitro 17:211 (Abstract).
54. FURUSAWA, I., TANAKA, K., THANUTONG, P., MIZUGUCHI, A., YAZAKI, M., ASADA, K. 1984. Paraquat resistant tobacco calluses with enhanced superoxide dismutase activity. Plant Cell Physiol. 25:1247–1254.
55. DAUB, M.E., BRIGGS, S.P. 1983. Changes in tobacco cell membrane composition and structure caused by the fungal toxin, cercosporin. Plant Physiol. 71:763–766.
56. DAUB, M.E. 1982. Peroxidation of tobacco membrane lipids by the photosensitizing toxin, cercosporin. Plant Physiol. 69:1361–1364.
57. MACRI, F., VIANELLO, A. 1979. Photodynamic activity of cerosporin on plant tissues. Plant Cell Environ. 2:267–271.
58. CAVALLINI, A., BINDOLI, A., MACRI, F., VIANELLO, A. 1979. Lipid peroxidation induced by cercosporin as a possible determinant of its toxicity. Chem. Biol. Interact. 28:139–146.
59. FUKUZAWA, K., CHIDA, H., TOKUMURA, A., TSUKATANI, H. 1981. Antioxidant effect of α-tocopherol incorporation into lecithin liposomes on ascorbic acid-Fe^{+2}-induced lipid peroxidation. Arch. Biochem. Biophys. 206:173–180.
60. PAULS, K.P., THOMPSON, J.E. 1981. Effects of *in vitro* treatment with ozone on the physical and chemical properties of membranes. Physiol. Plant 53:255–262.
61. STEINKAMP, M.P., MARTIN, S.S., HOEFERT, L.L., RUPPEL, E.G. 1979. Ultrastructure of lesions produced by *Cercospora beticola* in leaves of *Beta vulgaris*. Physiol. Plant Pathol. 15:13–16.
62. STEINKAMP, M.P., MARTIN, S.S., HOEFERT, L.L., RUPPEL, F.G. 1981. Ultrastructure of lesions produced in leaves of *Beta vulgaris* by cercosporin, a toxin from *Cercospora beticola*. Phytopathology 71:1272–1281.
63. ECHANDI, E. 1959. La chasparria de los cafetos causada por el hongo *Cercospora coffeicola* Berk and Cooke. Turrialba 9:54–67.
64. CALPOUZOS, L., STALKNECHT, G.F. 1967. Symptoms of Cercospora leaf spot of sugar beets influenced by light intensity. Phytopathology 57:799–800.
65. CALPOUZOS, L. 1966. Action of oil in the control of plant disease. Annu. Rev. Phytopathol. 4:369–390.
66. THOROLD, C.A. 1940. Cultivation of bananas under shade for the control of leaf spot disease. Trop. Agric. Trin. 17:213–214.

67. ROTEM, J., WENDT, U., DRANZ, J. 1988. The effect of sunlight on symptom expression of *Alternaria alternata* on cotton. Plant Pathology 37:12–15.
68. TOWERS, G.H.N. 1984. Interactions of light with phytochemicals in some natural and novel systems. Can. J. Bot. 62:2900–2911.
69. BERENBAUM, M.R. 1987. Charge of the light-activated brigade: Photoxicity as a defense against insects In: Light-Activated Pesticides, (J.R. Heitz, K.R. Downum, eds.), American Chemical Society, Washington, DC, pp. 206–216.
70. BELLUS, D. 1979. Physical quenchers of singlet molecular oxygen. Adv. Photochem. 11:105–205.
71. FOOTE, C.S., DENNY, R.W., WEAVER, L., CHANG, Y., PETERS, J. 1970. Quenching of singlet oxygen. Ann. N.Y. Acad. Sci. 171:139–148.
72. KRINSKY, N.I. 1979. Carotenoid protection against oxidation. Pure Appl. Chem. 51:649–660.
73. TRUSCOTT, T.G. 1990. New trends in photobiology: The photophysics and photochemistry of the carotenoids. J. Photochem. Photobiol. B. 6:359–371.
74. YOUNG, A.J. 1991. The photoprotective role of carotenoids in higher plants. Physiol. Plant. 83:702–708.
75. HARTMAN, P.E. 1990. Ergothioneine as an antioxidant. Meth. Enzymol. 186:310–318.
76. LINDIG, B.A., ROGERS, M.A.J. 1981. Rate parameters for the quenching of singlet oxygen by water-soluble and lipid-soluble substrates in aqueous and micellar systems. Photochem. Photobiol. 33:627–634.
77. ROUGEE, M., BENSASSON, R.V., LAND, E.J., PARIENTE, R. 1988. Deactivation of singlet molecular oxygen by thiols and related compounds, possible protectors against skin photosensitivity. Photochem. Photobiol. 47:485–489.
78. BORS, W., SARAN, M., TAIT, D. 1984. Oxygen Radicals in Chemistry and Biology. Walter de Gruyter, Berlin.
79. HARPER, D.B, HARVEY, B.M.R. 1978. Mechanism of paraquat tolerance in perennial ryegrass. Plant Cell Environ.1:211–215.
80. LEE, E.H., BENNETT, J.W. 1982. Superoxide dismutase. A possible enzyme against ozone injury in snapbeans (*Phaseolus vulgaris* L.). Plant Physiol. 69:1444–1449.
81. BATCHVAROVA, R.B., REDDY, V.S., BENNETT, J. 1992. Cellular resistance in rice to cercosporin, a toxin of *Cercospora*. Phytopathology 82:642–646.
82. DAUB, M.E. 1987. Resistance of fungi to the photosensitizing toxin cercosporin. Phytopathology 77:1515–1520.
83. TAMAOKI, T., NAKANO, H. 1990. Potent and specific inhibitors of protein kinase C of microbial origin. Bio/Technology 8:732–735.
84. ROLLINS, J.A., EHRENSHAFT, M., UPCHURCH, R.G. 1993. Effects of light and altered-cercosporin phenotypes on gene expression in *Cercospora kikuchii*. Can. J. Microbiol. 39:118–124.
85. DAUB, M.E., LEISMAN, G.B., CLARK, R.A, BOWDEN, E.F. 1992. Reductive detoxification as a mechanism of fungal resistance to singlet-oxygen-generating photosensitizers. Proc. Natl. Acad. Sci. USA 89:9588–9592.
86. JENNS, A.E., DAUB, M.E. 1995. Characterization of mutants of *Cercospora nicotianae* sensitive to the toxin cercosporin. Phytopathology 85:906–912.
87. JENNS, A.E., SCOTT, D.L., BOWDEN, E.F., DAUB, M.E. 1995. Isolation of mutants of the fungus *Cercospora nicotianae* altered in their response to singlet-oxygen-generating photosensitizers. Photochem. Photobiol. 61:488–493.
88. SOLLOD, C.C., JENNS, A.E., DAUB, M.E. 1992. Cell surface redox potential as a mechanism of defense against photosensitizers in fungi. Appl. Environ. Microbiol. 58:444–449.
89. DAUB, M.E., PAYNE, G.A. 1989. The role of carotenoids in resistance of fungi to cercosporin. Phytopathology 79:180–185.

90. EHRENSHAFT, M., DAUB, M.E. 1994. Isolation, sequence and characterization of the *Cercospora nicotianae* phytoene dehydrogenase gene. Appl. Environ. Microbiol. 60:2766–2771.
91. EHRENSHAFT, M., JENNS, A.E., DAUB, M.E. 1995. Targeted gene disruption of carotenoid biosynthesis in *Cercospora nicotianae* reveals no role for carotenoids in photosensitizer resistance. Molec. Plant Microbe Interact. 8:569–575.
92. RUDDAT, M., GARBER, E.D. 1983. Biochemistry, physiology and genetics of carotenogenesis in fungi In: Secondary Metabolism and Differentiation in Fungi, (J.W. Bennett, A. Ciegler, eds.), Marcel Dekker, New York, pp. 95–151.
93. GWINN, K.D., DAUB, M.E. 1988. Regenerating protoplasts from *Cercospora* and *Neurospora* differ in their response to cercosporin. Phytopathology 78: 414–418.
94. GWINN, K.D., DAUB, M.E., HUANG, P. 1989. Cytological comparison of early stages of wall regeneration of *Cercospora nicotianae* and *Neurospora crassa* protoplasts. Can. J. Bot. 67: 1938–1943.
95. CLARK, R.A., STEPHENS, T.R., BOWDEN, E.F., DAUB, M.E. 1995. Electrochemical reduction of the phytotoxin cercosporin. J. Electroanal. Chem. 389:205–208.
96. EHRENSHAFT, M., JENNS, A.E., CHUNG, K.R., DAUB, M.E. 1998. *SOR1*, a gene required for photosensitizer and singlet oxygen resistance in the fungus *Cercospora nicotianae* is highly conserved in divergent organisms. Molecular Cell (In press).
97. STRAUBINGER, B., STRAUBINGER, E., WIRSEL, S., TURGEON, G., YODER, O.C. 1992. Versatile fungal transformation vectors carrying the selectable *bar* gene of *Streptomyces hygroscopicus*. Fungal Genetics Newsletter 39:82–83.
98. BRAUN, E.L., FUGE, E.K., PADILLA P.A., WERNER-WASHBURNE, M. 1996. A stationary-phase gene in *Saccharomyces cerevisiae* is a member of a novel, highly conserved gene family. J. Bacteriol. 178:6865–6872.
99. SIVASUBRAMANIAM, S., VANNIASHINGHAM, V.M., TAN, C.T., CHUA, N.H. 1995. Characterization of HEVER, a novel stress-induced gene from *Hevea brasiliensis*. Plant Mol. Biol. 29:173–178.
100. TSUGE, T., KOBAYASHI, H., NISHIMURA, S. 1989. Organization of ribosomal RNA genes in *Alternaria alternata* Japanese pear pathotype, a host-selective AK-producing fungus. Curr. Genet. 16:267–272.
101. JOHAL, G.S., GRAY, J., GRUIS, D., BRIGGS, S.P. 1995. Convergent insights into mechanisms determining disease and resistance responses in plant-fungal interactions. Can. J. Bot. 73:S468–S474.
102. SANFORD, J.C, JOHNSTON, S.A. 1985. The concept of parasite-derived resistance—deriving resistance genes from the parasites' own genome. J. Theoretical Biol. 113:395–495.
103. WILSON, T.M.A. 1993. Strategies to protect crop plants against viruses: Pathogen-derived resistance blossoms. Proc. Natl. Acad. Sci. 90:3134–3141.
104. NAKAYA, K., OMATA, K., OKAHASHI, I., NAKAMURA, Y., KOLKENBROCK, H., ULBRICH, N. 1990. Amino acid sequence and disulfide bridges of an antifungal protein isolated from *Aspergillus giganteus*. Eur. J. Biochem. 193:31–38.
105. CORNELISSEN, B.J.C. MELCHERS, L.S. 1993. Strategies for control of fungal diseases with transgenic plants. Plant Physiol. 101:709–712.
106. YONEYAMA, K., ANZAI, H. 1993. Transgenic plants resistant to diseases by the detoxification of toxins. In: Biotechnology in Plant Disease Control, (I. Chet, ed.), Wiley-Liss, Inc. New York, pp. 115–137.
107. ANZAI, J., YONEYAMA, K. YAMAGUCHI, I. 1989. Transgenic tobacco resistant to bacterial disease by the detoxification of a pathogenic toxin. Mol. Gen. Genet. 219:492–494.

Chapter Four

TREE–FUNGUS INTERACTIONS IN ECTOMYCORRHIZAL SYMBIOSIS

Roger T. Koide, Laura Suomi, and Robert Berghage

Department of Horticulture
Pennsylvania State University
University Park, Pennsylvania 16802

Introduction to Mycorrhizal Symbioses 57
The Potential Roles of Phenolic Compounds in Ectomycorrhizal
    Symbiosis 59
Fungal Succession and Phenolic Compounds 60
Regulation of Mycorrhizal Infection by Phenolic Compounds 61
Host Specificity and Phenolic Compounds 62
Conclusion 67

INTRODUCTION TO MYCORRHIZAL SYMBIOSES

The mycorrhiza is probably the most common symbiosis in which higher plants engage. More than 80% of higher terrestrial plants may be mycorrhizal.[1] This may make the mycorrhiza the most common symbiosis in all of nature! Mycorrhizal fungi, those fungi that participate in the mycorrhizal symbiosis, are important and sometimes essential to their hosts, primarily because they can provide an avenue for nutrient (mostly N and P) acquisition in addition to the roots.

In an examination of broad latitudinal or altitudinal gradients, one may discern corresponding gradients in soil properties and in the kinds of mycorrhizas present.[2] At low latitudes, mineral soils may prevail. The limiting plant nutrient is often P, and the most abundant mycorrhiza is the so-called vesicular-arbuscular mycorrhiza. At higher latitudes, the rate of litter decomposition may decline and organic matter in the soil may increase. Nitrogen may become increasingly limiting while ectomycorrhizas become dominant. At still higher latitudes, soil

Phytochemical Signals and Plant–Microbe Interactions, edited by Romeo *et al.*
Plenum Press, New York, 1998.

organic matter content increases further, soil N occurs almost entirely as organic compounds, and ericoid mycorrhizas, those involving members of the Ericaceae, may be the most abundant.

Vesicular-arbuscular (VA) mycorrhizal fungi are useful in scavenging inorganic phosphate which, in soil, is a poorly diffusing plant nutrient. The surface area of the hyphae of these mycorrhizal fungi adds to that provided by the roots and so increases the absorption capacity of the root system. These fungi are largely incapable of hydrolyzing organic compounds containing N or P, compounds that are found only in small quantities in VA-dominated ecosystems due to the inherently rapid rates at which they are mineralized. Only some 150 or so species of VA mycorrhizal fungi have been described despite that fact that most mycorrhizal plants are associated with VA mycorrhizal fungi.[3]

At the other end of the gradient, and in contrast to vesicular-arbuscular mycorrhizal fungi, ericoid mycorrhizal fungi are capable of hydrolyzing some complex organic materials including protein,[4] chitin,[5] and tannins,[6] substances that are found in abundance in ecosystems dominated by ericoid mycorrhizas. Through the production of enzymes, such as proteases, chitinases, and polyphenol oxidases,[4-6] ericoid mycorrhizal fungi make available to their hosts sources of N that would otherwise be unavailable. As for the vesicular-arbuscular mycorrhizal fungi, there are relatively few described species of ericoid mycorrhizal fungi.[3]

We devote most of our discussion to the middle of the gradient, the inhabitants of which are primarily ectomycorrhizal fungi. In contrast to the VA mycorrhizal symbiosis, there are relatively few host species that are ectomycorrhizal. Most of these are members of the Pinaceae, Fagaceae, Myrtaceae, Dipterocarpaceae, and the subfamily Caesalpinioideae of the Fabaceae.[3] Despite this relatively limited set of hosts, there are literally thousands of fungal species that form ectomycorrhizas, most of which are basidiomycetes.[7] That there is such great taxonomic diversity among ectomycorrhizal fungi should not be surprising. The capacity to form ectomycorrhizas has arisen in several different fungal lineages.[8] Moreover, as occupants of the middle of the gradient, ectomycorrhizal fungi occur in a broad range of soils, from those that are largely mineral to those with well-developed layers (horizons) of partially decomposed organic materials.[2]

In addition to great taxonomic diversity, there exists a great range of functional diversity among ectomycorrhizal fungi. Species and even isolates of the same species may differ in such ecologically relevant characters as protease production, preference for N source, rhizomorph formation, pH optimum, temperature optimum, and tolerance to drought and freezing.[3,9-16] Different mycorrhizal fungi may thus have different effects on plant vigor because the physiological and morphological capacities of the fungi in large measure determine the capacities of their hosts to acquire resources such as N, P, and water from the soil. In this contribution, we discuss a possible way for plants to control the range of ectomycorrhizal fungal species infecting their roots.

THE POTENTIAL ROLES OF PHENOLIC COMPOUNDS IN ECTOMYCORRHIZAL SYMBIOSIS

As will be discussed below, chemicals produced by host plants may play roles in the regulation of mycorrhizal infection, in the determination of host specificity, and in the orderly succession of ectomycorrhizal fungal species. Host chemicals can interact with mycorrhizal fungi in at least two ways. First, compounds produced by roots can exert control at the point of attempted infection. Second, because mycorrhizal fungi make intimate contact with soil organic matter, compounds that are natural components of forest litters (contributors include primarily roots, leaves, and bark) may affect mycorrhizal fungi even before they come in contact with the roots.

Phenolic compounds are among the most frequently utilized chemicals of plants to communicate with microorganisms. They are involved in many chemically-mediated interactions between plants and root-infecting microorganisms, including bacteria and fungi involved in both mutualistic and parasitic associations.[17,18] For example, acetosyringone and related compounds, produced upon wounding of plant tissues, induce the expression of virulence genes of *Agrobacterium tumefaciens*.[19,20] Flavonoids may be involved in the vesicular-arbuscular mycorrhizal symbiosis. Some, such as naringenin[21] and quercetin or myricetin,[22] may increase hyphal growth. Host-specific flavonoids also induce nodulation genes in *Rhizobium spp.*[23,24] Toluate may serve as a chemoattractant for *Bradyrhizobium*.[25]

It is not surprising that phenolic compounds are utilized as means of communication between plants and microorganisms. They are capable of conveying a great deal of information as a consequence of the nearly infinite variety of structures made possible by variation in folding, the number and placement of hydroxyl groups and sugars, and the level of oxidation.[26,27]

In addition to serving as specific signal molecules, a general effect of phenolics stems from their ready ability to form hydrogen bonds with carbohydrates and proteins. Hydrogen bonds are formed between hydroxyl groups of the phenolic rings and, for example, the carboxyl groups of amino acids. The inhibition of many microorganisms by phenolic compounds results from their binding to vital proteins such as enzymes.[28] Hydrogen bonding may be particularly relevant to ectomycorrhizal fungi found in association with soil organic matter where they hydrolyze proteins into amino acids and liberate phosphate from organic phosphorus sources via the activities of proteases and phosphatases.[29] For example, some ectomycorrhizal fungi have little capacity to degrade tannic acid, a common phenolic compound of the general class of hydrolyzable tannins.[6] As a result, only some ectomycorrhizal fungi can mobilize nitrogen from proteins that have been precipitated by tannic acid.[30] In addition to the indirect effects of phenolic compounds on mycorrhizal fungi via their effects on N mobilization, phenolic molecules may directly affect the activities of ectomycorrhizal fungi and, in so doing, influence the rapidity and extent of infection, determine which fungi successfully infect a root, and alter the course of fungal succession.

FUNGAL SUCCESSION AND PHENOLIC COMPOUNDS

As a stand of trees ages, the composition of the mycorrhizal fungal community can change.[31] For example, Last et al.[32] showed that *Thelephora* occurred exclusively in young pine stands, *Scleroderma* exclusively in the older stands, and *Laccaria* and *Rhizopogon* in stands of all ages. In other research in pine stands, Visser[33] showed that clearly there are early successional and late successional mycorrhizal fungal species. Some of this succession may be attributable to different rates at which species of fungi propagate themselves[32,34] or to variation in fungal preferences for N source (*e.g.* NH_4 vs. protein).[35] Succession also may be driven by the change in phytochemical composition associated with the age of the stand.

Can phenolic compounds control succession of ectomycorrhizal fungi? If we consider the process of organic matter accretion through succession, one moves from a largely mineral soil to one whose surface is dominated by organic material. The presence of well-developed organic horizons in ecosystems, such as those dominated by ectomycorrhizal and ericoid mycorrhizal plants, is caused by slow rates of decomposition. Slow rates of decomposition in pine forests, where we primarily work, may be caused by high C:N ratios, high lignin content, and low pH. In addition, phenolic compounds, such as condensed tannins and tannin precursors, contribute to the resistance of the material to decomposition by binding to substrates and hydrolytic enzymes, slowing the rate of hydrolysis significantly.[30] Phenolic compounds can be found in high concentrations in litter and in the organic soil horizons, precisely where the abundance of mycorrhizal roots is greatest (Fig. 1).

As a forest ages (as succession proceeds), an increasingly complex set of organic soil horizons may develop. As litter ages, both the concentration and the

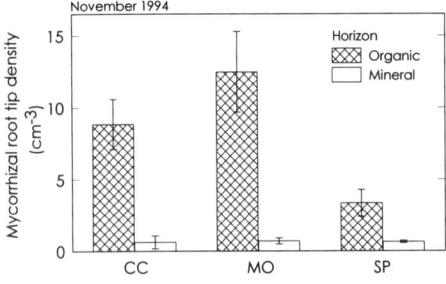

Figure 1. Mean (± se) density of ectomycorrhizal root tips of red pine (*Pinus resinosa*) at three different 50 year-old stands in central Pennsylvania (Clear Creek, CC; Moshanon, MO; Sproul, SP). There were significantly more tips located in the organic horizon than in the underlying mineral horizon in all three stands.

quality of phenolics ("tannins") changes[36] through leaching, decomposition, chemical immobilization, polymerization, and oxidation.[18] The surface of the organic horizon contains the newest litter which differs in physical and chemical composition from deeper layers. Organic matter accretion and change with time in the quality and quantity of phenolics contained therein would be expected to contribute to the succession of ectomycorrhizal fungi that is commonly seen.[18] This hypothesis, however, remains virtually untested by direct means. Neither has there been much indirect testing of the hypothesis, but Baar and associates[37,38] demonstrated that the diversity of the mycorrhizal fungal community can be altered by the removal of the organic soil horizon.

It is clear that water-soluble substances leached from litter can strongly influence mycorrhizal fungi.[37,39–42] For example, Rose et al.[40] showed that the relative proportion of two major morphotypes of mycorrhizas on Douglas fir was strongly affected by the type of litter applied. Persidsky et al.[41] showed that water soluble extracts of prairie soils inhibit the metabolisms of *Boletus spp.* mycorrhizas of *Pinus radiata*. Pellissier[42] demonstrated that simple phenols such as catechol, protocatechuic acid, p-hydroxyacetophenone, and p-hydroxybenzoic acid produced by *Vaccinium myrtillus* and *Picea abies* inhibited respiration of mycorrhizal fungi associated with *Picea abies* including *Cenococcum graniforme* and *Laccaria laccata*. Clearly, both roots and plant litter contain substances that can influence the physiology of soil-inhabiting fungi, including ectomycorrhizal fungi.

In *in vitro* cultures containing sugar, saprophytic fungi are almost always faster growing than mycorrhizal fungi. In the field, however, mycorrhizal fungi are able to exist and even thrive, presumably because they are readily supplied with simple carbohydrates by their hosts and thus do not compete with saprophytes for a source of carbon. In a carbon-limited habitat, such as in many temperate forests, one might predict that mycorrhizal fungi would outcompete saprophytic fungi. That this does not occur may be due, in part, to the presence of phenolic compounds in the soil. Aqueous extracts of aspen leaves containing benzoic acid and catechol, for example, are more inhibitory to mycorrhizal fungi than to litter-decomposing fungi.[39] In some cases, phenolic compounds actually stimulate the growth of saprophytes.[39,43]

REGULATION OF MYCORRHIZAL INFECTION BY PHENOLIC COMPOUNDS

Certain features of roots suggest a role for phenolic compounds in the regulation of both the species of fungi that can infect, and the degree to which they infect the root. Ling-Lee et al.[44] observed darkly staining cells in cross sections of *Eucalyptus* roots. These contained high concentrations of phenolic molecules. Similar structures occur in, for example, pine roots.[45,46] Many phenolic compounds are inhibitory or toxic to ectomycorrhizal fungi.[39,42] Phenolic compounds released from a layer of tannin-containing root cells may form a barrier, potentially limiting mycorrhizal fungal development.[47-49] This is why some believe that these defense molecules may be used to control the development of mycorrhizal fungi (personal communication, M. Weiss), preventing them from becoming parasitic or pathogenic.[44] If different mycorrhizal fungi have different tolerances for phenolic compounds, their presence would influence the structure of the mycorrhizal fungal community.

In some cases, mycorrhizal infection appears to reduce the concentration of several potentially toxic phenolic compounds including p-hydroxybenzoyl-glucose, p-hydroxybenzoic acid glucoside, picein, catechin, and epicatechin.[50] This suggests that while phenolic compounds inhibit mycorrhizal fungi, the symbiosis has evolved to the point that recognition of compatible fungi by the host results in an active reduction in this chemical defense. Alternatively, the result may simply have been a consequence of comparing old mycorrhizal roots to younger nonmycorrhizal roots. Indeed, some phenolic compounds, such as catechin and epicatechin, actually increase in concentration in the root as a consequence of mycorrhizal infection before a decline in concentration is observed (M. Weiss, personal communication).

HOST SPECIFICITY AND PHENOLIC COMPOUNDS

Can a host plant control which fungal species infect its roots? Possibly. For example, some ectomycorrhizal fungi have broad host ranges and tolerate a wide range of environmental conditions. This appears to be true for species in the genus *Amanita*.[7] Species of the genus *Suillus*, however, may have specific relationships with their hosts.[7] Host specificity implies that recognition of some kind occurs between host and fungus. This recognition may be mediated by phytochemicals.

Is there any evidence that phenolic compounds are involved in host specificity? In order for there to be specificity, host species must differ from one another in some relevant way. We determined the phenolic profiles of a number of different leaf litters. The results suggest that there is a high degree of variability among plant species in the kinds of phenolic compounds produced, probably sufficient variability to account for some ectomycorrhizal fungus/host specificity (Figs. 2, 3). For example, in a comparison of four species within the Pinaceae, we found both qualitative and quantitative variation in phenolic compounds (Fig. 2). Among species of the genus *Pinus,* there also was significant variation (Fig. 3). Such variability might account for differences in the kinds of fungi normally found associated with particular hosts.

We have pointed out that the genus *Suillus* contains species with narrower host ranges than the genus *Amanita*. In particular, *Suillus intermedius* (Smith & Thiers) Smith & Thiers is a species that has a narrowly defined host range, being

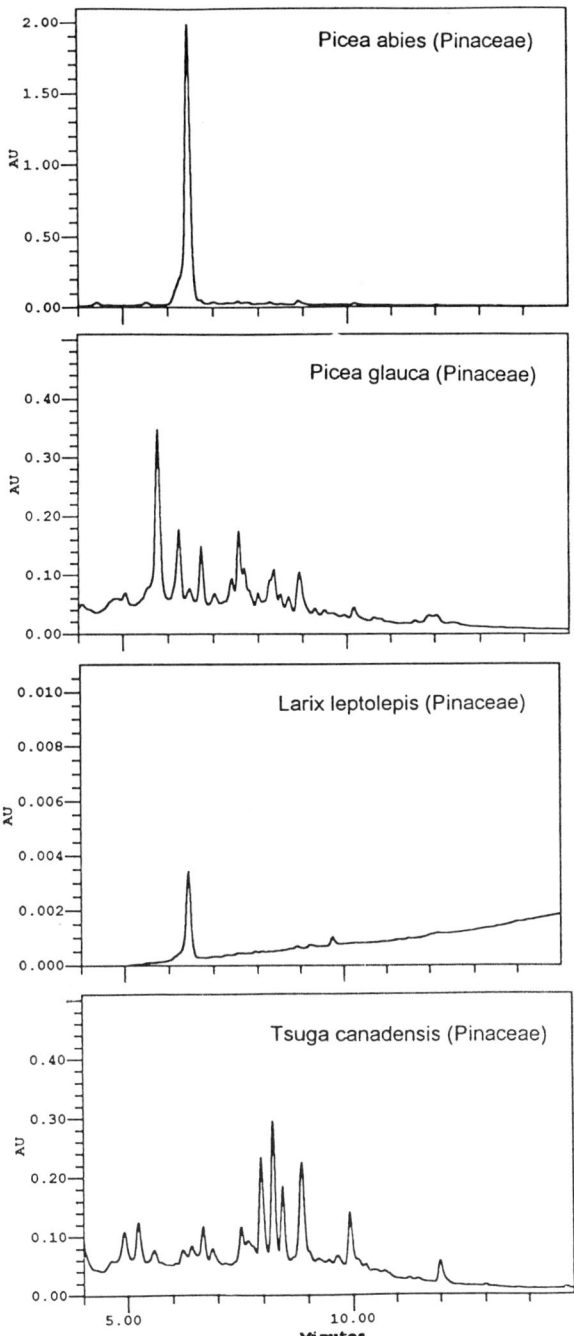

Figure 2. HPLC chromatograms from water extracts of leaves of four species in the Pinaceae. Absorbance at 280 nm. Method: Waters HPLC equipped with diode array detector monitoring 220–400 nm, water:acetonitrile gradient elution from 90:10 to 40:60, 1.5 mL min^{-1}, 30 min., Waters Nova-Pak C-18 column (3.9 × 50 mm).

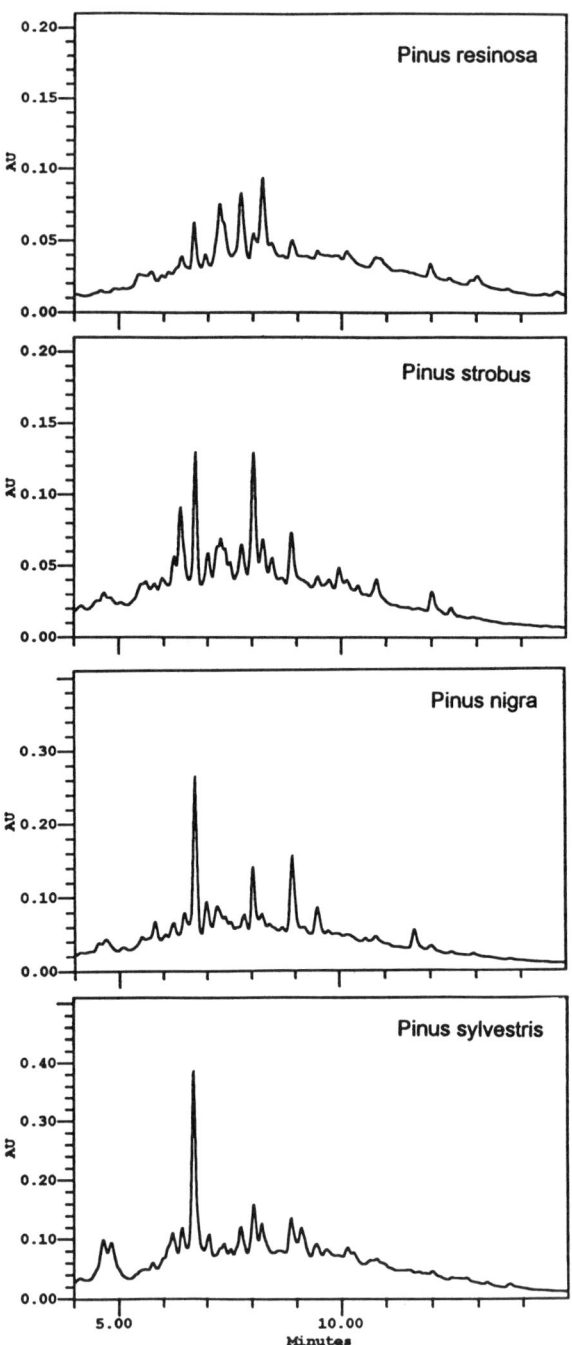

Figure 3. HPLC chromatograms from water extracts of leaves of four species in the genus *Pinus*. Absorbance at 280 nm. Method: see Figure 2 caption.

frequently associated with red pine (*Pinus resinosa* Ait.).[51] We have shown that red pine needles added to the growth medium stimulate the growth of *Suillus intermedius*. In contrast, the growth of *Amanita rubescens* Pers., a species with a broader host range, is negatively affected by the presence of pine needles.

Red pine needles contain both soluble and volatile components, either of which could be important in the effects on these two ectomycorrhizal fungi. The most abundant volatile components of red pine needles are α- and β-pinene. To determine the effect of these on fungal growth, we grew fungal cultures on split plates with gel growth medium on one side of the plates. On the other side, equal quantities of α- and β-pinene were slowly and steadily released through several layers of adhesive tape to control the pinene concentration at a level similar to that provided by fresh pine needles. *Amanita rubescens* was inhibited by the pinenes. *Suillus intermedius* also was inhibited by the same concentration of pinenes, but to a lesser extent.

Soluble substances, essentially exclusive of volatile components, were tested using water extracts of pine needles. Soluble components of pine needles had no significant effect on the growth of *Amanita rubescens* but stimulated the growth of *Suillus intermedius* (Fig. 4). We analyzed water-soluble extracts of red pine needles using HPLC. This revealed many substances that absorb strongly at 280 nm and that can be removed by passing through polyvinylpolypyrrolidone (PVPP). We have tentatively identified some of them as catechin, epicatechin, catechin gallate, epicatechin gallate, and taxifolin (Fig. 5), based on comparisons to reference compounds (using retention times and spectral profiles).

When we tested some of these individual phenolic components (5.0 mg L^{-1}), we found that catechin and epicatechin gallate were both slightly stimulatory, as were the pine needles and the pine needle water-extract (Fig. 6). This suggests that

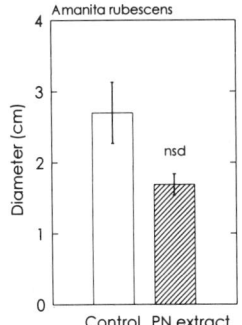
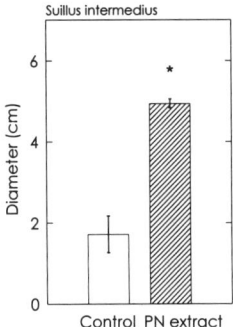

Figure 4. Mean (± se) colony diameter of two ectomycorrhizal fungal species growing on control medium and on medium to which was added a water-soluble extract of *Pinus resinosa* needles. The concentration of the extract was the equivalent of 0.5% ground pine needles. The asterisk indicates a significant ($p < 0.05$) difference.

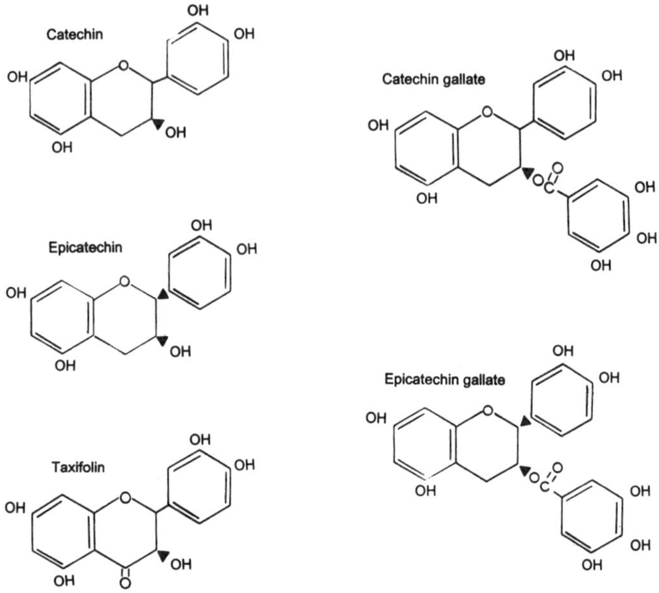

Figure 5. Structures of phenolic compounds extracted from needles of *Pinus resinosa*.

phenolic molecules may be responsible for the effects of pine needles on mycorrhizal fungal growth seen in some of our earlier experiments. The increase in catechin and epicatechin concentrations in recently infected roots (M. Weiss, personal communication) suggests a possible mechanism for promoting the growth of some ectomycorrhizal fungi over others.

In a broader survey of ectomycorrhizal fungal species collected from a red pine stand, we found that different species react to red pine needles differently (Fig. 7). Growth of *Leccinum* sp., *Amanita rubescens*, and *Pisolithus tinctorius* were strongly inhibited by pine needles. *Lactarius affinis, Suillus intermedius,* and

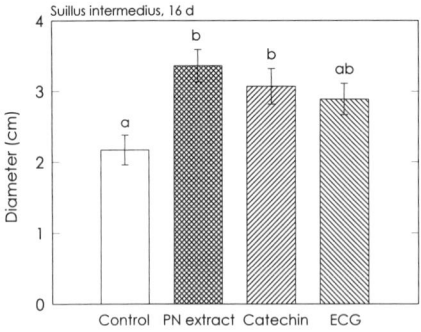

Figure 6. Mean (± se) colony diameter of *Suillus intermedius* growing on control medium, and on media to which was added either a water-soluble extract of *Pinus resinosa* needles (equivalent to 0.5% ground pine needles), catechin (5.0 $\mu g\ mL^{-1}$) or epicatechin gallate (5.0 $\mu g\ mL^{-1}$). Different letters indicate a significant ($p < 0.05$) difference between means.

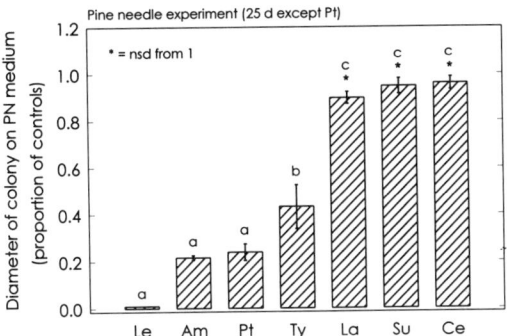

Figure 7. The proportion of fungal colony diameter on ground pine needle medium (0.5%) relative to that on control medium. Different letters indicate a significant ($p < 0.05$) difference between means. Species are *Leccinum sp.* (Le), *Amanita rubescens* (Am), *Pisolithus tinctorius* (Pt), *Tylopilus felleus* (Ty), *Lactarius affinis* (La), *Suillus intermedius* (Su) *and Cenococcum geophilum* (Ce).

Cenococcum geophilum were not significantly affected. Thus, variability occurs among ectomycorrhizal fungal species occupying red pine stands in their responses to red pine needles. If our measure of response (diameter growth) is ecologically relevant, then the presence of the litter could influence the composition of the fungal community. As mentioned, this is not a trivial fact because such a tremendous degree of functional variability among mycorrhizal fungi exists.

CONCLUSION

Our studies are preliminary. However, taken in conjunction with those of others, we hypothesize that phenolic compounds released by host roots and litter are important in communicating with ectomycorrhizal fungi and play a role in structuring the ectomycorrhizal fungal community.

ACKNOWLEDGMENTS

We thank the A. W. Mellon Foundation and the School of Forest Resources of Pennsylvania State University for providing funding for this project.

REFERENCES

1. SMITH, S.E., GIANINAZZI-PEARSON, V. 1988. Physiological interactions between symbionts in vesicular-arbuscular mycorrhizal plants. Ann. Rev. Plant Physiol. Plant Mol. Biol. 39: 221–244.

2. READ, D.J. 1984. The structure and function of the vegetative mycelium of mycorrhizal roots. In: The Ecology and Physiology of the Fungal Mycelium, (D.H. Jennings, A.D.M. Rayner, eds.), Cambridge University Press, Cambridge, pp. 215–240.
3. SMITH, S.E., READ, D.J. 1997. Mycorrhizal Symbiosis, Second Edition, Academic Press, San Diego, 605 pp.
4. READ, D.J., LEAKE, J.R., LANGDALE, A.R. 1989. The nitrogen nutrition of mycorrhizal fungi and their host plants. In: Nitrogen, Phosphorus and Sulphur Utilization by Fungi. (L. Boddy, R. Marchant, D.J. Read, eds.), Cambridge University Press, Cambridge, pp. 181–204.
5. LEAKE, J.R., READE, D.J. 1990. Chitin as a nitrogen source for mycorrhizal fungi. Mycol. Res. 94: 993–1008.
6. BENDING, G.D., READ, D.J. 1996a. Effects of the soluble polyphenol tannic acid on the activities of ericoid and ectomycorrhizal fungi. Soil. Biol. Biochem. 28: 1595–1602.
7. MOLINA, R., MASSICOTTE, H.B., TRAPPE J.M. 1992. Ecological role of specificity phenomena in ectomycorrhizal plant communities: Potentials for interplant linkages and guild development. In: Mycorrhizas in Ecosystems, (D.J. Read, D.H. Lewis, A.H. Fitter, I.J. Alexander, eds.), C.A.B. International, Oxon, pp. 106–112.
8. BRUNS, T.D. 1995. Thoughts on the processes that maintain local species diversity of ectomycorrhizal fungi. Plant and Soil 170: 63–73.
9. HACSKAYLO, E., PALMER, J.G., VOZZO, J.A. 1965. Effect of temperature on growth and respiration of ectotrophic mycorrhizal fungi. Mycologia 57: 748–756.
10. FRANCE, R.C., CLINE, M.L., REID, C.P.P. 1979. Recovery of ectomycorrhizal fungi after exposure to subfreezing temperatures. Can. J. Bot. 57: 1845–1848.
11. RAMSTEDT, M., SODERHALL, K. 1983. Protease, phenoloxidase and pectinase activities in mycorrhizal fungi. Trans. Brit. Mycol. Soc. 81: 157–161.
12. ABUZINADAH, R.A., READ, D.J. 1986. The role of proteins in the nitrogen nutrition of ectomycorrhizal plants. I. Utilization of peptides and proteins by ectomycorrhizal fungi. New Phytol. 103: 481–493.
13. CLINE, M.L., FRANCE, R.C., REID, C.P.P. 1987. Intraspecific and interspecific growth variation of ectomycorrhizal fungi at different temperatures. Can. J. Bot. 65: 869–875.
14. COLEMAN, M.D., BLEDSOE, C.S., LOPUSHINSKY, W. 1989. Pure culture response of ectomycorrhizal fungi to imposed water stress. Can. J. Bot. 67: 29–39.
15. WILLENBORG, A., SCHMITZ, D., LELLEY, J. 1990. Effects of environmental stress factors on ectomycorrhizal fungi in vitro. Can. J. Bot. 68: 1741–1746.
16. ANTIBUS, R.K., SINSABAUGH, R.L., LINKINS, A.E. 1992. Phosphatase activities and phosphorus uptake from inositol phosphate by ectomycorrhizal fungi. Can. J. Bot. 70: 794–801.
17. PETERS, N.K., VERMA, D.P.S. 1990. Phenolic compounds as regulators of gene expression in plant-microbe interactions. Molec. Plant-Microbe Interactions 3: 4–8.
18. SIQUEIRA, J.O., NAIR, M.G., HAMMERSCHMIDT, R., SAFIR, G.R. 1991. Significance of phenolic compounds in plant-soil-microbial systems. Crit. Rev. in Plant Sciences 10: 63–121.
19. STACHEL, S.E., MESSENS, E., VAN MONTAGU, M., ZAMBRYSKI, P. 1985. Identification of the signal molecules produced by wounded plant cells that activate T-DNA transfer in *Agrobacterium tumefaciens*. Nature 318: 624–629.
20. BOLTON, G.W., NESTER, E.W., GORDON, M.P. 1986. Plant phenolic compounds induce expression of the *Agrobacterium tumefaciens* loci needed for virulence. Science 232: 983–985.
21. GIANINAZZI-PEARSON, V., BRANZANTI, B., GIANINAZZI, S. 1989. *In vitro* enhancement of spore germination and early hyphal growth of a vesicular-arbuscular mycorrhizal fungus by host root exudates and plant flavonoids. Symbiosis 7: 243–255.
22. BECARD, G., DOUDS, D.D., PFEFFER, P.E. 1992. Extensive *in vitro* hyphal growth of vesicular-arbuscular mycorrhizal fungi in the presence of CO_2 and flavonols. Appl. Environ. Microbiol. 58: 821–825.

23. MULLIGAN, J.T., LONG, S.R. 1985. Induction of *Rhizobium nodC* expression by plant exudate required *nodD*. Proc. Natl. Acad. Sci. USA 82: 6609–6613.
24. ROSSEN, L., DAVIS, E.O., JOHNSTON, A.W.B. 1987. Plant-induced expression of *Rhizobium* genes involved in host specificity and early stages of nodulation. TIBS 12: 430–433.
25. HARWOOD, C.S., RIVELLI, M., ORNSTON, L.N. 1984. Aromatic acids are chemoattractants for *Pseudomonas putida*. J. Bacteriol. 160: 622–628.
26. GOTTLIEB, O.R. 1992. Plant phenolics as expressions of biological diversity. In: Plant Polyphenols, Synthesis, Properties, Significance, (R.W. Hemingway, P.E. Laks, eds.), Plenum Press, NY, pp. 523–538.
27. APPEL, H.M. 1993. Phenolics in ecological interactions: The importance of oxidation. J. Chem. Ecol. 19: 1521–1552.
28. FIELD, J.A., LETTINGA, G. 1992. Toxicity of tannic compounds to microorganisms. In: Plant Polyphenols, Synthesis, Properties, Significance, (R.W. Hemingway, P.E. Laks, eds.), Plenum Press, New York, pp. 673–692.
29. BENDING, G.D., READ, D.J. 1995. The structure and function of the vegetative mycelium of ectomycorrhizal plants. V. Foraging behaviour and translocation of nutrients from exploited litter. New Phytol. 130: 401–409.
30. BENDING, G.D., READ, D.J. 1996b. Nitrogen mobilization from protein-polyphenol complex by ericoid and ectomycorrhizal fungi. Soil. Biol. Biochem. 28: 1603–1612.
31. MASON, P.A., WILSON, J., LAST, F.T., WALKER, C. 1983. The concept of succession in relation to the spread of sheathing mycorrhizal fungi on inoculated tree seedlings growing in unsterile soils. Plant and Soil 71: 247–256.
32. LAST, F.T., NATARAJAN, K., MOHAN, V., MASON, P.A. 1992. Sequences of sheathing ecto-mycorrhizal fungi associated with man-made forests, temperate and tropical. In: Mycorrhizas in Ecosystems, (D.J. Read, D.H. Lewis, A.H. Fitter, I.J. Alexander, eds.), C.A.B. International, Oxon, pp. 214–219.
33. VISSER, S. 1995. Ectomycorrhizal fungal succession in jack pine stands following wildfire. New Phytol. 129: 389–401.
34. STENSTROM, E. 1989. The importance of infection methods and root environment on mycorrhiza formation. Agriculture, Ecosystems and Environment 28: 479–482.
35. DIGHTON, J., BODDY, L. 1989. Role of fungi in nitrogen, phosphorus and sulphur cycling in temperate forest ecosystems. In: Nitrogen, Phosphorus and Sulphur Utilisation by fungi, (L. Boddy, R, Marchant, D.J. Read, eds.), Cambridge University Press, Cambridge, pp. 269–298.
36. TIARKS, A.E., MEIER, C.E., FLAGLER, F.R.B., STEYNBERG, E.C. 1992. Sequential extraction of condensed tannins from pine litter at different stages of decomposition. In: Plant Polyphenols, Synthesis, Properties, Significance, (R.W. Hemingway, P.E. Laks, eds.), Plenum Press, New York, pp. 597–608.
37. BAAR, J., OZINGA, W.A., SWEERS, I.L., KUYPER, T.W. 1994. Stimulatory and inhibitory effects of needle litter and grass extracts on the growth of some ectomycorrhizal fungi. Soil Biol. Biochem. 26: 1073–1079.
38. BAAR, J., DE VRIES, F.W. 1995. Effects of manipulation of litter and humus layers on ectomycorrhizal colonization potential in Scots pine stands of different age. Mycorrhiza 5: 267–272.
39. OLSEN, R.A., ODHAM, G., LINDEBERG, G. 1971. Aromatic substances in leaves of *Populus tremula* as inhibitors of mycorrhizal fungi. Physiol. Plant. 25: 122–129.
40. ROSE, S.L., PERRY, D.A., PILZ, D., SCHOENEBERGER, M.M. 1983. Allelopathic effects of litter on the growth and colonization of mycorrhizal fungi. J. Chem. Ecol. 9: 1153–1162.
41. PERSIDSKY, D.J., LOEWENSTEIN, H., WILDE, S.A. 1965. Effect of extracts of prairie soils and prairie grass roots on the respiration of ectotrophic mycorrhizae. Agronomy J. 57: 311–312.
42. PELLISSIER, F. 1993. Allelopathic effect of phenolic acids from humic solutions on two spruce mycorrhizal fungi: *Cenococcum graniforme* and *Laccaria laccata*. J. Chem. Ecol. 19: 2105–2114.

43. LINDEBERG, G., LINDEBERG, M. 1980. Stimulation of litter-decomposing basidiomycetes by flavonoids. Trans. Brit. Mycol. Soc. 75:455–459.
44. LING-LEE, M., CHILVERS, G.A., ASHFORD, A.E. 1977. A histochemical study of phenolic materials in mycorrhizal and uninfected roots of *Eucalyptus fastigata* Deane and Maiden. New Phytol. 78: 313–328.
45. FOSTER, R.C., MARKS, G.C. 1967. Observations on the mycorrhizas of forest trees II. The rhizosphere of *Pinus radiata* D. Don. Aust. J. Biol. Sci 20: 915–926.
46. PICHE, Y., FORTIN, J.A., LAFONTAINE, J.G. 1981. Cytoplasmic phenols and polysaccharides in ectomycorrhizal and non-mycorrhizal short roots of pine. New Phytol. 88: 695–703.
47. FOSTER, R.C., MARKS, G.C. 1966. The fine structure of the mycorrhizas of *Pinus radiata* D. Don. Aust. J. Biol. Sci. 19: 1027–1038.
48. HILLIS, W.E., ISHIKURA, N., FOSTER, R.C., MARKS, G.C. 1968. The role of extractives in the formation of ectotrophic mycorrhizae. Phytochemistry 7: 409–410.
49. HILLIS, W.E., ISHIKURA, N. 1969. The extractives of the mycorrhizas and roots of *Pinus radiata* and *Pseudotsuga menziesii*. Aust. J. Biol. Sci. 22: 1425–1436.
50. MUNZENBERGER, B., KOTTKE, I., OBERWINKLER, F. 1995. Reduction of phenolics in mycorrhizas of *Larix decidua* Mill. Tree Physiology 15: 191–196.
51. PHILLIPS, R. 1991. Mushrooms of North America. Little Brown & Company, Boston. 319 pp.

Chapter Five

ALLELOCHEMICALS IN ROOT EXUDATES OF MAIZE

Effects on Root Lesion Nematode *Pratylenchus zeae*

Annette Friebe,[1,*] Wilma Klever,[1] Richard Sikora,[2] and Heide Schnabl[1]

[1] Institute of Agricultural Botany
University of Bonn
Meckenheimer Allee 176, 53115 Bonn, Germany
[2] Institute for Plant Diseases
University of Bonn
Nußallee 9, 53115 Bonn, Germany

Introduction 72
Cyclic Hydroxamic Acids—Allelochemicals of Maize 73
　Structural Diversity and Natural Occurrence 73
　Biological Activity 74
　Mode of Action 77
Analysis of Root Exudates from *Zea mays* 77
　Collection and Analytical Procedures 77
　Identification of Allelochemicals 78
Effects on Root Lesion Nematode *Pratylenchus zeae* 80
　Bioassay 80
　Influence of Host Root Exudate 82
　Influence of Selected Allelochemicals 82
Future Research Directions 87
Conclusion 88

---

* Author to whom correspondence should be addressed.

Phytochemical Signals and Plant–Microbe Interactions, edited by Romeo *et al.*
Plenum Press, New York, 1998.

INTRODUCTION

Root exudation of biologically active compounds plays an important role in the interaction of higher plants with the soil ecosystem. This includes, for instance, the exudation of chemical signals for the interaction of plants with symbiotic microorganisms as well as the exudation of antimicrobial substances for plant defense against soil-borne pathogens. Attraction, penetration, and feeding behavior of plant parasitic nematodes also involve molecular communication between the nematode and respective host plants. Chemotaxis of plant parasitic nematodes began to receive attention as early as 1925.[1] The influence of root exudates on such phenomena as attraction, repellence, inhibiton, and hatching stimulation have been described for a wide range of nematodes.[2-15]

The dependence of species such as the potato cyst nematode, *Globodera rostochiensis*, on host root exudates for hatching is well documented, and the stimulation of hatching has been the subject of extensive research.[2-8] Even very short exposures to root exudates of host plants are sufficient to induce hatching of juveniles from eggs.[7] The hatching stimulus is important for nematode survival in that it ensures the presence of actively growing roots needed for juvenile infection during the initial stages of host development. Stimulation of hatching is based on a spectrum of physiological changes in nematode eggs induced by host root exudates. This includes an alteration in cAMP levels[3,4] as well as activation of phosphoinositide signal transduction.[2] Movement of *Globodera rostochiensis* also is stimulated by potato root exudates.[9]

The range of responses of nematodes to roots and different chemicals with respect to their movement behavior has been reviewed by Prot.[10] The existence of kairomones for *Heterodera avenae* (Woll.) in root exudates of different cereal host plants such as oats, wheat, and barley is known.[11] In addition, attraction of soybean cyst nematode, *Heterodera glycines*, to soybean root leachates has been observed.[12] Nevertheless, the chemotactic behavior of plant parasitic nematodes to root exudates is controversial. Inhibitory and repellent effects of host exudates have been described.[13,14] Following fractionation of cucumber root exudate, fractions with both repellent and attractant activity to *Meloidogyne incognita* were isolated.[15]

Although many observations concerning the influence of root exudates of different plant species on the behavior of plant parasitic nematodes have been reported, little is known about the molecular structure of the respective signal compounds involved in these interactions. This paper deals with the activity of root exudates of *Zea mays* on the root lesion nematode, *Pratylenchus zeae*. On the basis of a comprehensive chemical analysis of low molecular allelochemicals exuded by maize cv. Mutin, the effects of selected compounds on nematode activity *in vitro* were investigated. Special attention was paid to cyclic hydroxamic acids and related compounds. The role of these substances in chemotaxis and host recognition of *Pratylenchus zeae* is discussed.

CYCLIC HYDROXAMIC ACIDS—ALLELOCHEMICALS OF MAIZE

Structural Diversity and Natural Occurrence

Heterocyclic 1,4-benzoxazin-3-ones and related benzoxazolinones are important natural products of cereal plants. The cyclic hydroxamic acids and respective derivatives have been described as constitutive compounds of a wide variety of gramineous plants including the cultivated crops maize, wheat, and rye, but not barley, oat, and rice. These substances have been found only in two dicot families—Acanthaceae and Scrophulariaceae.[16,17] During biosynthesis, 2,4-dihydroxy-2H-1,4-benzoxazin-3(4H)-one compounds are derived from their lactam precursors **1a–c** (Fig. 1).[18,19]

In intact plant cells, cyclic hydroxamic acids are sequestered and stabilized in the form of their (2R)-2-β-D-glucosides, **2a–c** (Fig. 2). In response to tissue damage or pathogen attack, the more toxic aglucones are released by vacuolar β-glucosidase catalyzed hydrolysis. The 2,4-dihydroxy-2H-1,4-benzoxazin-

Figure 1. Chemical structure of 2-hydroxy-2H-1,4-benzoxazin-3(4H)-ones (**1a–c**) as lactam precursors of cyclic hydroxamic acids.

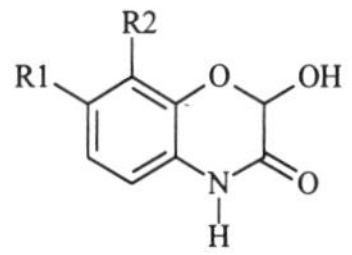

R_1 = H	R_2 = H	**1a** HBOA
R_1 = OMe	R_2 = H	**1b** HMBOA
R_1 = OMe	R_2 = OMe	**1c** HM$_2$BOA

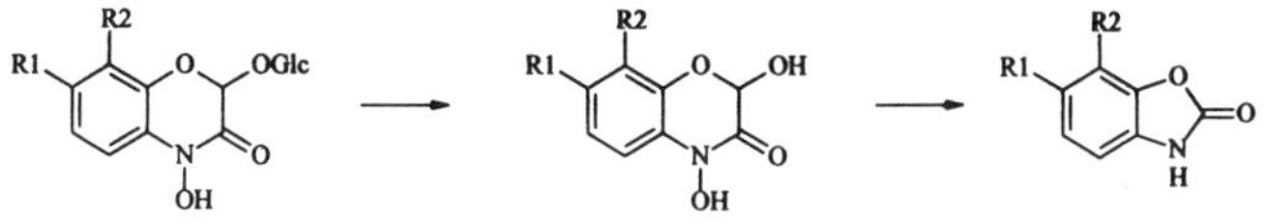

R_1 = H	R_2 = H	**2a** GDIBOA	**3a** DIBOA	**4a** BOA
R_1 = OMe	R_2 = H	**2b** GDIMBOA	**3b** DIMBOA	**4b** MBOA
R_1 = OMe	R_2 = OMe	**2c** GDIM$_2$BOA	**3c** DIM$_2$BOA	**4c** M$_2$BOA

Figure 2. Hydrolysis of (2R)-β-D-glucopyranosyloxy-4-hydroxy-2H-1,4-benzoxazin-3(4H)-ones (**2a–c**) to 2,4-dihydroxy-2H-1,4-benzoxazin-3(4H)-ones (**3a–c**) and further decomposition to benzoxazolinones (**4a–c**).

3(4H)-ones, **3a–c**, (Fig.2) subsequently decompose to give the respective ring-contracted benzoxazolinones, **4a–c** (Fig. 2).

The methoxylated hydroxamic acid, DIMBOA (**3b**), is the most abundant derivative in maize. Strong cultivar-dependent differences in the hydroxamic acid content have been described. Levels of DIMBOA in roots of various geographic origin reportedly range from 0.2 to 4.1 mmol/ kg fresh weight.[20] In *Zea mays*, DIMBOA (**3b**) content is much higher than DIM_2BOA (**3c**) content, and DIBOA (**3a**), a nonmethoxylated cyclic hydroxamic acid, has the lowest concentration.[21,22] Intense effort has gone into investigating the structural diversity of benzoxazinone related compounds, especially in maize. Glucosides of 7-hydroxylated cyclic hydroxamic acids and lactams have been described for maize and also for other plant species.[23–25] Additionally, the highly active and labile 4-methoxy-1,4-benzoxazin-3(4H)-one is found in maize.[26–29] Structural diversity has been demonstrated not only for the cyclic hydroxamic acids, but also for related benzoxazolinones. The allelochemical, 6-acetylbenzoxazolin-2-one,[30] and a chlorinated benzoxazolinone derivative[31] have been detected.

Hydroxamic acids are not present in seeds of cereal plants. After germination, the levels of 1,4-benzoxazin-3(4H)-ones increase and maximum levels are reached in young seedlings a few days after germination. Although cyclic hydroxamates are found in all parts of plants, the concentrations are generally higher in shoots than in roots.[32] Concentrations of these compounds in plants are highly dependent on environmental growth conditions. A number of studies have revealed increases in levels of hydroxamic acids caused by UV-light[33] and water deficiencies.[34]

Biological Activity

The secondary metabolites of cereal plants discussed above show a great variety of biological activity. Because of their phytotoxic properties, they have been implicated in allelopathic interferences of higher plants. Dicotylodonous species show a higher sensitivity to hydroxamic acids than monocotyledonous species.[32,35–37]

Much research has focused on the role of cyclic hydroxamic acids as defensive agents against plant diseases and pests. DIMBOA, the predominant benzoxazinone in wheat and maize, exhibits antibacterial and antifungal activity. Antifungal effects have been reported against a variety of pathogens including *Helminthosporium turcicum*,[38] *Septoria nodorum*,[39] and *Fusarium moniliforme*.[40] Inhibition of spore germination and mycelial growth have been demonstrated *in vitro*. Attempts to correlate the levels of hydroxamates in cultivated maize and wheat with resistance to fungal plant diseases have been made. Early reports demonstrated a linear relationship between the levels of hydroxamic acids and resistance of maize to stalk rot *Diploida maydis*[41] and wheat to

Puccinia graminis.[42] Observations that support a role of constitutive benzoxazinones in differences in plant defense regarding tolerance to host specific pathogens have been made. There is effective detoxification of benzoxazolinones by the maize pathogen *Fusarium moniliforme*.[40] Virulent strains of different wheat pathogenic fungi also degrade these compounds (Friebe et al., unpublished). Additionally, benzoxazinone compounds have been described as phytoalexins. Accumulation of 2-hydroxy-4,7-dimethoxy-2H-1,4-benzoxazin-3(4H)-one in wheat after infection with *Puccinia graminis* Pers. f. sp. *tritici* Ericss & Henn, and an increase in DIMBOA content in resistant maize after infection with *Exserohicum turcicum* have been reported.[26,43]

Cyclic hydroxamic acids also are directly associated with plant defense against herbivores. Among the best studied are interactions of cereal plants with aphids. Growth of aphids on wheat and maize negatively correlates with DIMBOA levels.[44,45] The influence of these allelochemicals on feeding behavior was studied by Givovich and Niemeyer[46] and Mayoral et al.[47] Aphid species were subdivided as sensitive or insensitive to the feeding deterrent effect. Advances in understanding the interactions came from experimental studies of glutathione S-transferases. The activity of the insect detoxification enzymes in the grain aphid, *Sitobion avenae*, was associated with concentration of hydroxamic acids in the host plant.[48]

The Southwestern corn borer (*Diatraea grandiosella*), European corn borer (*Ostrinia nubilalis*), and Asian corn borer (*Ostrinia furnacalis*) are major insect pests of maize, attacking the plant in the vegetative and reproductive stages of growth. DIMBOA is recognized as a major leaf feeding resistance factor.[33,49–52] An investigation of the effects of DIMBOA on several insect enzymatic systems revealed the allelochemical's ability to induce higher detoxification enzyme activity of cytochrome P-450 monooxygenase and glutathione S-transferase.[52] Nevertheless, the role of cyclic hydroxamic acids as defense chemicals is controversial. For instance, no correlation was found between DIMBOA levels in Peruvian maize cultivars and resistance to *Ostrinia nubialis*.[53]

Larval development of Western corn rootworm, including survival, weight, and head capsule width, was shown to be significantly and negatively correlated with hydroxamic acid content in roots. Resistant inbreds were distinguished by a remarkably high concentration of these compounds.[20] A degradation product, MBOA, was isolated and identified as an important semiochemical of maize roots for Western corn rootworm larvae. Attraction of larvae to MBOA purified from maize root extracts and synthetic MBOA was shown, suggesting the allelochemical is involved in host finding by this specialist insect.[54,55]

In soil, the microbial transformation of benzoxazinones and related benzoxazolinones into biologically active compounds has to be considered. The biotransformation of cyclic hydroxamic acids (**3a–c**) and related benzoxazoli-

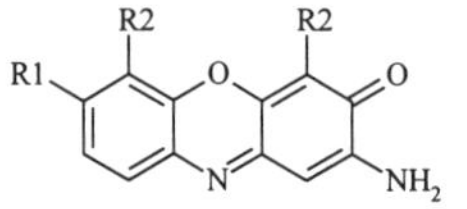

R1 = H	R2 = H	**5a** 2-amino-3H-phenoxazin-3one
R1 = OMe	R2 = H	**5b** 2-amino-7-methoxy-3H-phenoxazin-3-one
R1 = OMe	R2 = OMe	**5c** 2-amino-4,6,7-trimethoxy-3H-phenoxazin-3-one

Figure 3. Chemical structure of 2-amino-3H-phenoxazin-3-ones (**5a–c**), soil microbial metabolites of benzoxazolinones (**4a–c**).

nones (**4a–c**) to 2-amino-3H-phenoxazin-3-ones, **5a–c** (Fig. 3) by soil bacteria has been reported.[56,57]

Microbial transformation of BOA (**4a**) by rhizosphere bacteria of different cereal crops also has been observed.[58] Further acetylation of 2-amino-3H-phenoxazin-3-one can occur. 2-Amino-phenoxazin-3-one and its acetyl derivative are known as naturally occurring antibiotics, questiomycin and N-acetylquestiomycin. These microbial products belong to the important group of actinomycin analogs. The antibiotic effects of questiomycin and its acetyl derivative are of special interest in view of the development of plant species-dependent bacteria populations in the rhizosphere and their biological importance for the respective plant. Plant-associated microorganisms live in permanent competition for nutrients of limited availability at the plant surface. The ability to produce substances with inhibitory effects against competitors is of major importance for the development of the rhizosphere microflora.

Recently, N-(2-hydroxyphenyl)-malonamic acid, **6a**, and N-(2-hydroxy-4-methoxyphenyl)-malonamic acid, **6b**, were identified as metabolites of BOA and MBOA of soil-borne pathogenic fungi (Fig. 4) (Friebe et al., unpublished). These compounds must be considered detoxification products. No distinct antifungal and phytotoxic activities have been observed to date.

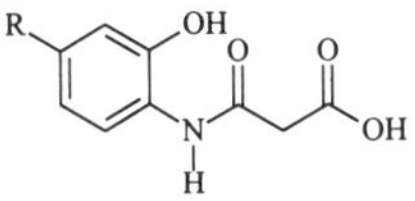

| R = H | **6a** N-(2-hydroxyphenyl)-malonamic acid |
| R = OMe | **6b** N-(2-hydroxy-4-methoxyphenyl)-malonamic acid |

Figure 4. Chemical structure of N-(2-hydroxyphenyl)-malonamic acid (**6a**) and N-(2-hydroxy-4-methoxyphenyl)-malonamic acid (**6b**) fungal metabolites of benzoxazolinones (**4a–c**).

Mode of Action

To understand molecular effects that can account for the biological activities discussed above, the influence of cyclic hydroxamic acids and related benzoxazolinones on the activity of different enzymatic systems was investigated. Inhibitory effects on the metabolism of chloroplasts were described, e.g. the chloroplast ATPase coupling factor was inhibited by cyclic hydroxamic acids.[59-61] Inhibition of mitrochondrial energy-linked reactions by the cyclic hydroxamic acid DIMBOA as well as the benzoxazolinone BOA also were found.[62,63] Significant inhibitory effects of 2,4-dihydroxy-1,4-benzoxazin-3-ones and corresponding benzoxazolinones also were shown for plasma-membrane H^+-ATPase of higher plants.[64] At lower effector concentrations, stimulation of this enzyme was observed.

Insect digestive enzymes have been suggested as possible targets of plant defense chemicals. For example, inactivation of thiol and serine proteinases, was shown for α-chymotrypsin and papain.[65,66] The reaction of hydroxamic acids with nucleophilic sites in enzymes, such as cysteine residues at the active site, has been directly related to enzyme inhibition. Investigation of structure-activity relationships of benzoxazinoid compounds have revealed that the alkylating action of these substances seems to be the chemical basis for their biological activity, at least in part.[67] Evidence for this was supported by mechanistic and kinetic investigations of reactions of hydroxamates with thiols and amines in aqueous solution.[68,69,78] Reaction of benzoxazinoide compounds with DNA and nucleotides also has been demonstrated.[70]

For the hydroxamate related benzoxazolinones, an inhibition of auxin induced growth has been observed. MBOA has been identified as an auxin inhibiting substance of maize.[71] Modification of auxin-receptor interaction by competitive effects has been implicated as the main reason for the inhibitory activity of benzoxazolinones.[72,73]

ANALYSIS OF ROOT EXUDATES FROM *ZEA MAYS*

Collection and Analytical Procedures

The effects of maize root exudates on the economically important lesion nematode, *Pratylenchus zeae*, recently were investigated. Experiments showed that damage by this migratory endoparasitic nematode is pronounced in the first weeks of plant growth. The mean number of nematodes in roots was highest three weeks after planting and decreased by almost 50% eleven weeks after planting.[74] For this reason, the exudation of allelochemicals by young seedlings was investigated in detail for a possible role in the regulation of parasitic nematode populations.

A modified continuous root exudate trapping system (CRET-system), as described by Tang and Young,[75] was used to collect allelochemicals from root exudates of actively growing *Zea mays* cv. Mutin plants. Thirty to 40 germinated seeds were planted in 100 g of autoclaved glass pearls in 2.5 l pots which were fitted with teflon sieve bottoms. The cultivation of the plants in the CRET-system was performed under controlled environmental conditions,[35] and the system was continuously watered by recirculation.

A chemical trap consisting of a glass column with a sintered disk bottom was filled with amberlite XAD-4 resin. Special attention was paid to purification of the absorber material. An adequate amount was purified via soxhlet extraction procedure (36 hr acetonitrile, 36 hr ethylacetate, 36 hr methanol) and then intensively washed with water. The chemical trap was arranged below the donor pot. After a collection time of 7 days, the trap was removed and replaced with a new one. After removing the trap from the CRET-system, the resin was washed with distilled water and eluted with methanol. The eluate was concentrated under vacuum, and the resulting root exudate was analyzed by HPLC and gas chromatography - mass spectrometry (GC-MS).

HPLC separation of root exudates and reference compounds was performed on a Beckman model 126 chromatograph equipped with a diode array detector 168 with a 4.6 × 250 mm ultrasphere ODS RP 18 column. Chromatographic conditions consisted of a mobile phase of 0.1% TFA v/v (A) and methanol (B) at a flow rate of 1 ml/min. The column was eluted using a solvent gradient profile. (0–1 min, 100% A; 1–9 min, 0–40% B; 9–15 min, 40% B; 15–27 min, 40–100% B; 27–35 min, 100% B). Eluents were observed at 275 and 430 nm. Compounds were analyzed using GC-MS in the form of their trimethylsilyl (TMS) derivatives. For derivatization, collected HPLC fractions were evaporated and supplemented with BSTFA (N,O-bis(trimethylsilyl)-trifluoroacetamide).[24]

For chemical identification and quantification of chemicals, reference substances were isolated and purified from seedlings. 2,4-Dihydroxy-7-methoxy-2H-1,4-benzoxazin-3(4H)-one (DIMBOA) was isolated from 7 day old corn seedlings by a procedure described by Woodward et al.[76] The procedure used for isolation of the respective 2-β-D-glucoside GDIMBOA was that of Hartenstein et al.[77] BOA and MBOA were obtained from ALDRICH and SIGMA.

Identification of Allelochemicals

After collection of exudates from the undisturbed root systems in CRET-systems, the compounds were separated using reverse phase HPLC. A representative chromatogram is shown in Fig. 5. After separation, fractionated compounds were analyzed by GC-MS. Derivatives of 1,4-benzoxazinone were identified on the basis of their UV and mass spectral data and comparison with respective reference compounds. The data are in good agreement with those in the literature.[22,24,78]

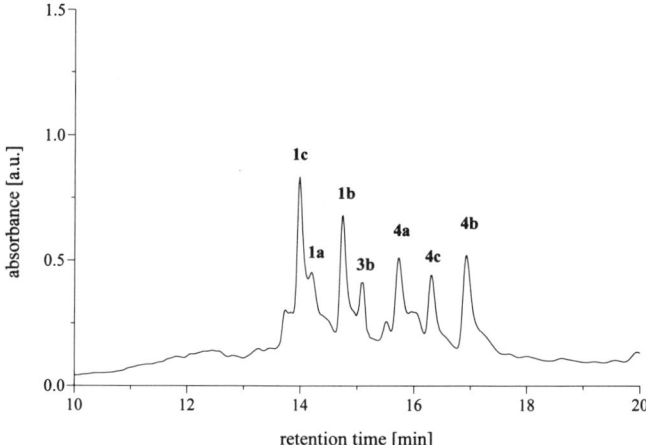

Figure 5. Reverse phase chromatogram of a root exudate of *Zea mays* cv. Mutin collected using a modified CRET-system. Allelochemicals were detected at 275 nm.

In addition to the cyclic hydroxamic acid, DIMBOA, a variety of lactam precursors (**1a–c**) and the corresponding benzoxazolinone decomposition products (**4a–c**) also were released by root exudation. The occurrence of lactams in exudates of maize is described here for the first time. Although the secretion of DIBOA and DIM_2BOA were not observed, it is likely that these compounds decomposed to BOA and M_2BOA.

Cyclic hydroxamic acids are well known natural products of different cereal plants, but there are only a few reports of the root exudation of these compounds. Differences in exudation have been found between wheat and rye cultivars.[79] Root exudation of 25 µmol DIBOA/kg fresh weight up to the first leaf stage for the rye cultivar Forrajero Baer has been found. In contrast, hydroxamic acids have not been detected in exudates of wheat cultivars. The exudation of both hydroxamic acids DIBOA and DIMBOA also has been detected in barnyard grass (*Echinochloa crusgalli*)[80] and quackgrass (*Agropyron repens*).[35,81] The glucoside bound hydroxamic acids, GDIBOA (**2a**) and GDIMBOA (**2b**) have been isolated from the rhizosphere of barnyard grass.[80]

Exudation of DIMBOA from roots of *Zea mays* was previously measured and analyzed.[82] Supplying iron significantly increased the exudation rate. Application of 0.5–5 µmol/l iron increased by 2–3 fold the amount of DIMBOA released by the roots of plants deficient in iron. Because of their formation of complexes with Fe(III)-ions,[82,83] it seems plausible that cyclic hydroxamic acids play a role in Fe-uptake and metabolism as phytosiderophores.

Using synthetic and isolated reference compounds, various allelochemicals were quantitatively analyzed by HPLC. The results for 2 and 3 week old

Table 1. Amounts of allelochemicals exuded by 2- and 3-week-old seedlings of *Zea mays* cv. Mutin[a]

Substance		Amount of allelochemicals exuded by roots [nmol per seedling per day] (mean ± SD)	
		Second week	Third week
HMBOA	(1b)	0.28 ± 0.12	0.31 ± 0.10
DIMBOA	(3b)	0.03 ± 0.05	0.05 ± 0.05
MBOA	(4b)	2.18 ± 0.93	2.10 ± 0.62
BOA	(4a)	0.09 ± 0.07	0.32 ± 0.25

[a]Values are means of two experiments with 3 replicates.

maize seedlings are shown (Table 1). The experimental data demonstrate that the majority of DIMBOA is decomposed to the related methoxybenzoxazolinone. No significant differences in exudation rates were observed between the 2nd and 3rd week of plant growth.

Root exudates of maize were collected in hydroponic culture for bioassays with *Pratylenchus zeae* in order to avoid matrix effects of XAD 4-absorber resin. Mean exudation rates of 0.05 nmol DIMBOA and 1.2 nmol MBOA per seedling per day were determined. In contrast to allelochemicals collected by the CRET-system, GDIMBOA also was detected in hydroponic root secretions. Exudation rates up to 10.1 nmol per seedling per day were observed.

EFFECTS ON ROOT LESION NEMATODE *PRATYLENCHUS ZEAE*

Bioassay

The effects of exudates of *Zea mays* as well as effects of selected isolated chemicals on the migratory endoparasitic nematode, *Pratylenchus zeae,* were investigated *in vitro* using a modified bioassay designed by Castro et al.[15] The influence of concentration gradients on the movement of nematodes was studied. In order to restrict nematode movement to one direction, a bioassay device was designed consisting of a plexiglass block with ten parallel tracks. All surfaces and tracks (90° grooves, 3 mm deep, 7.5 cm length) were polished using dental instruments. To insure optimal and reproducible conditions for nematode mobility, each track was filled with 340 μl of 1.5% agarose with a transparent plexi-glass cover placed 4 mm above the track to maintain constant moisture conditions. A 1 μl sample of the test solution was placed at one end of each track and 1 μl of distilled water at the other as a control. After a 2 hour waiting period to allow the formation of a concentration gradient, 50–180 nematodes (adults and juveniles) were placed in the center of each track, equidistant from the two test solutions. The device was stored at 26°C in an incubator in the dark, and after 6 and 24 hours, the migration of nematodes was determined microscopically.

The tracks were subdivided into 14 segments of 0.5 cm length, and the number of nematodes in each segment was counted. To make quantitative assessments, an attractant/repellent ratio was calculated which was equal to the ratio of the nematode densities, weighted for the distance they moved towards or away from a given sample. For each track, summations S_{sample} and $S_{control}$ were calculated according to Eq. 1.[15]

$$S = \Sigma (N)_i (D)_i \tag{1}$$

i = number of the segment
N = number of nematodes in the segment
D = distance from the center of the segment to the starting point
S = sum of nematode densities weighted for the moved distances

The ratio $S_{sample}/S_{control}$, defined by Castro et al.,[15] is a quantitative value for the nematode response to a given concentration gradient. The limits for this value are zero for maximum repellence and ∞ for maximum attraction of a given sample. To achieve a statistical normal distribution of the quantitative index, this ratio was mathematically transformed by following equations Eq. 2a and Eq. 2b:

$$I = 1 - (S_{control}/S_{sample}) \tag{2a}$$

for $S_{sample} > S_{control}$ $0 < I < 1$ (attractance)

$$I = (S_{sample}/S_{control}) - 1 \tag{2b}$$

for $S_{control} > S_{sample}$ $-1 < I < 0$ (repellence)

The data were tested for normal distribution by the Kolmogoroff-Smirnoff-test.[84] Variances were examined by F-test, and mean values of experimental series were compared using the two-tailed t-test corrected after Sidak (significance level $\alpha = 0.003$, $P = 0.05$).[85,86] A large number of control experiments (n = 265) was performed. Indices of −0.010 for an incubation time of 6 hours and 0.007 for a time of 24 hours indicated that there was no preferable direction for nematode migration. A standard error of ±0.020 was determined. All data were normally distributed.

For bioassays, *Pratylenchus zeae* was cultivated on sterile roots of the host plant maize in 25 ml of sterile growth media in Petri dishes 9 cm diameter and stored at 26°C for 2–4 months in the dark. Nematodes were extracted from the culture medium with tap water, washed through 250 μm and 25 μm sieves, and collected on a 20 μm sieve.

Influence of Host Root Exudate

In order to study attractant and repellent effects of root exudate on root lesion nematodes *in vitro*, a standard exudate was collected from a non-sterile hydroponic culture of plants. This was analyzed quantitatively for GDIMBOA, DIMBOA, and MBOA, and then five different dilutions were prepared. The resulting concentrations and the experimental results of bioassays obtained are shown in Table 2.

Strong concentration-dependent effects of exudates on nematodes were found after an incubation time of 6 hours. At the highest concentrations, significant repellent activity was observed, as indicated by negative indices. The bioassay also revealed significant repellent activity at the highest concentration after 24 hours of exposure. Dilution of the exudate resulted in a decrease in repellent activity. At the lowest concentration investigated, nematode attraction increased significantly. The index value of 0.80 corresponds to an attraction of 77% of the nematodes. Similar concentration dependent effects (attraction and repellent) of host root exudates have been described previously for juveniles of *Heterodera avenae*.[11] In general, activity decreased with increase in exposure time. The mean nematode densities in the different segments of the tracks are shown for 6 and 24 hours exposure of root exudate (Fig. 6).

Influence of Selected Allelochemicals

Selected derivatives of cyclic hydroxamic acids, important constituents of maize root exudates, also were investigated for their effects on *Pratylenchus zeae*. Different concentrations of GDIMBOA, DIMBOA, and MBOA were

Table 2. Effects of host root exudate of *Zea mays* cv. Mutin on *Pratylenchus zeae* behavior[a]

Concentration [mmol/l]			Dilution level	n	I (mean ± SE)	
GDIMBOA (2b)	DIMBOA (3b)	MBOA (4b)			Incubation 6h	Incubation 24 h
2.0	0.2	0.04	0	24	−0.42 ± 0.11*	−0.49 ± 0.06*
0.5	0.05	0.1	1	24	−0.53 ± 0.04*	−0.10 ± 0.07
0.1	0.01	0.02	2	8	−0.25 ± 0.07*	−0.11 ± 0.12
0.05	0.005	0.01	3	32	0.19 ± 0.06*	0.23 ± 0.06*
0.01	0.001	0.002	4	15	0.80 ± 0.02*	−0.02 ± 0.09

[a] Values larger than 0 indicate attraction and those below 0 repellent activity; * indicates values significantly different from control.

tested for effects on lesion nematode behavior in the *in vitro* bioassay. Results are shown in Table 3.

After exposure to GDIMBOA for 6 hours, only slight differences in nematode migration were observed. Only at a concentration of 0.2 mmol/l did a significant increase in attraction occur. The index value of 0.22 corresponded to an attraction of 56.5% of the nematodes. No significant effects were detected in

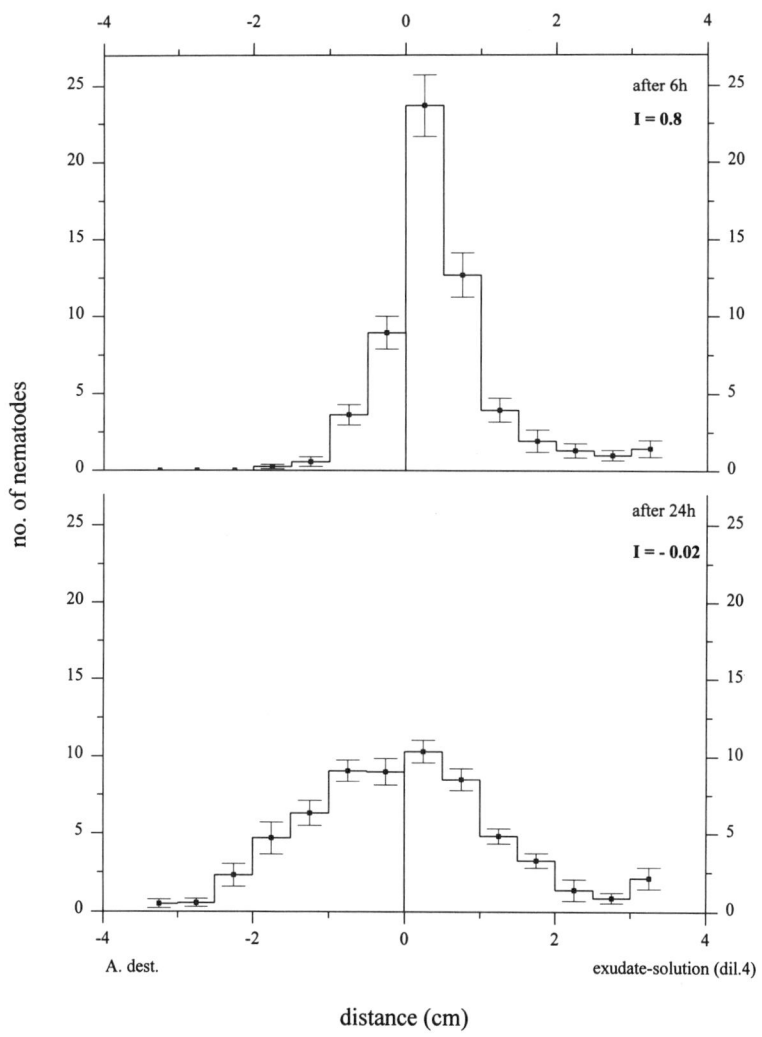

Figure 6. Effects of diluted host root exudate on root lesion nematode *P. zeae* behavior after 6 and 24 hours exposure time.

the treatments after 24 hours. The pronounced repellent activity at a concentration of 20 μmol/l GDIMBOA could not be explained. On the basis of these data, it seems unlikely that GDIMBOA exhibits chemotactic activity toward *P. zeae* in a concentration range relevant to normal maize root exudates.

A significant number of juveniles were attracted to the benzoxazinoide compound DIMBOA at a concentration of 2 mmol/l (Table 3). Attraction was greater after incubation for 24 hours than after 6 hours. The index value of 0.36 corresponds to 61.2% of attracted nematodes. Attractive activity of DIMBOA decreased with decreasing concentration, and at 20 μmol/l, strong repellent activity was seen. A slight decrease of activity was observed at after 24 hours. A further dilution lead to attractive effects. After 6 hours at a concentration of 2 μmol/l, an index of 0.26 was calculated, which corresponds to 58.8% attracted nematodes. After 24 hours, a more pronounced level of attraction was observed (Fig. 7). After 24 hours, an attractance/repellence index of 0.62 was obtained.

The increase in attraction of the juveniles to DIMBOA with time of incubation is illustrated by nematode densities shown in Figure 7. The chemical decomposition of DIMBOA to the related benzoxazolinone MBOA has to be considered. At an initial concentration of 0.1 mmol/l and a pH 6.75, a DIMBOA half-life time of 5.3 hr was determined previously by Woodward et al.[76] The effects of MBOA on *Pratylenchus zeae* are shown (Table 3). A significant number of juveniles were repelled by 0.2 mmol/l MBOA. At a lower concentration of 20 μmol/l, after 24 hours, a significant level of attraction was detected. Activity was lost after further dilution of the sample.

Table 3. Effects of 2,4-dihydroxy-7-methoxy-1,4-benzoxazin-3-one-(2R)-2-β-D-glucoside (2b),2,4-dihydroxy-7-methoxy-1,4-benzoxazin-3-one(3b), and 6-methoxy-benzoxazolin-2-one(4b) on *Pratylenchus zeae* behavior[a]

Substance	Concentration [mmol/l]	n	I (mean ± SE)	
			Incubation 6h	Incubation 24 h
GDIMBOA (2b)	4.0	21	−0.11 ± 0.10	−0.18 ± 0.09*
	2.0	20	0.17 ± 0.09	0.17 ± 0.09
	0.2	21	0.22 ± 0.08*	0.17 ± 0.09
	0.02	24	−0.12 ± 0.06	−0.57 ± 0.04*
	0.002	24	0.02 ± 0.08	0.08 ± 0.08
DIMBOA (3b)	2.0	45	0.24 ± 0.05*	0.36 ± 0.05*
	0.2	24	0.09 ± 0.08	0.04 ± 0.07
	0.02	24	−0.51 ± 0.06*	−0.38 ± 0.06*
	0.002	24	0.26 ± 0.08*	0.62 ± 0.05*
MBOA (4b)	2.0	23	−0.10 ± 0.10	−0.09 ± 0.10
	0.2	24	−0.22 ± 0.06*	−0.19 ± 0.06*
	0.02	18	0.11 ± 0.08	0.22 ± 0.07*
	0.002	32	0.09 ± 0.06	0.12 ± 0.07

[a]Values larger than 0 indicate attraction and those below 0 repellent activity; * indicates values significantly different from control.

We conclude that DIMBOA plays some role in *Pratylenchus zeae* host recognition. Comparison of the effects of host root exudates with the isolated DIMBOA are shown in (Fig. 8). Similarity in the form of the curves, in relation to DIMBOA concentration, can be seen up to the concentration of approximimately 0.02 mmol/l. Thereafter, the effects diverge with increasing concentration. Interestingly, effects of phytosiderophores released by root exudation of

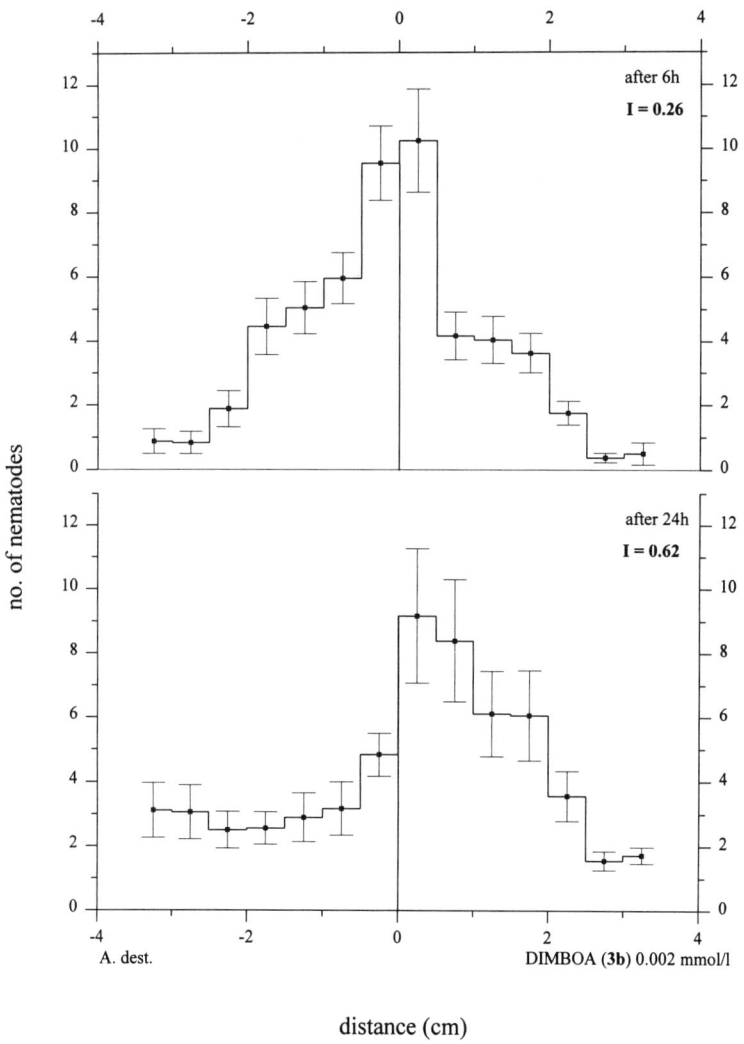

Figure 7. Effects of 2,4-dihydroxy-7-methoxy-*2H*-1,4-benzoxazin-3(4*H*)-one (**3b**) at concentration of 2 µmol/l on root lesion nematode behavior after 6 and 24 hours exposure time.

cereal plants also have been described for *Heterodera avena*. Lung tested activity of avenic acid, mugenic acid, and 2-desoxymugenic acid.[87] High attraction was found at concentrations below 0.047 μmol/l, whereas higher concentrations were repellent.

To further clarify the influence of hydroxamic acids on root lesion nematodes, we investigated DIMBOA concentrations in roots and their effects on nematode infection of seedlings by *Pratylenchus zeae* for 12 different cultivars of *Zea mays* (Table 4). The DIMBOA content of 8 day old seedling roots was determined by HPLC and expressed as the sum of GDIMBOA, DIMBOA, and MBOA concentrations. At the same time (4 days after inoculation), the penetration of roots by the nematodes was monitored microscopically.

The investigated maize cultivars significantly differed in root DIMBOA content up to about one order of magnitude. For the given range of concentrations, no direct correlation between DIMBOA in root tissue and nematode infection was found (Table 4). Nevertheless, cultivars with high levels of hydroxamic acids, such as cv. Mutin and cv. Marshall, exhibited high levels of root lesion nematode infection. Conversely, low levels of infection were observed with other cultivars having a lower DIMBOA content.

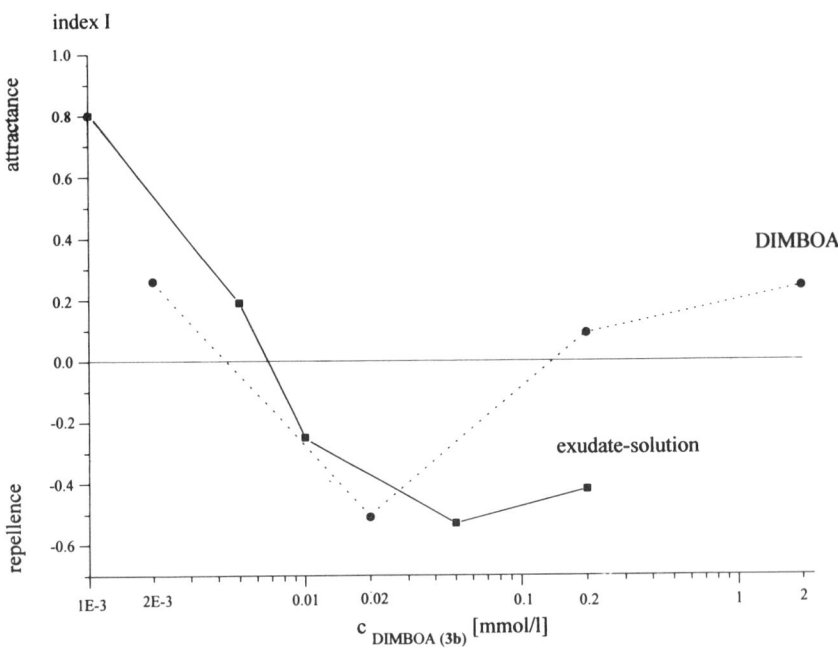

Figure 8. Comparison of the biological effect of different concentrations of the exudate-solution and DIMBOA after 6h exposure time.

Table 4. DIMBOA content and nematode attack of roots from 12 selected cultivars of *Zea mays*[a]

Cultivar	DIMBOA content (mean ± SE) [μmol/g fresh weight]	Infection
Mutin	9.98 ± 1.26	+++
Marshall	7.91 ± 1.89	+
Gracillima	5.52 ± 2.69	+
Osetiskaja	4.77 ± 1.34	+
Südkrim	4.27 ± 0.77	+
Extra early B.	3.86 ± 0.51	+
Marano	3.60 ± 0.35	+++
Albanien	3.36 ± 1.81	+
Jacque	3.12 ± 0.28	+
Canbreol	2.81 ± 2.27	–
Südtunesien	2.33 ± 1.01	+
White rice	1.83 ± 0.14	+

[a]Values are means of two experiments with two replicates; – no infection, + weak, ++ moderate and +++ strong infection.

After 4 days of infection, plants were analyzed for content of different hydroxamate related compounds in comparison to uninoculated controls. Results of these experiments demonstrated that infection does not influence the level of benzoxazinoide constituents in roots. Furthermore, induction of other cyclic hydroxamic acid related compounds was not detected. We conclude that there is no phytoalexin character to these allelochemicals.

As mentioned above, benzoxazinoide allelochemicals have been implicated in plant defense against different pathogens and herbivores. Surprisingly, and in contrast, an important role for these compounds as defense chemicals against *Pratylenchus zeae* seems unlikely. Our results, however, demonstrate clearly that these allelochemicals can act as signals for host recognition by nematodes.

FUTURE RESEARCH DIRECTIONS

The results of *in vitro* bioassays demonstrate that crude exudates of maize roots, which contain different derivatives of benzoxazinoide allelochemicals, influence the chemotactic behavior of the root lesion nematode, *P. zeae*. The attraction of the nematode at low concentrations may mimic allelochemical gradients along the root and/or the gradient extending into the rhizosphere. The concentration-dependent activity of the exudate may be due to sensitive chemotactic receptors on the nematode and to its related overall behavior in the rhizosphere. Future study might include, for instance, the investigation of stylet thrusting using video photographic technology. Physiological effects on specific enzymatic systems of nematodes would also be of special interest.

P. zeae is normally attracted not to the root tip zone, as many other plant parasites are, but rather penetrates the tissue at a distance behind the root tip. The strong concentration dependency may indicate that nematodes are attracted to root zones where optimal amounts of these substances are exuded, suggesting that they play some role in close quarter recognition of specific root zones.

An attractant activity of DIMBOA was observed at a low concentration, similar to concentrations expected in root exudates under natural conditions. The *in vitro* bioassay, however, produced only fragmentary information regarding the complex situation in the rhizosphere. Additive and synergistic effects of allelochemicals on root lesion nematodes may be possible, and studies of effects of microbial metabolites of the allelochemicals also are required.

Additionally, allelochemicals exuded by roots of maize may be important in the regulation of the behavior of other economically important plant parasitic neamtodes, *i.e.*, the ectoparasitic nematodes *Xiphinema, Trichodorus, Belonolaimus, Longidorus,* and *Tylenchorhynchus*. These genera feed directly at or near the root tip and could be affected by exudates of the host plant. Whether they are repelled or attracted to these substances at relevant concentrations needs study. Although there did not seem to be any correlation between cultivar content of cyclic hydroxamic acids and the level of *P. zeae* penetration, such studies could be relevant to cultivar attractiveness to cyst nematodes of maize (*Heterodera spp.*) and root-knot nematodes (*Meloidogyne spp.*) that penetrate through the root tip where allelochemicals may be found at higher concentrations.

CONCLUSION

It is generally accepted that root exudates play an important role in plant-nematode interaction, but little is known about the molecular nature of the signal compounds acting as attractants, repellents, or hatching stimuli to nematodes. We have provided new information about chemotactic effects of cyclic hydroxamic acids and related compounds exuded by maize. Biological activity of this important class of chemicals to plant parasitic nematodes is reported here for the first time.

Much research has focused on the importance of benzoxazinoide allelochemicals as defense agents of gramineous plants against diseases and pests. Antiherbivore and anti-pathogen effects are well documented in the literature as are allelopathic ones. The role of these natural products in interactions of plants with the environment is diverse. There are examples of effective detoxification strategies by herbivores and pathogens. The tolerance of organisms to cyclic hydroxamic acids may be considered an important feature of specific host-parasite and pathogen interactions. This study demonstrates attractant effects of

DIMBOA on the root lesion nematode *Pratylenchus zeae* at low concentrations, similar to concentrations in root exudates under natural conditions. The action of this compound as a signal for host recognition in nematodes is another example of the multiple roles that natural products play in ecological systems.

ACKNOWLEDGMENTS

We thank Daniela Hambrecht for careful attendance of maize CRET-systems and assistance in HPLC. We acknowledge the support in nematode cultivation of John Kimenju and in statistical analysis of data of Dr. T. Seelhorst (Mathematisches Seminar, Universität Bonn). Also, we thank especially Dr. D. Sicker (Institut für Organische Chemie, Universität Leipzig) for providing synthetic HMBOA and Prof. K. Hammer (Institut für Pflanzengenetik und Kulturpflanzenforschung Gatersleben) for seeds of different maize cultivars.

REFERENCES

1. STEINER, G. 1925. The problem of host selection and host specialization of certain plants infesting nemas and its application in the study of nemic pests. Phytopathology 15: 499–534.
2. ATKINSON, H.J., FOWLER, M. 1990. Changes in polyphosphoinositide metabolism in *Globodera rostochiensis* following stimulation to hatch by potato root diffusate. Nematologica 36: 417–423.
3. ATKINSON, H.J., FOWLER, M., ISAAC, R.E. 1987. Partial purification of hatching activity for *Globodera rostochiensis* from potato root diffusate. Ann. Appl. Biol. 110: 115–125.
4. ATKINSON, H.J., TAYLOR, J.D., FOWLER, M. 1987. Changes in the second stage juveniles of *Globodera rostochiensis* prior to hatching in response to potato root diffusate. Ann. Appl. Biol. 110: 105–114.
5. PERRY, R.N., GAUR, H.S. 1996. Host plant influences on the hatching of cyst nematodes. Fund. Appl. Nematol. 19: 505–510.
6. PERRY, R.N., ZUNKE, U., WYSS, U. 1989. Observations on the response of the dorsal and subventral oesophageal glands of *Globodera rostochiensis* to a hatching stimulus. Revue de Nematologie. 12: 91–96.
7. PERRY, R.N., BEANE, J. 1983. The hatching of *Heterodera goettingiana* in response to brief exposure to pea root diffusate. Nematologica 29: 34–38.
8. PERRY, R.N., HODGES, J.A., BEANE, J. 1981. Hatching of *Globodera rostochiensis* in response to potato root diffusate persisting in soil. Nematologia 27: 349–352.
9. CLARKE, A.J., HENNESY, J. 1984. Movement of *Globodera rostochiensis* (Wollenweber) juveniles stimulated by potato-root exudate. Nematologica 30: 206–212.
10. PROT, J.C. 1980. Migration of plant-parasitic nematodes towards plant roots. Revue de Nematologie 3: 305–318.
11. MOLTMANN, E. 1990. Kairomones in root exudate of cereals. Their importance in host finding of juveniles of the cereal cyst nematode, *Heterodera avenae* (Woll.), and their characterization. Z. Pflanzenkrankheiten Pflanzenschutz 97: 458–469.
12. PAPADEMETRIOU, M.K., BONE, L.W. 1983. Chemotaxis of larval soybean cyst nematode *Heterodera glycines* Race 3, to root leachates and ions. J. Chem Ecol. 9: 387–396.

13. DIEZ, J.A., DUSENBERY, D.B. 1989. Repellency of root-knot nematodes from exudate of host roots. J. Chem. Ecol. 15: 2445–2455.
14. TANDA, A.S., ATWAL, A.S., BAJAJ, Y.P.S. 1989. In vitro inhibition of root knot nematode, *Meloidogyne incognita*, by sesame root exudate and its amino acids. Nematologica 35: 115–124.
15. CASTRO, C.E., BELSER, N.O., McKINNEY, H.E., THOMASON, I.J. 1989. Quantitative bioassay for chemotaxis with plant parasitic nematodes. Attractant and repellent fractions for *Meloidogyne incognita* from cucumber roots. J. Chem. Ecol. 15: 1297–1309.
16. NIEMEYER, H.M. 1988. Hydroxamic acids (4-Hydroxy-1,4-benzoxazin-3-ones), defence chemicals in the gramineae. Phytochemistry 27: 3349–3358.
17. PRATT, K., KUMAR, P., CHILTON, W.S. 1995. Cyclic hydroxamic acids in dicotyledonous plants. Biochem. Syst. Ecol. 23: 781–785.
18. PENG, A.B., CHILTON W. S. 1994. Biosynthesis of DIMBOA in maize using deuterium oxid as a tracer. Phytochemistry 37: 167–171.
19. BAILEY, B.A., LARSON, R.L. 1991. Maize microsomal benzoxazinone-N-monooxygenase. Plant Physiol. 95: 792–796.
20. XIE, Y., ARNASON, J.T., PHILOGENE, B.J.R., OLECHOWSKI, H.T., HAMILTON, R.I. 1992. Variation of hydroxamic acid content in maize roots in relation to geographic origin of maize germplasm and resistance to western corn rootworm (Coleoptera: Chrysomelidae). J. Econ. Entomol. 85: 2478–2485.
21. XIE, Y., ARNASON, J.T., PHILOGENE, B.J.R., ATKINSON, J., MORAND, P. 1991. Distribution and variation of cyclic hydroxamic acids and related compounds in maize (*Zea mays*) root system. Can. J. Bot. 69: 677–681.
22. KLUN, J.A., TIPTON, C.L., ROBINSON, J.F. 1970. Isolation and identification of 6,7-dimethoxy-2-benzoxazolinone from dried tissues of *Zea mays* (L.) and evidence of its cyclic hydroxamic acid precursor. J. Agr. Food Chem. 18: 663–665.
23. NAGAO, T., OTSUKA, H., KOHDA, H., SATO, T., YAMASAKI, K. 1985. Benzoxazinones from *Coix Lachryma-Jobi* var. *Ma-Yuen*. Phytochemistry 24: 2959–2962.
24. WOODWARD, M.D., CORCUERA, L.J., SCHNOES, H.K., HELGESON, J.P., UPPER, C.D. 1979. Identification of 1,4-benzoxazin-3-ones in maize extracts by gas-liquid chromatography and mass spectrometry. Plant Physiol. 63: 9–13.
25. HOFMAN, J., HOFMANOVA, O., HANUS, V. 1969. 1,4-Benzoxazinone derivatives in plants. A new glucoside derivative from *Zea mays*. Tetrahedron Lett. No. 57: 5001–5002.
26. BÜCKER, C., GRAMBOW, H.J. 1990. Alterations in 1,4-benzoxazinone levels following inoculation with stem rust in wheat leaves carrying various alleles for resistance and their possible role as phytoalexins in moderately resistant leaves. Z. Naturforsch. 45c: 1151–1155.
27. GRAMBOW, H.J., LÜCKGE, J. 1986. Occurence of 2-(2-hydroxy-4,7-dimethoxy-$2H$-1,4-benzoxazin-3-one)-β-D-glucopyranoside in *Triticum aestivum* leaves and its conversion into 6-methoxy-benzoxazolinone. Z. Naturforsch. 41c: 684–690.
28. KLUGE, M., GRAMBOW, H.J., SICKER, D. 1997. (2R)-2-β-D-Glucopyranosyloxy-4,7-dimethoxy-$2H$-1,4-benzoxazin-3(4H)-one from *Triticum aestivum*. Phytochemistry 44: 639–641.
29. HEDIN, P.A., DAVIS, F.M., WILLIAMS, W.P. 1993. 2-Hydroxy-4,7-dimethoxy-1,4-benzoxazin-3-one (N-O-Me-DIMBOA), a possible toxic factor in corn, to the southwestern corn borer. J. Chem. Ecol. 19: 531–542.
30. FIELDER, D.A., COLLINS, F.W., BLACKWELL, B.A., BENSIMIN, C., APSIMON, J.W. 1994. Isolation and characterization of 4-acetyl-benzoxazolin-2-one (4-ABOA), a new benzoxazolinone from *Zea mays*. Tetrahedron Lett. 35: 521–524.
31. ANAI, T., AIZAWA, H., OHTAKE, N., KOSEMURA, S., YAMAMURA, S., HASEGAWA, K. 1996. A new auxin-inhibiting substance, 4-Cl-6,7-dimethoxy-2-benzoxazolinone, from light-grown maize shoots. Phytochemistry 42: 273–275.
32. BARNES, J.P., PUTNAM, A.R. 1987. Role of benzoxazinones in allelopathy by rye (*Secale cereale* L.). J. Chem. Ecol. 13: 889–906.

33. BERGVINSON, D.J., LARSEN, J.S., ARNASON, J.T. 1995. Effect of light on changes in maize resistance against the European corn borer, *Ostrinia nubililalis* (Hübner). Can. Entomol. 127: 111–122.
34. MORSE, S., WRATTEN, S.D., EDWARDS, P.J., NIEMEYER, H.M. 1991. Changes in hydroxamic acid content of maize leaves with time and after artificial damage: implications for insect attack. Ann. Appl. Biol. 119: 239–249.
35. SCHULZ, M., FRIEBE, A., KÜCK, P., SEIPEL, M., SCHNABL, H. 1994. Allelopathic effects of living quackgrass (*Agropyron repens* L.). Identification of inhibitory allelochemicals exuded from rhizome borne roots. Appl. Bot. 68: 195–200.
36. PEREZ, F.J., ORMENO-NUNEZ, J. 1993. Weed growth interference from temperate cereals: The effect of a hydroxamic-acids exuding rye (*Secale cereale* L.) cultivar. Weed Res. 33: 115–119.
37. PEREZ, F.J. 1991. Allelopathic effect of hydroxamic acids from cereals on *Avena sativa* and *Avena fatua*. Phytochemistry 29: 773–776.
38. COUTURE, R.M., ROUTLEY, D., DUNN, G.M. 1971. Role of cyclic hydroxamic acids in monogenic resistance of maize to *Helminthosporium turcicum*. Physiol. Pl. Path. 1: 515–521.
39. BAKER, E.A., SMITH, I.M. 1977. Antifungal compounds in winter wheat resistant and susceptible to *Septoria nodorum*. Ann. Appl. Biol. 87: 67–73.
40. RICHARDSON, M.D., BACON, C.W. 1995. Catabolism of 6-methoxy-benzoxazolinone and benzoxazolinone by *Fusarium moniliforme*. Mycologia 87: 510–517.
41. DABLER, J.M., PAPPELIS, A.J., BEMILLER, J.N. 1969. Effects of phenolic acids and corn extracts upon spore germination of *Diplodia zeae*. Phytopathology 69: 1098–1101.
42. EL NAGHY, M.A., SHAW, M. 1966. Correlation between resistance to stem rust and the concentration of a glucoside in wheat. Nature 23: 17–18.
43. ZHU, Y.L., HE, R., LIU, J.L. 1994. Changes of DIMBOA contents in resistant and susceptible near-isolines of maize during infection with *Exserohicum turcicum*. Acta Agron. Sinica 20: 653–657.
44. THACKRAY, D.J., WRATTEN, S.D., EDWARDS, P.J., NIEMEYER, H.M. 1990. Resistance to the aphids *Sitobion avenae* and *Rhopalosiphum padi* in Gramineae in relation to hydroxamic levels. Ann. Appl. Biol. 116: 573–582.
45. BOHIDAR, K., WRATTEN, S.D., NIEMEYER, H.M. 1986. Effect of hydroxamic acids on the resistance of wheat to the aphid *Sitobion avenae*. Ann. Appl. Biol. 109: 193–198.
46. GIVOVICH, A., NIEMEYER, H.M. 1995. Comparison of the effect of hydroxamic acids from wheat on five species of cereal aphids. Entomol. Exp. Appl. 74: 115–119.
47. MAYORAL, A.M., TJALLINGII, W.F., CASTANERA, P. 1996. Probing behavior of *Diuraphis noxia* on five cereal species with different hydroxamic levels. Entomol. Exp. Appl. 78: 341–348.
48. LESZCZYNSKI, B., DIXON, A.F.G. 1992. Resistance of cereals to aphids: The interaction between hydroxamic acids and glutathione S-transferases in the grain aphid *Sitobion avenae* (F.) (Homoptera: Aphididae). J. Appl. Ent. 113: 61–67.
49. ROBINSON, J.F., KLUN, J.A., BRINDLEY, T.A. 1978. European corn borer: A non preference mechanism of leaf feeding resistance and its relationship to 1,4-benzoxazin-3-one concentration in dent corn tissue. J. Econ. Entomol. 71: 461–465.
50. GUTHRIE, W.D. 1981 Maize whorl stage resistance to the first four instars of European corn borer larvae (Lepidoptera: Pyralidae). J. Kans. Entomol. Soc. 54: 737–740.
51. HOUSEMAN, J.G., CAMPOS, F., THIE, N.M.R., PHILOGENE, B.J.R., ATKINSON, J., NORAND, P., ARNASON, J.T. 1992. Effect of maize-derived compounds of European corn borer (Lepidoptera: Pyralidae). J. Econ. Ent. 85: 669–674.
52. FENGMING, Y., CHONGREN, X., SONGGANG, L., CHANGSHAN, L., JUHUAI, L. 1995. Effects of DIMBOA on several enzymatic systems in Asian corn borer, *Ostrinia furnacalis* (Guenee). J. Chem. Ecol. 21: 2047–2056.
53. ABEL, C.A., WILSON, R.C., ROBBINS, J.C. 1995. Evaluation of peruvian maize for resistance to European corn borer *Ostrinia nubialis*. J. Econ. Ent. 88: 1044–1048.

54. HIBBARD, B.E., BJOSTARD, L.B. 1990. Isolation of corn semiochemicals attractive and repellent to Western corn rootworm larvae. J. Chem. Ecol. 16: 3425–3439.
55. BJOSTARD, L.B., HIBBARD, B.E. 1992. 6-Methoxy-2-benzoxazolinone: A semiochemical for host location by Western corn rootworm larvae. J. Chem. Ecol. 18: 931–944.
56. GAGLIARDO, R.W., CHILTON, W.S. 1992. Soil transformation of 2(3H)-benzoxazolone of rye into phytotoxic 2-amino-3H-phenoxazin-3-one. J. Chem. Ecol. 18: 1683–1691.
57. KUMAR, P., GAGLIARDO, R.W., CHILTON, W.S. 1993. Soil transformation of wheat and corn metabolites MBOA and DIM$_2$BOA into aminophenoxazinones. J. Chem. Ecol. 19: 2453–2461.
58. FRIEBE, A., WIELAND, I., SCHULZ, M. 1996. Tolerance of *Avena sativa* to the allelochemical benzoxazolinone. Degradation of BOA by root-colonizing bacteria. Appl. Bot. 70: 150–154.
59. QUEIROLO, C.B., ANDREO, C.S., NIEMEYER, H.M., CORCUERA, L.J. 1983. Inhibition of ATPase from chloroplasts by a hydroxamic acid from the Gramineae. Phytochemistry 22: 2455–2458.
60. MASSARDO, F., ZUNIGA, G.E., PEREZ, L.M., CORCUERA, L.J. 1994. Effects of hydroxamic acids on electron transport and their cellular location in corn. Phytochemistry 35: 873–876.
61. QUEIROLO, C.B., ANDREO, C.S., VALLEJOS, R.H., NIEMEYER, H.M., CORCUERA, L.J. 1981. Effects of hydroxamic acids isolated from Gramineae on adenosine 5'-triphosphate synthesis in chloroplasts. Plant Physiol. 68: 941–943.
62. NIEMEYER, H.M., CALCATERRA, N.B., ROVERI, O.A. 1986. Inhibition of mitochondrial energy-linked reactions by 2,4-dihydroxy-7-methoxy-1,4-benzoxazin-3-one (DIMBOA), a hydroxamic acid from Gramineae. Biochem. Pharmacol. 35: 3909–3914.
63. NIEMEYER, H.M., CALCATERRA, N.B., ROVERI, O.A. 1987. Inhibition of energy metabolism by benzoxazolin-2-one. Comp. Biochem. Physiol. B 87: 35–39.
64. FRIEBE, A., ROTH, U., KÜCK, P., SCHNABL, H., SCHULZ, M. 1997. Effects of 2,4-dihydroxy-1,4-benzoxazin-3-ones on the activity of plasma membrane H$^+$-ATPase. Phytochemistry 44: 979–983.
65. CUEVAS, L., NIEMEYER, H.M., PEREZ, F.J. 1990. Reaction of DIMBOA, a resistance factor from cereals, with α-chymotrypsin. Phytochemistry 29: 1429–1432.
66. PEREZ, F.J., NIEMEYER, H.M. 1989. Reaction of DIMBOA, a resistance factor from cereals, with papain. Phytochemistry 28: 1597–1600.
67. HASHIMOTO, Y., SHUDO, K. 1996. Chemistry of biologically active benzoxazinoids. Phytochemistry 43: 551–559.
68. NIEMEYER, H.M, CORCUERA, L.J., PEREZ, F.J. 1982. Reaction of a cyclic hydroxamic acid from Gramineae with thiols. Phytochemistry 21: 2287–2289.
69. PEREZ, F.J., NIEMEYER, H.M. 1989. Reaction of DIMBOA with amines. Phytochemistry 28: 1831–1834.
70. ISHIZAKI, T., HASHIMOTO, Y., SHUDO, K., OKAMOTO, T. 1982. Reaction of 4-acetoxy-1,4-benzoxazin-3-one with DNA. A possible chemical mechanism for the antifungal and mutagenic activities. Tetrahedron Lett. 23: 4055–4056.
71. HASEGAWA, K., TOGO, S., URASHIMA, M., MIZUTANI, J., KOSEMURA, S., YAMAMURA, S. 1992. An auxin-inhibiting substance from light-grown maize shoots. Phytochemistry 31: 3673–3676.
72. VENIS, M.A., WATSON, P.J. 1978. Naturally occurring modifiers of auxin-receptor interaction in corn: Identification as benzoxazolinones. Planta 142: 103–107.
73. HOSHI-SAKODA, M., USUI, K., ISHIZUKA, K., KOSEMURA, S., YAMAMURA, S., HASEGAWA, K. 1994. Structure-activity relationships of benzoxazolinones with respect to auxin-induced growth and auxin-binding protein. Phytochemistry 37: 297–300.
74. DE WAELE, D., JORDAAN, E.M. 1988. Plant-parasitic nematodes on field crops in South africa. 1. Maize. Revue Nematol. 11: 65–74.

75. TANG, C. S., YOUNG, C.C. 1982. Collection and identification of allelopathic compounds from the undisturbed root system of bigalta limpograss (*Himarthria altissima*). Plant Physiol. 69: 155–160.
76. WOODWARD, D., CORCUERA, L.J., HELGESON, J.P., KELMAN, A., UPPER, C.D. 1978. Decomposition of 2,4-dihydroxy-7-methoxy-2H-1,4-benzoxazin-3(4H)-one in aqueous solution. Plant Physiol. 61: 796–802.
77. HARTENSTEIN, H., KLEIN, J., SICKER, D. 1993. Efficient procedure for (2R)-ß-D-glucopyranosyloxy-4-hydroxy-7methoxy-2H-1,4-benzoxazin-3(4H)-one from maize. Ind. J. Heterocycl. Chem. 2: 151–153.
78. ATKINSON, J., MORAND, P., ARNASON, J.T., NIEMEYER, H.M., BRAVO, V. 1991. Analogues of the cyclic hydroxamic acid 2,4-dihydroxy-7-methoxy-2H-1,4-benzoxazin-3-one: Decomposition to benzoxazolinones and reaction with mercaptoethanol. J. Org. Chem. 56: 1788–1800.
79. PEREZ, F.J., ORMENO-NUNEZ, J. 1991. Differences in hydroxamic acids content in roots and root exudates of wheat (*Triticum aestivum* L.) and rye (*Secale cereale* L.): Possible role in allelopathy. J. Chem. Ecol. 17: 1037–1043.
80. PETHO, M. 1993. Occurrence of cyclic hydroxamic acids in the tissues of barnyard grass (*Echinochloa crus-galli*/L./P. B.), and their role in allelopathy. Acta Agron. Hungarica 42: 197–202.
81. FRIEBE, A., SCHULZ, M., KÜCK, P., SCHNABL, H. 1995: Phytotoxins from shoot extracts and root exudates of *Agropyron repens* seedlings. Phytochemistry 38: 1157–1159.
82. PETHO, M. 1992. Occurrence and physiological role of benzoxazinones and their derivatives. III. Possible role of 7-methoxy-benzoxazinone in iron uptake of maize. Acta Agron. Hungarica 41: 57–64.
83. PETHO, M. 1993. Possible role of cyclic hydroxamic acids in the iron uptake by grasses. Acta Agron. Hungarica 42: 203–214.
84. SACHS, L. 1968. Statistische Auswertungsmethoden. Springer-Verlag Berlin, pp. 326–329.
85. SIDAK, Z. 1967. Rectangular confidence regions for the means of multivariate normal distributions. J. Am. Stat. Association 62: 626–633.
86. SOKAL, R.R., ROHLF, F.J. 1995. Biometry. The Principles and Practice of Statistics in Biological Research. 3rd Ed., W.H. Freeman & Comp., New York, pp. 239–240.
87. LUNG, G. 1993. The role of phytosiderophores as an attractive substances of root exudates from several cereals for second stage juvèniles of *Heterodera avenae*. Med. Fac. Landbouww. Univ. Gent. 58: 729–735.

Chapter Six

CHEMICAL SIGNALS IN THE PLANT–NEMATODE INTERACTION

A Complex System?

Godelieve Gheysen

Department of Genetics
Flanders Interuniversity Institute for Biotechnology (VIB)
Universiteit Gent
K.L. Ledeganckstraat 35
B-9000 Gent, Belgium

Introduction 95
Chemoreception in Nematodes 96
Hatching Factors 97
Host Finding 99
In Search of a Place to Settle 101
The Plant Hormone Debate 102
Resistant Plant Cultivars—Hypersensitive Response 103
Root Cell Development—Nematode Feeding Site 105
Sex Pheromones 107
Plant Host Influence on Nematode and Egg Development 107
Nematode Infection Influence on Host Defense Response 108
Conclusion 109

INTRODUCTION

Compared to bacterial and fungal phytopathology, and in contrast to semiochemical research related to insect behavior, the knowledge of plant–nematode interactions and the chemical signaling involved is limited and progressing slowly. Their complex life cycles and the small size of these obligate parasites are major stumbling blocks for experimental approaches.

Phytochemical Signals and Plant–Microbe Interactions, edited by Romeo *et al.*
Plenum Press, New York, 1998.

The sedentary endoparasitic nematodes especially have evolved a complex relationship with their plant hosts. After penetration of the plant root and migration to a suitable site, the second-stage juveniles induce specialized feeding cells. The nematodes become sedentary and then undergo three molts during their development to the adult stage. The formation and functioning of the permanent feeding sites is dependent on a continuous interaction between the nematode and the redifferentiated root cells.[1] Root-knot nematodes induce giant cells by mitosis uncoupled from cytokinesis, whereas cyst nematodes generate a syncytium through cell wall dissolution.[2] Both types of feeding cells contain a dense cytoplasm with many organelles and multiple enlarged nuclei. Root-knot nematodes are embedded in a gall or root knot originating from cell divisions and hypertrophy. Adult females lay their eggs outside of the root in a gelatinous matrix from which the next generation can immediately reinvade the host plant. In the case of cyst nematodes, the adult females become filled with hundreds of embryonated eggs. When the female dies, her body wall hardens to form a protective coat enclosing the eggs, and this cyst is an ideal survival structure to bridge adverse conditions until the next growing season.

Adaptations of nematodes to plant parasitism involved considerable changes to the esophageal glands and the development of the stylet, a hollow protrusible structure that is used to pierce plant cell walls. Secretions from two subventral glands and one dorsal gland are thought to be essential for many aspects of parasitism.[3] The subventral glands have been postulated to secrete substances important for penetration, migration, feeding cell induction, and food digestion. One of the functions of the dorsal gland is most probably the formation of feeding tubes in the cytoplasm of the feeding cells. Other possible signaling molecules of the nematode might originate from the cuticle surface coat or amphidial gland secretions.[4,5]

Because endoparasitic nematodes cannot feed outside the plant root, random finding of a food source or a sexual partner would often hamper survival due to unnecessary energy expenditures. Therefore, chemical signals and corresponding behavioral responses must play a major role in many aspects of the nematode's life cycle. Based mainly on ultrastructural studies, the amphids are considered to be the primary chemosensory organs in nematodes. Therefore, first an overview on the current knowledge of the structure and function of the amphids will be given. Then, the nematode's life cycle, with special attention to the exchange of signals between the parasite and its host plant, will be discussed.

CHEMORECEPTION IN NEMATODES

The detailed ultrastructure of the sense organs of nematodes has been reviewed by Coomans and De Grisse.[6] The amphids are the largest and most complex of the sense organs and their structure is remarkably conserved in a wide range of nematodes. The two amphids are situated laterally on either side of the mouth, each opening to the exterior via a prominent pore. The amphidial cavity contains a number of dendritic processes which are surrounded by secretions produced by the glandular sheath cell.

Chemical signals initially come into contact with the secretions filling the amphidial duct. Several studies indicate that at least some of the components in the amphidial secretions are glycoproteins, and the composition of these secretions is different among nematode species.[7–10] This is illustrated by the differential immunoreactivity of an antiserum towards a 32-kDa glycoprotein (gp32), specifically present in the amphidial secretions of five *Meloidogyne* species, but not found in representatives of eight other genera, including *Globodera* and *Heterodera*.[11] Gp32 is produced in all stages of the *Meloidogyne* life cycle, except in the sedentary adult female.[12] Differences among *Globodera* species in the reactivity of a monoclonal antibody to amphidial secretions have also been observed.[13]

In *Caenorhabditis elegans*, structure–function relationships for the role of amphids in chemoreception have been demonstrated by the occurrence of behavioral mutants with structurally altered amphids[14] and, more recently, by laser microbeam ablation of individual amphidial neurons.[15] In plant-parasitic species, evidence comes from nematicide research, *e.g.* aldicarb disturbs nematode orientation and causes changes in the ultrastructure of the amphidial sheath cell in the affected nematodes.[16] A more direct proof is the finding that incubation of infective *M. javanica* juveniles with gp32 antiserum significantly retards attraction of the nematode to host roots.[12]

HATCHING FACTORS

The first step in the nematode's life cycle towards parasitism is hatching. For most phytoparasitic nematodes, hatching occurs once development to second-stage juveniles has been completed and the environmental conditions, such as temperature and moisture, are favorable. In many species, however, this spontaneous hatch rate is enhanced by specific plant root diffusates. Generally, nematodes with a narrow host range require host root diffusate for efficient hatching. The most extreme example is the potato cyst nematode that almost completely depends on root diffusates for hatching.[17] This dependence, in the case of *Globodera rostochiensis* and *G. pallida*, is an obvious target for specific control strategies, such as hatching induction in the absence of a host, resulting in starvation of the nematode. Therefore, and because of the specificity of the stimulus, many details about hatching mechanisms have been provided by studies with these particular nematode species.

Initial studies indicated that at least four different hatching factors were present in potato root diffusate affecting different aspects of the hatching,[18,19] but later work revealed the presence of at least 25 hatching factors.[20] The molecular mass of these hatching factors in potato root diffusate was reported to be 350–437[18] and 498.[21] Two factors have been identified as the potato glycoalkaloids, α-chaconine and solanine (Fig. 1a).[22] Natural hatching factors for *Heterodera glycines* have also been identified and named glycinoeclepin A, B, and C (Fig. 1b).[23–25] Because these chemicals are effective at very low doses (glycinoeclepin A induces hatching at a concentration of 10^{-12} g/ml), they are interesting candidates for agrochemic use. Analogs and precursors of glycinoe-

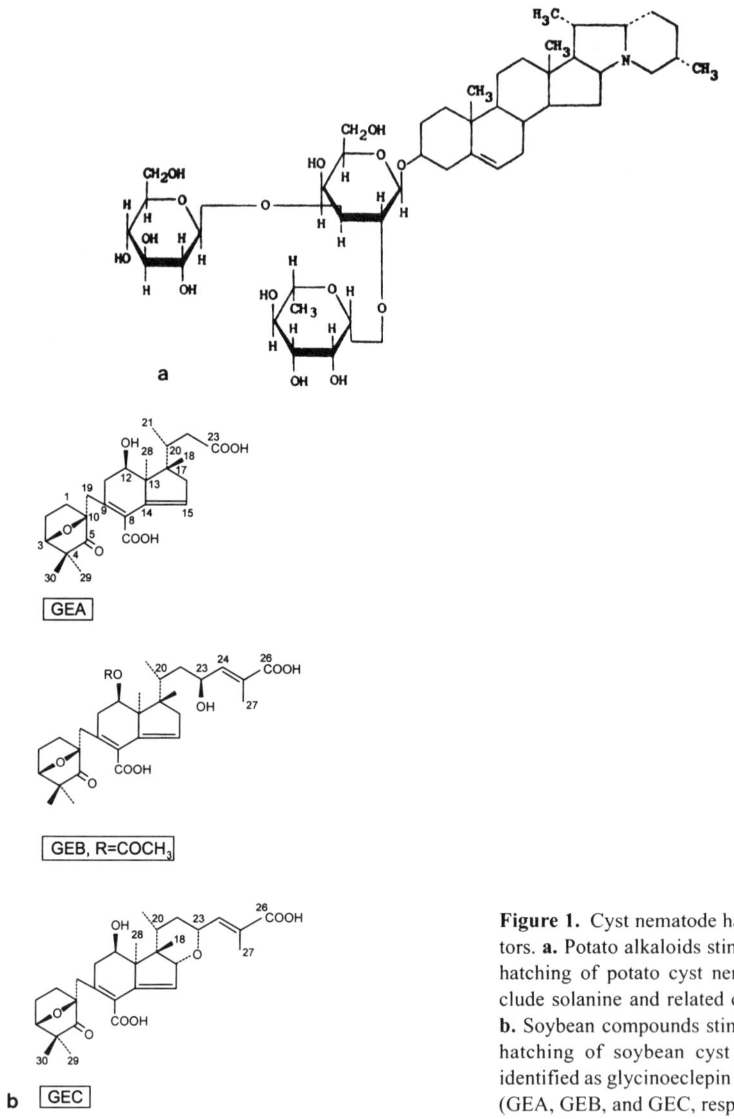

Figure 1. Cyst nematode hatching factors. **a.** Potato alkaloids stimulating the hatching of potato cyst nematodes include solanine and related compounds. **b.** Soybean compounds stimulating the hatching of soybean cyst nematodes identified as glycinoeclepin A, B, and C (GEA, GEB, and GEC, respectively).

clepin A have been synthesized and tested for effects on hatching of *H. glycines*, and some of them did not stimulate but rather inhibited hatching.[26]

The activity of plant root diffusates is not constant during the plant's life.[27,28] At the end of the season, aging plants produce diffusates with lower hatching activity, ensuring a "carry over" population after host senescence or crop harvest. Besides this adaptation for surviving dry or cold seasons, there generally is also a correlation between the suitability of a plant to serve as a host for a nematode species and the corresponding hatching activity in its root exudates. For example, black nightshade stimulates more hatch of *G. tabacum* than tobacco and seems also to be a better host.[29] Poor hosts have been reported not to stimulate hatch of *H. glycines*.[30] However, no correlation has been found between the hatching activity of root diffusates and host resistance or susceptibility. For example, potato cultivars carrying resistance from *Solanum tuberosum* ssp. *andigena* produce hatching factors, and some of the *ex-vernei* resistant potato genotypes induce even more hatching than the susceptible control.[31,32] Cultivars of barley, oat, and wheat resistant to *H. avenae* also stimulate hatching of this nematode species.[33]

The absence of secretions and the shrunken state of the sheath cell in unhatched potato cyst nematodes indicates that the amphids may not be functional before hatching and, thus, play no role in the detection of the hatching stimuli present in potato root diffusate.[34] How then is the signal perceived? The hatching factors apparently have a direct effect on eggshell permeability through a Ca^{2+}-mediated change of the inner lipid layer of the eggshell.[35,36] As a result, trehalose escapes from the perivitelline fluid through the eggshell, thus reducing the osmotic pressure on the unhatched juvenile.[37] As a consequence, the second-stage juvenile becomes hydrated and active, cutting its way through the eggshell. The substantial hatch of *H. schachtii* in water could, at least partially, be explained by the lower osmotic pressure of the perivitelline fluid and the fact that these juveniles are unaffected by osmotic pressures high enough to inhibit movement of *G. rostochiensis*.[38]

The eggshell remains rigid during hatching of *G. rostochiensis*, and no lipase or chitinase activity can be detected.[39] In contrast, hatching of *Meloidogyne* is clearly associated with increased flexibility of the eggshell before eclosion.[40] Several enzymes can be detected, including chitinase, collagenase, and lipase, the activity level of the latter being correlated with the cumulative hatching percentage of *M. incognita*.[39] These enzymes could be produced from the activated juvenile or they could be present in an inactive state in the perivitelline fluid until hydration.

HOST FINDING

Because most plant-parasitic nematodes lay their eggs outside the roots, infective juveniles that have emerged from the eggs are unprotected. They are vulnerable to environmental conditions and dispose of a limited energy reserve to locate and penetrate roots of a host plant. As stated above, the perception of plant roots is most probably accomplished through the amphids. The nature of the signal(s) is, however, rather controversial, not only because it might be an interplay of general and more specific molecules, but also because such a variety of experimental conditions have been used, making it difficult to compare results among studies. Nematodes normally move through a three-dimensional soil environment responding to interfering gradients of different stimuli. In contrast, many bioassays have employed a simplified system of radial two-dimensional attraction gradients in agar[41,42] and are, thus, only a fragmentary simulation of the rhizosphere.

It is widely accepted that nematodes are attracted to plant roots and that this attraction is generally non-specific.[43] Physiologically active roots generate several gradients of non-specific attractants in the rhizosphere. In experimental set-ups, nematodes are capable of orientation in (electro)chemical gradients,[43] in CO_2 gradients with a concentration difference of 0.08% per cm,[44] and in temperature gradients of only 0.033°C per cm.[45] Despite some contradictions among observations, it appears that CO_2 is one of the attracting factors for migration of plant-parasitic nematodes toward roots.[43] Additional attractants probably are amino acids, sugars, and other metabolites in the root exudates. Initial penetration of resistant cultivars is, in general, similar to that of susceptible plants.[46]

Nevertheless, in particular cases, a specific interaction has been recognized. For example, Viglierchio[47] found that for *H. schachtii*, attractiveness was correlated with efficiency as a host. A similar relation was observed for *Aphelenchoides besseyi* and different rice cultivars.[48] Less attraction can be the result of a repelling chemical (Fig. 2), *e.g.* cucumber plants carrying a single dominant bitter gene produce triterpenoid compounds that repel nematodes. They attract significantly fewer *M. incognita* juveniles than do nonbitter isogenic lines.[49]

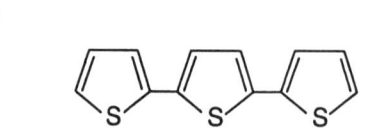

Figure 2. Plant secondary metabolites with nematode-repelling activity. a. Cucurbitacin A, a bitter compound from cucumber. b. α-Terthienyl, repellent and nematotoxic compound from *Tagetes erecta* L. (marigold).

Nematodes preferentially aggregate around specific root zones, with the most important area being the zone of cellular elongation close to the root tip. This region of the root is metabolically very active, strongly exudating, and creating electrical potential gradients.[50–52] Other focal points are lateral root primordia, areas of previous penetration, and other injured root regions.[43]

Interestingly, root exudates not only attract the infective juveniles, but can also induce the characteristic behavior necessary for root penetration. *Heterodera schachtii* juveniles that aggregated around agarose discs containing *Sinapis alba* root exudates showed pre-infection exploratory behavior, including stylet thrusting.[41]

IN SEARCH OF A PLACE TO SETTLE

The initial stages of root-knot nematode infection of plant roots, described in the early forties,[53,54] have been monitored in detail by video-enhanced microscopy inside the thin transparent roots of *Arabidopsis thaliana*.[55] The juveniles penetrate the root in the elongation zone and then migrate intercellularly between cortical cells down to the root tip. There they reverse direction and enter the vascular cylinder where they select parenchymatous cells to induce a feeding site. It has been postulated that the endodermis forms a physical barrier. An accidental breaching of the endodermis, however, does not influence the migration path of the nematode.[56] Although the nature of the signal is unknown,[57] the nematode perceives and follows a gradient that leads it to the root apex. This hypothesis is consistent with the observation that the juveniles, upon entering the root, normally, immediately choose the right direction, toward the root tip.[55] When they arrive in the root meristematic region, most turn around above the apical initials at the quiescent center[56] and move upwards into the developing vascular cylinder. That this second part of their journey is dependent on specific plant signals can be deduced from the behavior of root-knot nematodes in the resistant alfalfa cultivar Moapa 69.[58] Although no physical barrier appeared to be present in the root tips of this resistant cultivar, no juveniles were observed entering the vascular cylinder. Instead they remained clumped and apparently disoriented at the root apex.

Cyst nematode behavior during early stages of the infection process differs in some important aspects from that of root-knot nematodes.[59,60] Second-stage juveniles penetrate the epidermis with a preference for the elongation zone of the root tip. Then, they migrate, cutting through successive cells, through the cortex towards the vascular cylinder. The cellular damage usually causes necrosis along the migration path of the nematode, which completely alters its behavior when it arrives at suitable cells for feeding-site induction.[60] The destructive cutting is then switched to a gentle piercing of the initial syncytial cell, and saliva is injected into the cytoplasm. Again, there is no information on how the juvenile recognizes its target cells.[61]

The migration process seems to involve both mechanical force and enzymatic secretions from the juvenile. Cellulase activity has been detected in several nematodes, such as *H. schachtii*, *M. incognita*, and *M. javanica*.[62,63] Microscopy analysis and immunolabeling confirmed that the root-knot nematodes progress intercellularly by enzymatic separation of contiguous cells at the middle lamella.[55,56] Some of the immunodominant proteins recognized by a panel of monoclonal antibodies are detected both in the secretory granules of the subventral glands of *G. rostochiensis* and in the saliva of preparasitic juveniles. Presence of these proteins is developmentally regulated, as they are found only in (pre)parasitic second-stage juveniles and adult males. These developmental stages have, in common, the ability to migrate within root tissue. It is postulated that these antigens are related to this ability.[64]

THE PLANT HORMONE DEBATE

Phytohormones are plant root components that might be informative for orientation of juveniles. The involvement of these growth regulators in the plant–nematode interaction has long been postulated.[65] The recent finding that *Globodera pallida* secretes a putative auxin-binding protein, possibly associated with the amphids,[66] is a starting point for further analysis.

The cellular changes occurring when the nematodes become sedentary can be mimicked partially by application of indole-3-acetic acid (IAA), *i.e.* hypertrophy, hyperplasia, adventitious root formation, nuclear division without cytokinesis, and the breakdown of cell walls.[67–69] There are several reports of the presence of indole compounds in nematode-induced galls.[1,70–72] The question then arises as to the source of these auxins. Auxin determination by paper chromatography indicates that the type and level of auxins in the gall are characteristic of the parasitizing nematode but can be moderated somewhat by the plant host.[71,72] This phenomenon could be due to the presence of auxins in nematode saliva or to the secretion of nematode enzymes that release free auxins from conjugates in the plant root.[73] However, other observations suggest that the higher auxin levels in galls might rather be the result of a retention of auxins that otherwise would be transported to other plant parts.[74,75]

A different approach for investigating the importance of phytohormones in the establishment of a successful infection is to compare or to modify the levels of specific hormones in compatible versus incompatible plant–nematode interactions. In peach rootstocks resistant to *M. javanica*, nematodes are not capable of completing their life cycle. When 1-naphthylacetic acid is applied, however, roots develop vigorously, galls are formed, and nematodes grow to maturity.[76] Dropkin *et al.*, however, could not see an effect of IAA on tomatoes resistant to *M. incognita*.[77] This apparent contradiction could be due to the lower stability of IAA or to a differential sensitivity of the different plant–nematode interactions to specific auxins. For example, IAA was detected in galls induced by *M.*

javanica but not by *M. incognita* in tomato.[71] Several reports claim the presence of auxin-degrading enzymes in resistant cultivars, *e.g.* in Solanaceae resistant to *G. rostochiensis*[65] and in cotton resistant to *M. incognita*.[78]

In tomato, cytokinin activity was lower in resistant than in susceptible cultivars,[79] and exogenously supplied cytokinins shifted the response of resistant tomato plants toward susceptibility.[77] Similarly, Kochba and Samish[76,80] found that resistance to *M. javanica* in peach rootstocks was correlated with lower endogenous cytokinin levels, and that, although no significant effect on plant growth was observed, the application of kinetin abolished resistance. In contrast, Huettel and Hammerschlag[81] applied 6-benzylaminopurine to the same nematode-resistant peach rootstock Nemaguard and observed initial galling, but a failure of *M. incognita* to develop to maturity. It is difficult to compare these two contradictory observations because different nematodes and hormones were used, and Huettel and Hammerschlag[81] used an *in vitro* system.

There are additional but fragmentary data on the possible role of other hormones in plant–nematode interactions.[82] Spraying 2,4-dichlorophenoxyacetic acid decreased the resistance of oat, but not of barley, to *Ditylenchus dipsaci*.[83] After *Heterodera avenae* penetration, suppressed growth of root axes and emerged laterals corresponded to an increase in abscisic acid and ethylene in oat roots,[84] whereas exogenous application of abscisic acid to susceptible potato plants appeared to suppress reproduction of *M. incognita*.[85]

To obtain clearer insight into the possible causal relationship between phytohormones and plant responses to nematode infection, defined *Arabidopsis thaliana* hormone mutants have been analyzed.[75,86] The mutants were defective in either the biosynthesis of or response to auxin, abscisic acid, gibberellins, and ethylene. Surprisingly, *H. schachtii* as well as *M. incognita* reproduced well on all mutants. No differences were found in syncytium and gall formation in any of the mutant lines tested.

It can be concluded that despite extensive work examining the role of plant hormones in plant–nematode interactions, little progress has been made in this area. Most probably, the reason is the complex nature of plant hormones, which is not even completely understood in normal plant growth. More sophisticated microtechniques, such as immunocytochemistry and electro-spray tandem mass spectroscopy,[87,88] might resolve some of the contradictions about hormonal levels and reveal their origin and exact location in the nematode-infected roots. The eventual causal role of phytohormones, however, will probably only be clarified once the molecular triggers that induce nematode-feeding cells are known.

RESISTANT PLANT CULTIVARS—HYPERSENSITIVE RESPONSE

Plant resistance to nematodes can be broad or only effective against specific pathotypes of a species. In many cases, monogenic dominant resistance genes have been identified, but polygenic resistance also occurs.[78] Although any step of the plant–nematode interaction can be hindered, leading to an unsuccessful infection, in many of the resistant plant genotypes, the juveniles penetrate the roots normally and attempt initiating a nematode-feeding site. It is only then that the plant reacts with a localized necrosis or hypersensitive response characteristic of pathogen resistance in general.[89] Indeed, the cloned nematode resistance gene, $HS1^{pro-1}$, is similar to previously isolated other resistance genes.[90,91]

One of the best characterized resistance traits is the Mi locus conferring resistance to several *Meloidogyne* species in tomato.[92] By 8 to 12 hours after nematode inoculation, a hypersensitive response is visible around the head of the nematode,[93] preventing the formation of giant cells, a process which requires 24 to 48 hours in susceptible plants. The Mi-mediated resistance is temperature sensitive, and temperature shifts have shown that later than 48 hours after inoculation, resistance is no longer triggered at the permissive temperature.[94]

The timing and the exact mechanism of the host-resistant reaction appears to be different depending on the particular resistance gene.[61] For example, in potato clones carrying resistance from different sources, the potato cyst nematode induces a feeding site and becomes sedentary. Sooner or later, depending on the particular resistance gene,[95–97] the initial syncytium becomes surrounded by necrosing tissues and deteriorates. This degeneration limits nematode development to the third or fourth stage, and the few nematodes that develop further become males, an indication of poor nutritional functioning of the syncytium. In the cells adjacent to the syncytia, necrosis is already visible in line PI 437654 two days after inoculation with *H. glycines*.[98] Five days after infection, the syncytium is completely collapsed in a process resembling apoptosis or programmed cell death. In the soybean cultivar Bedford, however, signs of degeneration are first visible inside the syncytium and only after five days of infection.[99] Resistance to *M. incognita* in the cotton cultivar Clevewilt was associated with greatly retarded nematode development, but not with increased necrosis.[100]

The products of resistance genes to a variety of pathogens have turned out to be receptors that trigger defense reactions when challenged by a specific pathogen compound.[90,101,102] In the case of nematodes, this could be a glycoprotein from the surface coat or an esophageal gland product injected into the root cells. A protein characteristic for avirulent *M. incognita* females has been identified in a comparison between near-isogenic virulent and avirulent females,[103] but the role of this protein in avirulence is unknown. Inbred populations of *G. rostochiensis* and *H. glycines* have been produced, and the genetic maps for these species are under development.[104–106] These will provide valuable tools for identification and isolation of the respective avirulence genes.[107]

ROOT CELL DEVELOPMENT—NEMATODE FEEDING SITE

In a compatible plant–nematode interaction, the plant seems to ignore the parasitic nature of the intruding nematode, despite the important necrosis along the migration path of the cyst nematodes. After selecting suitable cells in the root, cyst and root-knot nematodes establish feeding sites that are nematode specific regardless of the tissue and host in which they are induced. Therefore, the key to understanding the complex plant response lies in the signal(s) coming from the nematode. The secretions, that originate in the esophageal glands and are injected into the plant root through the stylet, are believed to be the source of these signals.[3] These secretions might interact with a receptor at the outer side of the plasma membrane, or they could be deposited directly in the cytoplasm of the target cell through a perforation of the plasma membrane at the stylet orifice. Either way, the activation of a receptor protein may trigger the signal transduction cascade leading to feeding site formation. Secretions may also provide specific transcription factors, directly switching on a developmental program.

The exact mechanism causing the cellular changes will become better understood with the identification of the molecules in the secretions. The small amount of material available has hindered easy access to this information. Currently, investigations concentrate on the use of monoclonal antibodies specific for secretory components.[108–111]

Because plants have not evolved for the purpose of feeding nematodes, parasites must exploit existing plant cell differentiation patterns. Many of the characteristics of giant cells and syncytia are similar to those of cells of nutritional organs: dense cytoplasm, genome multiplication, and many organelles. For example, analogies between the tapetum and nematode-feeding cells extend from cytological to molecular characteristics, such as the down-regulation of the CaMV 35S promoter in both organs.[112,113] One of the earliest visible changes during feeding cell initiation is cell cycle activation.[2] This can be seen at the cytological level and has been confirmed by using gene markers. Genes encoding cyclin-dependent kinases and cyclins are transcriptionally activated a few hours after the nematode has selected an initial feeding cell.[114]

The expression of many genes is altered in nematode-feeding sites.[115] Therefore, an alternative way of unraveling the signal transduction pathway from the nematode to the cellular changes is to approach it from the other end. Instead of trying to identify the nematode signal directly, the inducible plant promoters and the transcription factors involved can be characterized. The *TobRB7* promoter, expressed in young vascular root tissue and highly induced in giant cells but not in syncytia, has been dissected molecularly. A 300-bp fragment of this promoter turned out to be specific for the giant cells in infected roots.[116] Only a few nematode-inducible promoters have been analyzed in this way. Apparently, only one promoter, from the *lemmi9* gene (acronym for *Lycopersicon esculentum*/*Meloidogyne incognita*[117]), has been scanned for protein binding sites.[115] A

12-bp imperfect repeat close to the TATA box in the *lemmi9* promoter has been shown to bind specifically a nuclear protein from tomato galls. The motif CANNTG present in this repeat is also found in the *TobRB7* and *hmg1* promoters,[115] which have in common with *lemmi9* that they are highly upregulated in giant cells.

In animals, CANNTG boxes are recognized by proteins belonging to the superfamily of basic helix-loop-helix (bHLH) transcriptional regulators. Interestingly, *Trichinella spiralis*, a parasitic nematode of mammalian muscle, induces a nurse cell system similar to giant cells, apparently by shifting the normal balance of bHLH activator and repressor proteins that control myofiber differentiation.[118] Recently, plant bHLH proteins have been identified, such as a water-stress-related, root-specific protein.[119,120] To date, only one putative transcription factor has been reported to be transcribed in giant cells, a MYB-like protein corresponding to a cDNA identified in a subtraction screening.[121] It would be interesting to test whether this particular protein actually binds to any of the reported giant cell-induced promoters.

Because of the limited amount of material available at early infection stages, the molecular data currently available are mainly from genes induced relatively late in pathogenesis. Most of the dramatic changes in feeding site formation occur in the first days after inoculation. PCR-based RNA amplification methods and promoter-trapping approaches have successfully been applied to study these early stages. Differential display of control and root-knot infected *Arabidopsis* roots has generated many putatively induced clones,[122] and by sequencing analysis, various housekeeping and stress-related genes as well as other potentially interesting genes have been identified. In a joint effort, approximately 2000 transgenic lines tagged with a promoterless *gus* gene were screened for *gus* expression in nematode-feeding cells.[123] Several different lines were obtained containing tags that are rapidly activated in syncytia, giant cells, and/or *Xiphinema*-induced galls. These and other early induced genes or promoters will be useful to study further the plant–nematode interaction and to design genetically engineered nematode-resistant plants.[124]

Detailed GUS analysis of sectioned roots revealed different gene expression patterns inside galls. For example, promoter activity of the *A. thaliana* line Att0001 is high in giant cells, whereas another line shows strong GUS staining in the surrounding gall cells but is negative inside the giant cells.[123] In GUS analyses of whole roots, this difference can easily be overlooked. Another gall-positive *gus* expression pattern has been observed for *Lea14*.[124] Although this gene shows significant sequence similarity in the coding region with *lemmi9* (70% amino acid identity in exon 2), the expression pattern is completely different. *Lemmi9* is highly expressed in the giant cells, whereas *Lea14* is activated in the plant cells around the egg mass of mature females.[124] These differential gene expression patterns in *M. incognita*-induced galls illustrate the complexity of the signals that activate plant gene expression in nematode-infected roots.

SEX PHEROMONES

The sedentary adult female ruptures the root cortex and needs to attract the males which have escaped from the plant roots after their final molt. Little is known about the chemical nature of the sex pheromones of plant-parasitic nematodes. Vanillic acid (Fig. 3), however, has been identified as a substance with sex pheromone activity from *H. glycines* females,[125] attracting males of *H. glycines* but not *H. schachtii*[126] or *G. rostochiensis*.[4] An alternative to behavioral studies, especially when several HPLC fractions need to be tested for male response, is the electrophysiological recording of electrical activity within the cephalic region of individual males.[4] The small size of the plant-parasitic nematodes makes direct recordings from individual sense organs extremely difficult. Nevertheless, it is possible to detect significant increases in spike frequency after application of sex pheromones.[127] Using this method, it was shown that *G. rostochiensis* males exhibit specific mate recognition, whereas *G. pallida* respond to female extracts from both *Globodera* species.[127] It was also observed that root exudates do not play a role in attracting the males, rather they rely solely on the sex pheromones to locate the female.

PLANT HOST INFLUENCE ON NEMATODE AND EGG DEVELOPMENT

Ellenby reported in 1954 that the sex of *Globodera rostochiensis* is epigenetically determined.[128] The same was shown for *M. incognita* by Triantaphyllou.[129,130] Unfavorable conditions, such as high population densities, cause large proportions of males to develop in this parthenogenetic species. This sex reversal is thought to occur as a survival mechanism because males do not feed, leaving sufficient food for the remaining females to complete their reproduction cycle. No pheromones have been identified that could be involved in this sex shift. It appears to be the host response that determines the sex of the developing nematode. Indeed, smaller giant cells, or syncytia in the case of cyst nematodes, do not have the capacity of supplying the necessary nutrients for female development which consumes about 30-fold more food than males.[131] As the density

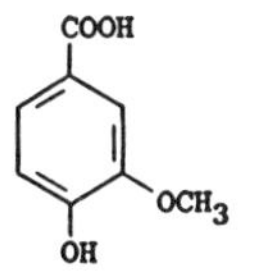

Figure 3. Vanillic acid, produced by *Heterodera glycines* females, that specifically attracts *H. glycines* males.

of the invading juveniles increases, fewer find a site capable of containing a high-capacity feeding cell, and an increasing proportion become males.[132] With only one developing nematode per root, excluding competition, root diameter influences sex determination: the smaller the root diameter, the higher the proportion of developing males (shown for *G. pallida*).[133] Therefore, most juveniles that invade the narrower lateral roots become males (observed for *G. rostochiensis* and for *H. schachtii* by Trudgill[134] and Sobczak,[135] respectively). The role of host nutrition in sex determination has been directly shown by McClure and Viglierchio.[136]

Using distilled water or standard root exudates, the hatching response of cysts has been shown to change with the age of the host plant on which they were produced.[137–139] As the host plant starts to senesce, a greater proportion of the eggs in the cysts depend on root diffusate to hatch. This has been observed for *H. cajani* on cowpea,[137] *H. sacchari* on rice,[138] and *H. sorghi* on sorghum,[139] but not for *H. oryzicola* on rice.[138] The presence of certain compounds or changes in the syncytium associated with plant senescence may be the signal for possible changes, e.g. in eggshell structure. Also, diapause, a state in which the juveniles are unresponsive to hatching stimuli, appears to be initiated by signals passed to the nematode from the plant during the growing season. The photoperiod, acting on the host plant, affects developing females and influences subsequent hatching of juveniles.[140]

NEMATODE INFECTION INFLUENCE ON HOST DEFENSE RESPONSE

When plants are infected by pathogens to which they are resistant, they normally respond by expressing diverse defense mechanisms that inhibit pathogen growth. These defense reactions can be induced systemically throughout the plant and limit subsequent infections with compatible pathogens.[141] This systemic acquired resistance has been well-documented for viral, fungal, and bacterial infections, but data on nematode infections are scarce.

Ogallo and McClure[142,143] demonstrated that prior inoculation of the tomato cultivar 'Celebrity' with the incompatible *M. incognita* significantly suppressed reproduction of the compatible *M. hapla* inoculated 5 or more days later. Prior inoculation with *M. hapla*, on the other hand, significantly enhanced reproduction of *M. incognita*. Competition for space and infection sites was excluded by employing a split-root assay, which also demonstrated the systemic nature of the factors associated with changes of host suitability. Ibrahim and Lewis[144] reported a similar resistance induction on soybean using root-knot nematodes.

One class of defense genes that is induced in systemic acquired resistance is that encoding pathogenesis-related proteins. For example, glucanases and chitinases have been postulated to be important for induced resistance against

Figure 4. Phytoalexins produced by some plants as a defense mechanism against nematodes. **a.** Glyceollin from *Glycine max*. **b.** Gossypol from *Gossypium hirsutum*.

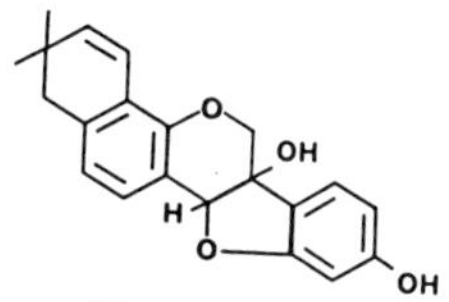

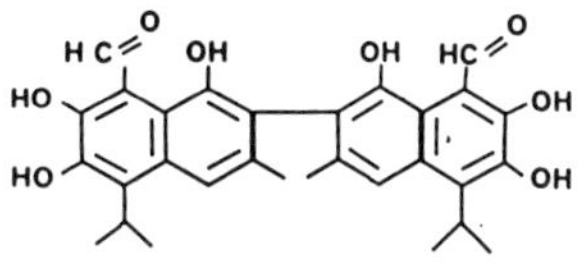

certain fungal infections. Nematodes also induce pathogenesis-related proteins, such as β-1,3-glucanases,[145] but only in leaves and not in roots of infected plants.

Other host-defensive substances are proteinase inhibitors and phytoalexins.[146] For instance, the soybean phytoalexin, glyceollin (Fig. 4a), is nematistatic to *M. incognita*, and a clear correlation has been found between the accumulation of glyceollin and incompatibility to *M. incognita* in soybean.[147] Other examples of phytoalexins associated with plant resistance to nematodes are known, *e.g.* gossypol in coton (Fig. 4b).[148,149] To my knowledge, no literature is available on the effect of incompatible nematode infection on subsequent attack by a different pathogen, *e.g.* in the leaf, or the reverse, *e.g.* effect of an incompatible leaf pathogen on subsequent nematode infection of plant roots.

CONCLUSION

In comparison to other fields in phytopathology, plant–nematode research has been hampered by the specific characteristics of this plant–pathogen interaction: a long life cycle, obligate biotrophic parasites, only a few plant cells directly affected by the infection. The availability of a common host system (*Arabidopsis thaliana*) for use in research on different nematodes and the wealth of molecular data that has been generated from this plant have certainly increased interest and scientific output of research on plant–nematode interactions. Nevertheless, it remains a challenge to investigate and understand the particular characteristics of different nematode–host interactions. It is the diversity of the possible chemical signals and their effects that underlies the natural variety of plant–nematode interactions.

ACKNOWLEDGMENTS

The author thanks C. Fenoll, F.W. Grundler, H. Helder, R.N. Perry, and W. Robertson for providing unpublished information and preprints of articles and M. De Cock for help in preparing the manuscript. G.G. is a Postdoctoral fellow of the Fund for Scientific Research (Flanders).

REFERENCES

1. BIRD, A.F. 1962. The inducement of giant cells by *Meloidogyne javanica*. Nematologica 8:1–10.
2. GHEYSEN, G., DE ALMEIDA ENGLER, J., VAN MONTAGU, M. 1997. Cell cycle regulation in nematode feeding sites. In: Cellular and Molecular Basis of Plant–Nematode Interactions, F. Grundler, S. Ohl, and C. Fenoll, (eds.), Kluwer Academic Publishers, Dordrecht, pp. 120–132.
3. HUSSEY, R.S., DAVIS, E.L., RAY, C. 1994. *Meloidogyne* stylet secretions. In: Advances in Molecular Plant Nematology, (NATO ASI Series A: Life Sciences, Vol. 268), F. Lamberti, C. De Giorgi, and D. McK. Bird, (eds.), Plenum Press, New York, pp. 233–249.
4. PERRY, R.N. 1996. Chemoreception in plant parasitic nematodes. Ann. Rev. Phytopathol. 34:181–199.
5. JONES, J.T., ROBERTSON, W.M. 1997. Nematode secretion. In: Cellular and Molecular Basis of Plant–Nematode Interactions, F. Grundler, S. Ohl, and C. Fenoll, (eds.), Kluwer Academic Publishers, Dordrecht, pp. 98–106.
6. COOMANS, A., DE GRISSE, A. 1981. Sensory structures. In Plant Parasitic Nematodes, Vol. 3, B.M. Zuckerman, R.A. Rohde, (eds.), Academic Press, New York, pp. 127–174.
7. FORREST, J.M.S., ROBERTSON, W.M. 1986. Characterization and localization of saccharides on the head region of four populations of the potato cyst nematode *Globodera rostochiensis* and *G. pallida*. J. Nematol. 18:23–26.
8. PREMACHANDRAN, D., VON MENDE, N., HUSSEY, R.S., McCLURE, M.A. 1988. A method for staining nematode secretions and structures. J. Nematol. 20:70–78.
9. AUMANN, J. 1989. Enzymatic effects on lectin binding to *Heterodera schachtii* (Nematoda: Heteroderidae) males. Nematologica 35:461–468.
10. AUMANN, J., WYSS, U. 1989. Histochemical studies on exudates of *Heterodera schachtii* (Nematoda: Heteroderidae) males. Rev. Nématol. 12:309–315.
11. STEWART, G.R., PERRY, R.N., ALEXANDER, J., WRIGHT, D.J. 1993. A glycoprotein specific to the amphids of *Meloidogyne* species. Parasitology 106:405–412.
12. STEWART, G.R., PERRY, R.N., WRIGHT, D.J. 1993. Studies on the amphid specific glycoprotein gp32 in different life-cycle stages of *Meloidogyne* species. Parasitology 107:573–578.
13. CURTIS, R.H.C., SEGERS, I., EVANS, K. 1996. Observations on the specificity of *Globodera pallida* and *G. rostochiensis* diagnostic proteins to amphids of *Globodera* species. Nematologica 42:466–480.
14. LEWIS, J.A., HODGKIN, J.A. 1975. Specific neuroanatomical changes in chemosensory mutants of the nematode *Caenorhabditis elegans*. J. Comp. Neurol. 172:489–510.
15. BARGMANN, C.I., THOMAS, J.H., HORVITZ, H.R. 1990. Chemosensory cell function in the behaviour and development of *Caenorhabditis elegans*. Cold Spring Harbor Symp. Quant. Biol. 55:529–538.
16. TRETT, M.W., PERRY, R.N. 1985. Functional and evolutionary implications of the anterior sensory anatomy of species of root-lesion nematodes (genus *Pratylenchus*). Rev. Nématol. 8:341–355.

17. PERRY, R.N. 1997. Physiology and sensory perception of potato cyst-nematodes. In: Potato Cyst Nematodes, R.J. Marks, and B.B. Brodie, (eds.), CAB International, Wallingford, In Press.
18. ATKINSON, H.J., FOWLER, M., ISAAC, R.E. 1987. Partial purification of hatching activity for *Globodera rostochiensis* from potato diffusate. Ann. Appl. Biol. 110:115–125.
19. PERRY, R.N. 1989. Dormancy and hatching of nematode eggs. Parasitology Today 5:377–383.
20. BYRNE, J., WALSH, D., DEVINE, K., JONES, P. 1996. Investigations into the cause of the delayed hatch of *Globodera pallida* compared to *G. rostochiensis*. Nematropica 26:247.
21. MULDER, J.G., DIEPENHORST, P., PLIEGER, P., BRÜGGEMANN-ROTGANS, I.E.M. 1993. Hatching agent for the potato cyst nematode. Patent Application, Int. Appl. No. PCT/NL92/00126, Int. Publ. No. WO 93/02083.
22. DEVINE, K.J., BYRNE, J., MAHER, N., JONES, P.W. 1996. Resolution of natural hatching factors for the golden potato cyst nematode, *Globodera rostochiensis*. Ann. Appl. Biol. 129:323–334.
23. MASAMUNE, T., ANETAI, M., TAKASUGI, M., KATSUI, N. 1982. Isolation of a natural hatching stimulus, glycinoeclepin A, for the soybean cyst nematode. Nature 297:495–496.
24. FUKUSAWA, A., FURUSAKI, A., IKURA, M., MASAMUNE, R. 1985. Glycinoeclepin A, a natural hatching stimulus for the soybean cyst nematode. J. Chem. Soc. 4:222–224.
25. FUKUSAWA, A., MATSUE, H., IKURA, M., MASAMUNE, R. 1985. Glycinoeclepins B and C, nortriterpenes related to glycinoeclepin A. Tetrahedron Lett. 26:5539–5542.
26. KRAUS, G.A., VANDER LOUW, S.J., TYLKA, G.L., SOH, D.H. 1996. Synthesis and testing of compounds that inhibit soybean cyst nematode egg hatch. J. Agric. Food Chem. 44:1548–1550.
27. PERRY, R.N., CLARKE, A.J., BEANE, J. 1980. Hatching of *Heterodera goettingiana* in vitro. Nematologica 26:493–495.
28. GRECO, N., BRANDONISIO, A. 1986. The biology of *Heterodera carotae*. Nematologica 32:447–460.
29. LAMONDIA, J.A. 1995. Hatch and reproduction of *Globodera tabacum tabacum* in response to tobacco, tomato, or black nightshade. J. Nematol. 27:382–386.
30. SCHMITT, D.P., RIGGS, R.D. 1991. Influence of selected plant species on hatching of eggs and development of juveniles of *Heterodera glycines*. J. Nematol. 23:1–6.
31. TURNER, S.J., STONE, A.R. 1981. Hatching of potato cyst-nematodes (*Globodera rostochiensis*, *G. pallida*) in root exudates of *Solanum vernei* hybrids. Nematologica 27:315–318.
32. MUGNIERY, D., BALANDRAS, C., ROUSSELLE, P. 1994. Existence chez la pomme de terre de résistance induite vis-à-vis de *Globodera rostochiensis*. Fundam. Appl. Nematol. 17:383–388.
33. WILLIAMS, T.D., BEANE, J. 1979. Temperature and root exudates on the cereal cyst-nematode *Heterodera avenae*. Nematologica 25:397–405.
34. JONES, J.T., PERRY, R.N., JOHNSTON, M.R.L. 1994. Changes in the ultrastructure of the amphids of the potato cyst nematode, *Globodera rostochiensis*, during development and infection. Fundam. Appl. Nematol. 17:369–382.
35. CLARKE, A.J., PERRY, R.N., HENNESSY, J. 1978. Osmotic stress and the hatching of *Globodera rostochiensis*. Nematologica 24:384–392.
36. CLARKE, A.J., HENNESSY, J. 1983. The role of calcium in the hatching of *Globodera rostochiensis*. Rev. Nématol. 6:247–255.
37. ELLENBY, C., PERRY, R.N. 1976. The influence of the hatching factor on the water uptake of the second stage larva of the potato cyst nematode *Heterodera rostochiensis*. J. Exp. Biol. 64:141–147.
38. PERRY, R.N., CLARKE, A.J., HENNESSY, J. 1980. The influence of osmotic pressure on the hatching of *Heterodera schachtii*. Rev. Nématol. 3:3–9.
39. PERRY, R.N., KNOX, D.P., BEANE, J. 1992. Enzymes released during hatching of *Globodera rostochiensis* and *Meloidogyne incognita*. Fundam. Appl. Nematol. 15:283–288.

40. BIRD, A.F. 1968. Changes associated with parasitism in nematodes. III. Ultrastructure of the egg shell, larval cuticle, and contents of the subventral esophageal glands in *Meloidogyne javanica*, with some observations on hatching. J. Parasitol. 54:475–489.
41. GRUNDLER, F., SCHNIBBE, L., WYSS, U. 1991. In vitro studies on the behaviour of second-stage juveniles of *Heterodera schachtii* (Nematoda: Heteroderidae) in response to host plant root exudates. Parasitology 103:149–155.
42. CLEMENS, C.D., AUMANN, J., SPIEGEL, Y., WYSS, U. 1994. Attractant-mediated behaviour of mobile stages of *Heterodera schachtii*. Fundam. Appl. Nematol. 17:569–574.
43. PROT, J.-C. 1980. Migration of plant-parasitic nematodes towards plant roots. Rev. Nématol. 3:305–318.
44. KLINGLER, J. 1963. Die Orientierung von *Ditylenchus dipsaci* in gemessenen künstlichen und biologischen CO_2-Gradienten. Nematologica 9:185–199.
45. EL-SHERIF, M., MAI, W.F. 1969. Thermotactic response of some plant parasitic nematodes. J. Nematol. 1:43–48.
46. GRIFFIN, G.D., ELGIN, J.H. Jr. 1977. Penetration and development of *Meloidogyne hapla* in resistant and susceptible alfalfa under differing temperatures. J. Nematol. 9:51–56.
47. VIGLIERCHIO, D.R. 1961. Attraction of parasitic nematodes by plant root emanations. Phytopathology 51:136–142.
48. LEE, Y.B., EVANS, A.A.F. 1973. Correlation between attractions and susceptibilities of rice varieties to *Aphelenchoides besseyi* Christie, 1942. Kor. J. Pl. Prot. 12:147–151.
49. HAYNES, R.L., JONES, C.M. 1976. Effects of the *Bi* locus in cucumber on reproduction, attraction, and response of the plant to infection by the southern root-knot nematode. J. Am. Soc. Hort. Sci. 101:422–424.
50. RAWSTHORNE, D., BRODIE, B.B. 1986. Relationship between root growth of potato, root diffusate production, and hatching of *Globodera rostochiensis*. J. Nematol. 18:379–384.
51. ROBERTSON, W.M., FORREST, J.M.S. 1989. Factors involved in host recognition by plant-parasitic nematodes. Aspects App. Biol. 22:129–133.
52. BALHADÈRE, P., EVANS, A.A.F. 1994. Characterization of attractiveness of excised root tips of resistant and susceptible plants for *Meloidogyne naasi*. Fundam. Appl. Nematol. 17:527–536.
53. LINDFORD, M.B. 1941. Parasitism of the root-knot nematode in leaves and stems. Phytopathology 31:634–648.
54. LINDFORD, M.B. 1942. The transient feeding of root-knot nematode larvae. Phytopathology 32:580–589.
55. WYSS, U., GRUNDLER, F.M.W., MÜNCH, A. 1992. The parasitic behaviour of second-stage juveniles of *Meloidogyne incognita* in roots of *Arabidopsis thaliana*. Nematologica 38:98–111.
56. GRAVATO NOBRE, M.J., VON MENDE, N., DOLAN, L., SCHMIDT, K.P., EVANS, K., MULLIGAN, B. 1995. Immunolabelling of cell surfaces of *Arabidopsis thaliana* roots following infection by *Meloidogyne incognita* (Nematoda). J. Exp. Bot. 46:1711–1720.
57. VON MENDE, N. 1997. Invasion and migration behaviour of sedentary nematodes. In: Cellular and Molecular Basis of Plant–Nematode Interactions, F. Grundler, S. Ohl, and C. Fenoll, (eds.), Kluwer Academic Publishers, Dordrecht, pp. 51–64.
58. POTENZA, C.L., THOMAS, S.H., HIGGINS, E.A., SENGUPTA-GOPALAN, C. 1996. Early root response to *Meloidogyne incognita* in resistant and susceptible alfalfa cultivars. J. Nematol. 28:475–484.
59. WYSS, U., ZUNKE, U. 1986. Observations on the behaviour of second stage juveniles of *Heterodera schachtii* inside host roots. Rev. Nématol. 9:153–165.
60. WYSS, U., GRUNDLER, F.M.W. 1992. Seminar: *Heterodera schachtii* and *Arabidopsis thaliana*, a model host-parasite interaction. Nematologica 38:488–493.
61. GOLINOWSKI, W., SOBCZAK, M., KUREK, W., GRYMASZEWSKA, G. 1997. The structure of syncytia. In: Cellular and Molecular Basis of Plant–Nematode Interactions, F. Grundler, S. Ohl, and C. Fenoll, (eds.), Kluwer Academic Publishers, Dordrecht, pp. 80–97.

62. DROPKIN, V.H. 1963. Cellulase in phytoparasitic nematodes. Nematologica 9:444–454.
63. BIRD, A.F., DOWNTOWN, W.J.S, AND HAWKER, J.S. 1975. Cellulase secretion by the second-stage juvenile of the root-knot nematode (*Meloidogyne javanica*). Marcellia 38:165–169.
64. SMANT, G., GOVERSE, A., STOKKERMANS, J.W.P.G., DE BOER, J.M., POMP, H.R., ZILVERENTANT, J.F., OVERMARS, H.A., GOMMERS, F.J., HELDER, J., SCHOTS, A., BAKKER, J. 1997. Analysis of subventral easophageal gland proteins detected in stylet secretions of potato cyst nematode second-stage juveniles. Phytopathology: In press.
65. GIEBEL, J. 1974. Biochemical mechanisms of plant resistance to nematodes: A review. J. Nematol. 6:175–184.
66. DUNCAN, L.H., ROBERTSON, W.M., KUSEL, J.R., PHILLIPS, M.S. 1996. A putative nematode auxin binding protein from the potato cyst nematode *Globodera pallida*. Nematropica 26:259.
67. NAYLOR, J., SANDER, G., SKOOG, F. 1954. Mitosis and cell enlargement without cell division in excised tobacco pith tissue. Physiol. Plant. 7:25–29.
68. FAN, D.-F., MACLACHLAN, G.A. 1967. Massive synthesis of ribonucleic acid and cellulase in the pea epicotyl in response to indoleacetic acid, with and without concurrent cell division. Plant Physiol. 42:1114–1122.
69. KACZMAREK, U., GIEBEL, J. 1980. Disturbances of plant cell mitosis caused by *Globodera rostochiensis* and some plant tissue substances. Bull. Acad. Pol. Sci. Ser. Sci. Biol. 27:969–974.
70. BALASUBRAMANIAN, M., RANGARWAMI, G. 1962. Presence of indole compounds in nematode galls. Nature 194:714–715.
71. YU, P.K., VIGLIERCHIO, D.R. 1964. Plant growth substances and parasitic nematodes. I. Root knot nematodes and tomato. Exp. Parasitol. 15:242–248.
72. VIGLIERCHIO, D.R., YU, P.K. 1968. Plant growth substances and plant parasitic nematodes. II. Host influence on auxin content. Exp. Parasitol. 23:88–95.
73. GIEBEL, J., PIEGAT, M., WILSKI, A. 1966. The influence of some exogenic enzymes on root tissues of potatoes susceptible and resistant to the golden nematode (Heterodera rostochiensis Woll.). Pr. Nauk. Inst. Ochr. Rośl. Warszawa 8:205–211 (in Polish with English summary).
74. SANSTEDT, R., SCHUSTER, M.L. 1966. The role of auxins in root-knot nematode-induced growth on excised tobacco stem segments. Physiol. Plant. 19:960–967.
75. VERCAUTEREN, I., NIEBEL, A., VAN MONTAGU, M., GHEYSEN, G. 1995. *Arabidopsis thaliana* roots show an increased auxin concentration upon nematode infection. Med. Fac. Landbouww. Univ Gent 60/4a:1661–1664.
76. KOCHBA, J., SAMISH, R.M. 1971. Effect of kinetin and 1-naphthylacetic acid on root-knot nematodes in resistant and susceptible peach rootstocks. J. Am. Soc. Hort. Sci. 96:458–461.
77. DROPKIN, V.H., HELGESON, J.P., UPPER, C.D. 1969. The hypersensitivity reaction of tomatoes resistant to *Meloidogyne incognita*: Reversal by cytokinins. J. Nematol. 1:55–61.
78. BINGEFORS, S. 1982. Nature of inherited nematode resistance in plants. In: Pathogens, Vectors, and Plant Diseases: Approaches to Control, K.F. Harris, K. Maramorosch, (eds.), Academic Press, New York, pp. 187–219.
79. VAN STADEN, J., DIMALLA, G.G. 1977. A comparison of the endogenous cytokinins in the roots and xylem exudate of nematode-resistant and susceptible tomato cultivars. J. Exp. Bot. 28:1351–1356.
80. KOCHBA, J., SAMISH, R.M. 1972. Level of endogenous cytokinins and auxin in roots of nematode-resistant and susceptible peach rootstocks. J. Am. Soc. Hort. Sci. 97:115–119.
81. HUETTEL, R.N., HAMMERSCHLAG, F.A. 1986. Influence of cytokinin on in vitro screening of peaches for resistance to nematodes. Plant Disease 70:1141–1144.
82. GLAZER, I., EPSTEIN, E., ORION, D., APELBAUM, A. 1986. Interactions between auxin and ethylene in root-knot nematode (*Meloidogyne javanica*) infected tomato roots. Physiol. Mol. Plant Pathol. 28:171–179.

83. WEBSTER, J.M. 1967. Some effects of 2,4-dichlorophenoxyacetic acid herbicides on nematode-infested cereals. Plant Pathol. 16:23–26.
84. VOLKMAR, K.M. 1991. Abscisic acid and ethylene increase in *Heterodera avenae*-infected tolerant or intolerant oat cultivars. J. Nematol. 23:425–431.
85. KARIMI, M., VAN MONTAGU, M., GHEYSEN, G. 1995. Exogenous application of abscisic acid to potato plants suppresses reproduction of *Meloidogyne incognita*. Med. Fac. Landbouwwet. Univ. Gent 60/3b:1033–1035.
86. SIJMONS, P.C., VON MENDE, N., GRUNDLER, F.M.W. 1994. Plant-parasitic nematodes. In: Arabidopsis, E.M. Meyerowitz, and C.R. Somerville, (eds.), Cold Spring Harbor Laboratory Press, Cold Spring Harbor, pp. 749–767.
87. PRINSEN, E., REDIG, P., VAN DONGEN, W., ESMANS, E.L., VAN ONCKELEN, H. 1995. Quantitative analysis of cytokinins by electrospray tandem mass spectrometry. Rapid Commun. Mass Spectrom. 9:948–953.
88. PRINSEN, E., VAN DONGEN, W., ESMANS, E.L., VAN ONCKELEN, H. 1997. HPLC linked electrospray tandem mass spectrometry: A rapid and reliable method to analyse indole-3-acetic acid metabolism in bacteria. J. Mass Spectrom. 32:12–22.
89. HAMMOND-KOSACK, K.E., JONES, J.D.G. 1996. Resistance gene-dependent plant defense responses. Plant Cell 8:1773–1791.
90. BENT, A.F. 1996. Plant disease resistance genes: Function meets structure. Plant Cell 8:1757–1771.
91. CAI, D., KLEINE, M., KIFLE, S., HARLOFF, H.-J., SANDAL, N.N., MARCKER, K.A., KLEIN-LANKHORST, R.M., SALENTIJN, E.M.J., LANGE, W., STIEKEMA, W.J., WYSS, U., GRUNDLER, F.M.W., JUNG, C. 1997. Positional cloning of a gene for nematode resistance in sugar beet. Science 275:832–834.
92. WILLIAMSON, V.M., LAMBERT, K.N., KALOSHIAN, I. 1994. Molecular biology of nematode resistance in tomato. In: Advances in Molecular Plant Nematology, (NATO ASI Series A: Life Sciences, Vol. 268), F. Lamberti, C. De Giorgi, and D. McK. Bird, (eds.), Plenum Press, New York, pp. 211–219.
93. PAULSON, R.E., WEBSTER, J.M. 1972. Ultrastructure of the hypersensitive reaction in roots of tomato, *Lycopersicon esculentum* L. to infection by the root-knot nematode, *Meloidogyne incognita*. Physiol. Plant Pathol. 2:227–234.
94. DROPKIN, V.H. 1969. Cellular responses of plants to nematode infections. Ann. Rev. Phytopathol. 7:101–122.
95. TURNER, S.J., STONE, A.R. 1984. Development of potato cyst-nematodes in roots of resistant *Solanum tuberosum* ssp. *andigena* and *S. vernei* hybrids. Nematologica 30:324–332.
96. RICE, S.L., LEADBEATER, B.S.C., STONE, A.R. 1985. Changes in cell structure in roots of resistant potatoes parasitized by potato cyst-nematodes. I. Potatoes with resistance gene H_1 derived from *Solanum tuberosum* ssp. *andigena*. Physiol. Plant Pathol. 27:219–234.
97. RICE, S.L., STONE, A.R., LEADBEATER, B.S.C. 1987. Changes in cell structure in roots of resistant potatoes parasitized by potato cyst nematodes. 2. Potatoes with resistance from *Solanum vernei*. Physiol. Mol. Plant Pathol. 31:1–14.
98. MAHALINGAM, R., SKORUPSKA, H.T. 1996. Cytological expression of early response to infection by *Heterodera glycines* Ichinohe in resistant PI 437654 soybean. Genome 39:986–998.
99. KIM, Y.H., RIGGS, R.D., KIM, K.S. 1987. Structural changes associated with resistance of soybean to *Heterodera glycines*. J. Nematol. 19:177–187.
100. McCLURE, M.A., ELLIS, K.C., NIGH, E.L. 1974. Post-infection development and histopathology of *Meloidogyne incognita* in resistant cotton. J. Nematol. 6:21–26.
101. SCOFIELD, S.R., TOBIAS, C.M., RATHJEN, J.P., CHANG, J.H., LAVELLE, D.T., MICHELMORE, R.W., STASKAWICZ, B.J. 1996. Molecular basis of gene-for-gene specificity in bacterial speck disease of tomato. Science 274:2063–2065.

102. TANG, X., FREDERICK, R.D., ZHOU, J., HALTERMAN, D.A., JIA, Y., MARTIN, G.B. 1996. Initiation of plant disease resistance by physical interaction of AvrPto and Pto kinase. Science 274:2060–2063.
103. CASTAGNONE-SERENO, P., ROSSO, M.-N., BONGIOVANNI, M., DALMASSO, A. 1995. Electrophoretic analysis of near-isogenic avirulent and virulent lineages of the parthenogenetic root-knot nematode *Meloidogyne incognita*. Physiol. Mol. Plant Pathol. 47:293–302.
104. JANSSEN, R., BAKKER, J., GOMMERS, F.J. 1991. Mendelian proof for a gene-for-gene relationship between virulence of *Globodera rostochiensis* and the H_1 resistance gene in *Solanum tuberosum* ssp. *andigena* CPC 1673. Rev. Nématol. 14:207–212.
105. ROUPPE VAN DER VOORT, J.N.A.M., ROOSIEN, J., VAN ZANDVOORT, P.M., FOLKERTSMA, R.T., VAN ENCKEVORT, E.L.J.G., JANSSEN, R., GOMMERS, F.J., BAKKER, J. 1994. Linkage mapping in potato cyst nematodes. In: Advances in Molecular Plant Nematology, (NATO ASI Series A: Life Sciences, Vol. 268), F. Lamberti, C. De Giorgi, and D. McK. Bird, (eds.), Plenum Press, New York, pp. 57–63.
106. OPPERMAN, C.H., DONG, K., CHANG, S. 1994. Genetic analysis of the soybean-*Heterodera glycines* interaction. In: Advances in Molecular Plant Nematology, (NATO ASI Series A: Life Sciences, Vol. 268), F. Lamberti, C. De Giorgi, and D. McK. Bird, (eds.), Plenum Press, New York, pp. 65–75.
107. CASTAGNONE-SERENO, P., ABAD, P., BAKKER, J., WILLIAMSON, V.M., GOMMERS, F.J., DALMASSO, A. 1997. Genetic and molecular strategies for the cloning of (a)virulence genes in sedentary plant-parasitic nematodes. In: Cellular and Molecular Basis of Plant–Nematode Interactions, F. Grundler, S. Ohl, C. Fenoll, (eds.), Kluwer Academic Publishers, Dordrecht, pp. 167–175.
108. DAVIS, E.L., ALLEN, R., HUSSEY, R.S. 1994. Developmental expression of esophageal gland antigens and their detection in stylet secretions of *Meloidogyne incognita*. Fundam. Appl. Nematol. 17:255–262.
109. GOVERSE, A., DAVIS, E.L., HUSSEY, R.S. 1994. Monoclonal antibodies to the esophageal glands and stylet secretions of *Heterodera glycines*. J. Nematol. 26:251–259.
110. DE BOER, J.M., SMANT, G., GOVERSE, A., DAVIS, E.L., OVERMARS, H.A., POMP, H., VAN GENT-PELZER, M., ZILVERENTANT, J.F., STOKKERMANS, J.P.W.G., HUSSEY, R.S., GOMMERS, F.J., BAKKER, J., SCHOTS, A. 1996. Secretory granule proteins from the subventral esophageal glands of the potato cyst nematode identified by monoclonal antibodies to a protein fraction from second-stage juveniles. Mol. Plant-Microbe Interact. 9:39–46.
111. DE BOER, J.M., OVERMARS, H.A., POMP, H.R., DAVIS, E.L., ZILVERENTANT, J.F., GOVERSE, A., SMANT, G., STOKKERMANS, J.P.W.G., HUSSEY, R.S., GOMMERS, F.J., BAKKER, J., SCHOTS, A. 1996. Production and characterization of monoclonal antibodies to antigens from second stage juveniles of the potato cyst nematode, *Globodera rostochiensis*. Fundam. Appl. Nematol. 19:545–554.
112. PLEGT, L., BINO, R.J. 1989. β-Glucuronidase activity during development of the male gametophyte from transgenic and non-transgenic plants. Mol. Gen. Genet. 216:321–327.
113. GODDIJN, O.J.M., LINDSEY, K., VAN DER LEE, F.M., KLAP, J.C., SIJMONS, P.C. 1993. Differential gene expression in nematode-induced feeding structures of transgenic plants harbouring promoter-*gus*A fusion constructs. Plant J. 4:863–873.
114. NIEBEL, A., DE ALMEIDA ENGLER, J., HEMERLY, A., FERREIRA, P., INZÉ, D., VAN MONTAGU, M., GHEYSEN, G. 1996. Induction of *cdc2a* and *cyc1At* expression in *Arabidopsis thaliana* during early phases of nematode-induced feeding cell formation. Plant J. 10:1037–1043.
115. FENOLL, C., ARISTIZÁBAL, F.A., SANZ-ALFÉREZ, S., DEL CAMPO, F.F. 1997. Regulation of gene expression in feeding sites. In: Cellular and Molecular Basis of Plant–Nematode Interactions, F. Grundler, S. Ohl, and C. Fenoll, (eds.), Kluwer Academic Publishers, Dordrecht, pp. 133–149.

116. OPPERMAN, C.H., TAYLOR, C.G., CONKLING, M.A. 1994. Root-knot nematode-directed expression of a plant root-specific gene. Science 263:221–223.
117. VAN DER EYCKEN, W., DE ALMEIDA ENGLER, J., INZÉ, D., VAN MONTAGU, M., GHEYSEN, G. 1996. A molecular study of root-knot nematode-induced feeding sites. Plant J. 9:45–54.
118. JASMER, D.P. 1993. *Trichinella spiralis* infected skeletal muscle cells arrest in G_2/M and cease muscle gene expression. J. Cell Biol. 121:785–793.
119. URAO, T., YAMAGUCHI-SHINOZAKI, K., URAO, S., SHINOZAKI, K. 1993. An Arabidopsis *myb* homolog is induced by dehydration stress and its gene product binds to the conserved MYB recognition sequence. Plant Cell 5:1529–1539.
120. URAO, T., NOJI, M.-A., YAMAGUCHI-SHINOZAKI, K., SHINOZAKI, K. 1996. A transcriptional activation domain of ATMYB2, a drought-inducible *Arabidopsis* Myb-related protein. Plant J. 10:1145–1148.
121. McK. BIRD, D., WILSON, M.A. 1994. DNA sequence and expression analysis of root-knot nematode-elicited giant cell transcripts. Mol. Plant-Microbe Interact. 7:419–424.
122. VERCAUTEREN, I., VAN MONTAGU, M., GHEYSEN, G. 1996. Identification of mRNA species expressed upon nematode infection by the differential display technique. Nematropica 26:323.
123. BARTHELS, N., VAN DER LEE, F.M., KLAP, J., GODDIJN, O.J.M., OHL, S.A., KARIMI, M., PUZIO, P., GRUNDLER, F.M.W., LINDSEY, K., ROBERTSON, L., ROBERTSON, W.M., VAN MONTAGU, M., GHEYSEN, G., SIJMONS, P.C. 1997. Regulatory sequences of Arabidopsis drive reporter gene expression in nematode feeding structures. Plant Cell: In Press.
124. GHEYSEN, G., VAN DER EYCKEN, W., BARTHELS, N., KARIMI, M., VAN MONTAGU, M. 1996. The exploitation of nematode-responsive plant genes in novel nematode control methods. Pestic. Sci. 47:95–101.
125. JAFFE, H., HUETTEL, R.N., DEMILO, A.B., HAYES, D.K., REBOIS, R.V. 1989. Isolation and identification of a compound from soybean cyst nematode, *Heterodera glycines*, with sex pheromone activity. J. Chem. Ecol. 15:2031–2043.
126. AUMANN, J., HASHEM, M. 1993. Studies on substances with sex pheromone activity produced by *Heterodera schachtii* females. Fundam. Appl. Nematol. 16:43–46.
127. RIGA, E., PERRY, R.N., BARRETT, J. 1996. Electrophysiological analysis of the response of males of *Globodera rostochiensis* and *G. pallida* to their female sex pheromones and to potato root diffusate. Nematologica 42:493–498.
128. ELLENBY, C. 1954. Environmental determination of the sex ratio of a plant parasitic nematode. Nature 174:1016–1017.
129. TRIANTAPHYLLOU, A.C. 1960. Sex determination in *Meloidogyne incognita* Chitwood, 1949, and intersexuality in *M. javanica* Treub, 1885, Chitwood, 1949. Ann. Inst. Phytopathol. Benaki N.S. 3:12–31.
130. TRIANTAPHYLLOU, A.C. 1973. Environmental sex differentiation of nematodes in relation to pest management. Ann. Rev. Phytopathol. 11:441–462.
131. MÜLLER, J., REHBOCK, K., WYSS, U. 1981. Growth of *Heterodera schachtii* with remarks on amounts of food consumed. Rev. Nématol. 4:227–234.
132. MUGNIERY, D., FAYET, G. 1984. Détermination du sexe de *Globodera rostochiensis* Woll. et influence des niveaux d'infestation sur la pénétration, le développement et le sexe de ce nématode. Rev. Nématol. 7:233–238.
133. MUGNIERY, D., FAYET, G. 1981. Détermination du sexe chez *Globobera pallida* Stone. Rev. Nématol. 4:41–45.
134. TRUDGILL, D.L. 1967. The effect of environment on sex determination in *Heterodera rostochiensis*. Nematologica 13:263–272.
135. SOBCZAK, M. 1996. Investigations on the structure of syncytia in roots of *Arabidopsis thaliana* induced by the beet cyst nematode *Heterodera schachtii* and its relevance to the sex

of the nematode. Ph.D. thesis. Kiel, Institut für Phytopathologie des Christian-Albrechts-Universität.
136. McCLURE, M.A., VIGLIERCHIO, D.R. 1966. The influence of host nutrition and intensity of infection on the sex ratio and development of *Meloidogyne incognita* in sterile agar cultures of excised cucumber roots. Nematologica 12:248–258.
137. GAUR, H.S., PERRY, R.N., BEANE, J. 1992. Hatching behaviour of six successive generations of the pigeon-pea cyst nematode, *Heterodera cajani*, in relation to growth and senescence of cowpea, *Vigna unguiculata*. Nematologica 38:190–202.
138. IBRAHIM, S.K., PERRY, R.N., PLOWRIGHT, R.A., ROWE, J. 1993. Hatching behaviour of the rice cyst nematodes *Heterodera sacchari* and *H. oryzicola* in relation to age of host plant. Fundam. Appl. Nematol. 16:23–29.
139. GAUR, H.S., BEANE, J., PERRY, R.N. 1995. Hatching of four successive generations of *Heterodera sorghi* in relation to the age of sorghum, *Sorghum vulgare*. Fundam. Appl. Nematol. 18:599–601.
140. HOMINICK, W.M. 1986. Photoperiod and diapause in the potato cyst-nematode, *Globodera rostochiensis*. Nematologica 32:408–418.
141. RYALS, J.A., NEUENSCHWANDER, U.H., WILLITS, M.G., MOLINA, A., STEINER, H.-Y., HUNT, M.D. 1996. Systemic acquired resistance. Plant Cell 8:1809–1819.
142. OGALLO, J.L., McCLURE, M.A. 1995. Induced resistance to *Meloidogyne hapla* by other *Meloidogyne* species on tomato and pyrethrum plants. J. Nematol. 27:441–447.
143. OGALLO, J.L., McCLURE, M.A. 1996. Systemic acquired resistance and susceptibility to root-knot nematodes in tomato. Phytopathology 86:498–501.
144. IBRAHIM, I.K.A., LEWIS, S.A. 1986. Interrelationships of *Meloidogyne arenaria* and *M. incognita* on tolerant soybean. J. Nematol. 18:106–111.
145. RAHIMI, S., PERRY, R.N., WRIGHT, D.J. 1996. Identification of pathogenesis-related proteins induced in leaves of potato plants infected with potato cyst nematodes, *Globodera* species. Physiol. Mol. Plant Pathol. 49:49–59.
146. ZACHEO, G., BLEVE-ZACHEO, T., MELILLO, M.T. 1997. Biochemistry of plant defence responses to nematode infection. In: Cellular and Molecular Basis of Plant–Nematode Interactions, F. Grundler, S. Ohl, and C. Fenoll, (eds.), Kluwer Academic Publishers, Dordrecht, pp. 201–213.
147. KAPLAN, D.T., KEEN, N.T., THOMASON, I.J. 1980. Studies on the mode of action of glyceollin in soybean incompatibility to the root-knot nematode, *Meloidogyne incognita*. Physiol. Plant Pathol. 16:319–325.
148. VEECH, J.A. 1979. Histochemical localization and nematoxicity of terpenoid aldehydes in cotton. J. Nematol. 11:240–246.
149. VEECH, J.A. 1982. Phytoalexins and their role in the resistance of plants to nematodes. J. Nematol. 14:2–9.

Chapter Seven
SALICYLIC ACID-MEDIATED SIGNAL TRANSDUCTION IN PLANT DISEASE RESISTANCE

Daniel F. Klessig, Jörg Durner, Jyoti Shah, and Yinong Yang

Waksman Institute and Department of Molecular Biology and
 Biochemistry
Rutgers, The State University of New Jersey
P.O. Box 759
Piscataway, New Jersey 08855

Introduction 119
SA Metabolism 121
Is SA The Mobile Signal? 123
Mechanisms of Action for SA 124
SA-Mediated Gene Induction 127
Genetic Analysis of SA-Mediated Signal Transduction 129
Concluding Remarks 130

INTRODUCTION

During the past decade, extensive efforts have been made to unveil the molecular mechanisms of recognition, perception, and signal transduction in plant-pathogen interactions. The recent cloning of plant disease resistance genes, in particular, has greatly advanced our understanding of the signal recognition and perception mechanisms involved in race-specific disease resistance.[1-4] However, at present, still relatively little is known about the downstream signal transduction pathway(s) that leads to the activation of plant defense genes and disease resistance.

In plant-pathogen interactions, a susceptible plant cannot restrict pathogen growth and/or spread; thus, the pathogen often causes severe damage or even

Phytochemical Signals and Plant–Microbe Interactions, edited by Romeo *et al.*
Plenum Press, New York, 1998.

death of the entire plant. In contrast, a resistant plant is capable of deploying a variety of defense responses to prevent pathogen colonization. A key difference between resistant and susceptible plants is the timely recognition of the invading pathogen and the rapid and effective activation of host defenses. The activated defense response is frequently manifested, in part, as the so-called hypersensitive response (HR), which is characterized by necrosis at the sites of infection (resembling animal programmed cell death),[5,6] and restriction of pathogen growth and spread.[7] In addition to the localized HR, many plants respond to pathogen infection by activating defenses in uninfected parts of the plant. As a result, the entire plant is more resistant to a secondary infection. This systemic acquired resistance (SAR) is long lasting and often confers broad-based resistance to different pathogens.[8]

The development of HR and SAR is accompanied by the activation of many plant protectant and defense genes. Their products include glutathione S-transferases (GST), peroxidases, cell wall proteins, proteinase inhibitors, hydrolytic enzymes, and enzymes involved in phytoalexin biosynthesis, such as phenylalanine ammonia lyase (PAL) and chalcone synthase.[8,9] The synthesis of several families of pathogenesis-related (PR) proteins also correlates well with development of both HR and SAR, particularly in tobacco and *Arabidopsis*. Furthermore, many of these proteins exhibit antimicrobial activity *in vitro* and/or when overexpressed in transgenic plants. In general, the acidic extracellular PRs are dramatically induced after infection, which makes them excellent markers for resistance. In contrast, the basic, intracellular PRs have a higher basal level, are more modestly induced by infection, and are also induced by wounding and ethylene treatment.[10–12]

In general, a plant's reaction to biotic and abiotic stresses requires the perception of a stimulus by a receptor and the subsequent involvement of second messengers and effector proteins to trigger an appropriate response. The activation of plant defense mechanisms is initiated by host recognition of race-specific (*e.g.* avirulence gene products) or non-specific signals (*e.g.* microbial proteins, glycoproteins, small peptides and oligosaccharides, etc.). This signal perception process may be mediated through membrane-bound or cytosolic receptors. In some gene-for-gene interactions, for example, a bacterial avirulence gene product may enter plant cells and directly interact with a cytosolic receptor encoded by the corresponding disease resistance gene.[13,14] Plasma membrane receptors have also been identified that appear to participate in the perception of fungal elicitors.[15] Following the initial perception, signals may be transduced through G-proteins,[16,17] ion fluxes,[18,19] reactive oxygen species,[20,21] and/or phosphorylation cascades involving various kinases/phosphatases.[22–24] Subsequent transcriptional and/or posttranslational activation of transcription factors eventually leads to the induction of plant defense genes.

Both race-specific and non-specific signals may be amplified through the generation of secondary signal molecules such as salicylic acid (SA). In the late

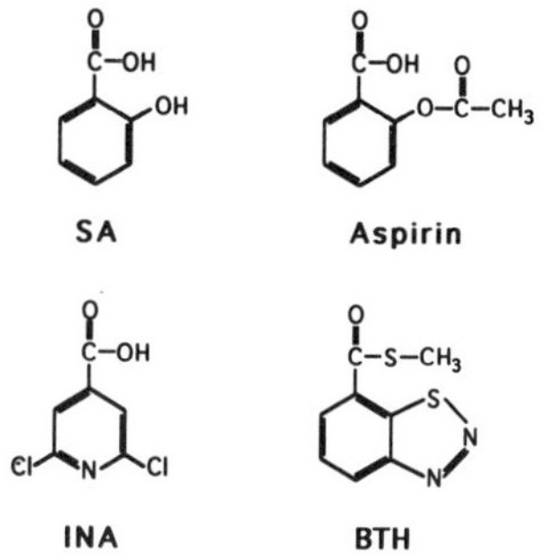

Figure 1. Salicylic acid (SA) and its functional analogues. The SA derivative, aspirin or acetyl salicylic acid, was the first of these four depicted compounds shown to induce *PR* genes and enhanced resistance to pathogens.[25,26] Aspirin is broken down to SA spontaneously or with the help of esterases and likely is active in plants only in its unacetylated form. More recently, two synthetic compounds, 2,6-dichloroisonicotinic acid (INA) and benzothiadiazole (BTH), have been identified which also induce *PR* genes and enhanced disease resistance. Both show structural similarities to SA and appear to be functional analogues of SA. BTH is being used commercially as a plant protecting agent.[68]

'70s and early '80s, White and colleagues[25,26] demonstrated that treatment of tobacco with SA or its derivative, aspirin (Fig. 1), induced *PR* gene expression and enhanced resistance to pathogens such as tobacco mosaic virus (TMV). However, not until the early '90s did it become apparent that SA is an endogenous signal in plant defense. It was found that SA accumulated to high levels at the site of TMV infection in resistant tobacco plants, with a subsequent, but much smaller, rise in the uninoculated tissues.[27-29] This increase paralleled or slightly preceded the induction of *PR* genes in both the inoculated and uninoculated leaves. Endogenous SA levels were also shown to rise in the phloem of cucumber plants infected with tobacco necrosis virus or *Colletotrichum lagenarium*.[30] This increase preceded the development of SAR. Furthermore, SA treatment activated the same set of nine genes that are induced systemically by TMV infection of tobacco.[31]

More recently, SA's involvement in plant defense has been demonstrated through analysis of transgenic tobacco and *Arabidopsis* plants that express the *Pseudomonas pudita nahG* gene encoding salicylate hydroxylase. Salicylate hydroxylase converts SA to catechol, a compound which does not induce *PR* genes or enhanced resistance. The transgenic plants accumulate little, if any, SA, exhibit reduced or no *PR* gene expression, are more susceptible to a primary infection, and fail to develop SAR.[32,33] In addition, specific inhibition of PAL, which catalyzes the first step in the phenylpropanoid pathway leading to SA biosynthesis, results in greater disease susceptibility in *Arabidopsis*.[34] The enhanced susceptibility is likely due to a block in SA production as resistance was restored in these plants by addition of exogenous SA.

SA METABOLISM

In plants, the phenylpropanoid pathway is rapidly induced after pathogen infection and is responsible for the synthesis of several defense-related compounds, such as phytoalexins and lignin, as well as SA. The first step in this

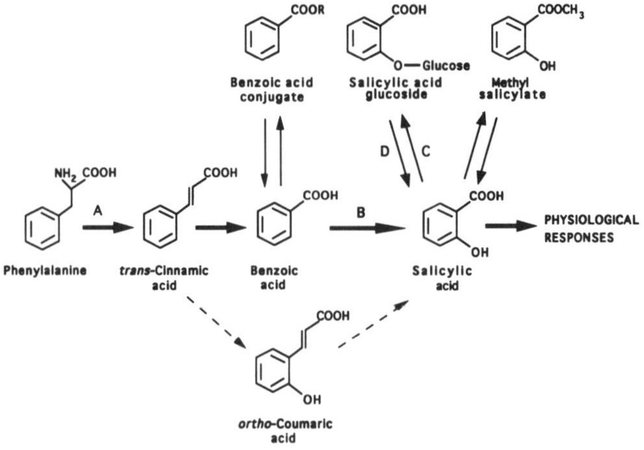

Figure 2. SA metabolism. SA can be synthesized from phenylalanine through *ortho*-coumaric acid or benzoic acid (BA). TMV-infected tobacco uses the BA intermediate (bold arrows). Both BA and SA can be converted to conjugated forms, which may serve as rapidly releasable sources of these compounds. The structure of the BA conjugate is unknown; an ester conjugate is shown as an example. After infection, SA levels rise and induce a variety of physiological responses, including induction of *PR* gene expression and enhanced resistance. Much of the newly synthesized SA after TMV infection of tobacco is converted to the glucose conjugate, SA β-glucoside, or to the volatile derivative, methyl salicylate. Methyl salicylate may be involved in long distance transmission of the SA signal. The enzymatic steps A through D are catalyzed by PAL, BA-2-hydroxylase, UDP-glucose:SA glucosyltransferase, and SA β-glucosidase, respectively.

pathway is the conversion of phenylalanine to *trans*-cinnamic acid by PAL. SA is subsequently synthesized from *trans*-cinnamic acid through one of two possible intermediates: *o*-coumaric acid or benzoic acid (BA; Fig. 2). Different plants use either or both of these intermediates for SA synthesis. In TMV-infected tobacco, SA is primarily synthesized through BA.[35] The activity of BA-2 hydroxylase, which converts BA to SA, is elevated upon TMV infection or BA treatment.[36] Since both BA-2 hydroxylase activity and SA levels are directly proportional to BA levels, the rate-limiting step for SA synthesis must occur further upstream, perhaps at BA formation.

Much of the SA synthesized after TMV infection in tobacco is conjugated to glucose to form SA β-glucoside (SAG).[28,37] The enzyme responsible for conjugating SA, UDP-glucose:SA glucosyltransferase, has been characterized in several plant species including tobacco.[38–40] The levels of both SA and SAG, which are very low in uninfected tobacco, increase substantially after TMV infection. With time, SAG becomes the predominant form in the infected leaves.[28,37] SAG levels also increase in the uninoculated leaves of tobacco,[41] cucumber,[42] and *Arabidopsis*[43] after pathogen attack.

While the function of SAG in resistance is still unresolved, its existence suggests additional complexity in the modulation of the SA signal, since many biologically active compounds in plants (*e.g.* auxin, cytokinins, and giberellins) are inactivated via glucose conjugation. Unexpectedly, Hennig et al.[44] found that injection of chemically synthesized SAG into tobacco leaves induced *PR-1* gene expression. However, this induction was preceded by a transient release of SA from SAG, which occurred in the extracellular spaces. The β-glucosidase responsible for this release is cell wall associated and exhibits very broad substrate specificity.[45] The presence of a mechanism that releases SA from SAG suggests that SAG might serve as an inactive storage form from which SA is rapidly released at the site of a challenging infection.[44] This rapidly released SA might then super induce defense responses. Interestingly, a conjugate of BA has been shown to accumulate to high levels in uninfected tobacco plants.[35] After TMV infection, the level of this conjugate drops concurrent with the rise of BA and SA. Based on these results, the BA conjugate appears to serve as a storage form that can be rapidly hydrolyzed after infection, thereby increasing the levels of BA and, subsequently, SA.[35]

IS SA THE MOBILE SIGNAL?

Development of SAR involves a mobile signal which is translocated from the infected tissue to other parts of the plant. For example, Gianinazzi and Ahl[46] demonstrated that SAR could be induced by grafting an infected leaf onto a healthy plant. Several studies have suggested that SA is the translocated signal. SA levels were found to increase prior to *PR* gene induction in the uninoculated leaves of TMV-infected tobacco.[27] Furthermore, elevated levels of SA were detected in the phloem of pathogen-infected cucumber and tobacco plants prior to development of SAR.[29,30] Recently, Shulaev et al.[47] used labeling studies with $^{18}O_2$ to demonstrate that as much as 70% of the SA present in upper uninoculated tobacco leaves was synthesized and transported from the TMV-inoculated leaf. Similarly, after labeling *C. lagenarium*-infected cucumber cotyledons with ^{14}C-BA, ^{14}C-labeled SA was detected in the upper uninoculated leaves before the development of SAR.[48] As ^{14}C-BA was not translocated to the upper leaves, this result also argues that SA is mobile.

In contrast, results from a number of studies have argued that SA is not the translocated signal necessary for SAR development. Application of exogenous SA to a subset of the plant's leaves does not induce *PR* gene expression or resistance in untreated leaves. Similarly, treatment of leaves with compounds that stimulate the production of SA induces defense responses only in the treated tissue.[49] Furthermore, it was found that the signal for SAR moved out of *Pseudomonas syringae*-infected cucumber leaves prior to any detectable increase in SA levels in the phloem sap.[50]

The strongest evidence that SA is not the translocated signal has come from grafting studies using transgenic plants. When NahG transgenic tobacco were infected with TMV, there was little increase in SA levels compared to those observed after infection of wild type (wt) tobacco. SAR also failed to develop in the systemic tissues of NahG plants. Nonetheless, SAR did develop in a wt scion grafted to a NahG rootstock after TMV infection of rootstock leaves.[51] Furthermore, it was found that transgenic tobacco expressing the gene for the A1 subunit of cholera toxin accumulate high levels of SA, express *PR* genes, develop spontaneous lesions, and exhibit enhanced resistance.[16] However, when a wt scion was grafted onto the transgenic rootstock, SAR failed to develop in the grafted tissue, despite the high level of SA within the rootstock.

In sum, many studies suggest that despite SA's ability to be translocated, another mobile signal compound must exist. Perhaps there is redundancy, and translocation of either SA or the yet to be identified compound is sufficient to induce SAR. Although it appears that SAR can develop in the absence of SA translocation, SA is still needed in the systemic tissue, where it presumably is synthesized. This conclusion is based on results from the NahG grafting experiments described above; SAR was not induced in a NahG scion after the wt rootstock leaves were infected with TMV.[51]

Recently, it was reported that a portion of the SA synthesized in TMV-infected tobacco is converted to methyl salicylate (Fig. 2).[52] As methyl salicylate is volatile, it may function as an airborne signal which activates defense responses in uninfected tissues and possibly even neighboring plants. At ambient temperature, methyl salicylate is a liquid and, therefore, could also be translocated through the vascular system of the plant. Once in the responding tissue, methyl salicylate is thought to act by being converted back to SA. Thus, the mobile signal could be methyl salicylate, which may utilize an unexpected route of translocation.

MECHANISMS OF ACTION FOR SA

While it is becoming increasingly clear that SA plays an important role(s) in both local and systemic resistance, its mechanism(s) of action has not yet been well defined. In an attempt to determine SA's mode of action, Klessig and coworkers have identified several proteins with which SA interacts.[53–55] The first SA binding protein (SABP) identified in tobacco[56,57] was later found to be a family of catalases[53,58] whose H_2O_2-degrading activity was inhibited by SA and those analogues of SA which are biologically active for induction of *PR* genes and enhanced disease resistance. The findings that SA treatment resulted in elevated levels of endogenous H_2O_2 and that treatment of plants with H_2O_2 or H_2O_2-generating agents induced *PR-1* gene expression led to the working hypothesis that one of SA's mechanisms of action is to inhibit catalase, with the resulting elevated

H_2O_2 perhaps serving as a second messenger to activate defense responses.[53] In support of this model, it was discovered that 2,6-dichloroisonicotinic acid (INA; a synthetic inducer of *PR* genes and enhanced resistance; Fig. 1) and its biologically active analogues also inhibited tobacco catalase *in vivo*.[59] Moreover, SA and INA were found to inhibit ascorbate peroxidase, the other major H_2O_2-scavenging enzyme.[55] In contrast, the guaiacol-utilizing peroxidases, which are involved in processes associated with disease resistance such as lignification and cross-linking of cell wall proteins, were not blocked by SA or INA.

The role of SA-mediated catalase inhibition and elevated H_2O_2 levels in plant defense responses is an area of active debate. While H_2O_2 and H_2O_2-inducing chemicals activated *PR-1* genes in wt tobacco, *PR-1* induction was strongly suppressed in NahG transgenic plants.[60,61] In addition, no detectable increase of H_2O_2 levels was found during the onset of SAR,[61] and Bi and coworkers[60] were unable to detect significant reductions in catalase activity after *Pseudomonas syringae* infection of tobacco or pretreatment of leaf discs with SA. Furthermore, H_2O_2 at very high concentrations (150–1000 mM) was found to stimulate accumulation of SA and activate BA-2 hydroxylase.[61-63]

A more direct test of the involvement of catalase and H_2O_2 in the activation of defense responses was performed using transgenic tobacco in which catalase activity was depressed either through the expression of antisense constructs[64] or through sense cosuppression.[65] Suppression of catalase activity was not accompanied by induction of *PR-1* genes in most of the transgenic plants. However, those which had the most severely reduced catalase levels developed necrosis and concurrently exhibited elevated levels of PR-1 protein and enhanced resistance to TMV. As SA and SAG levels were also elevated in the necrotic tissue, the simplest explanation for these results is that oxidative stress, resulting from severe catalase deficiency, induced necrosis and SA synthesis; SA in turn activated the defense responses.[64] When the catalase deficient antisense plants were crossed with NahG plants, the progeny still developed necrosis but *PR-1* gene expression and enhanced resistance were suppressed.[66] Thus, SA seems to be required for *PR-1* gene induction in the catalase deficient plants. Taken together, these results argue that H_2O_2 acts upstream of SA in the signal transduction pathway rather than, or in addition to, acting downstream of SA.

If the predominant mechanism by which SA induces defense responses is not through elevated H_2O_2 levels caused by the inhibition of catalase and ascorbate peroxidase, how is the SA signal transduced? One possibility is through the generation of free radicals. SA inhibits catalase by acting as a one electron (e^-)-donating substrate to siphon catalase from its extremely rapid catalatic (a) cycle into the much slower (~1000 fold) peroxidative (b) cycle (Fig. 3).[58] This results in not only the inhibition of catalase but also the formation of a SA free radical. Since free radicals are known to cause lipid peroxidation, Anderson et al.[67] tested the effects of SA and its analogues on the *in vivo* accumulation of lipid peroxides and related by-products. SA and its biologically active analogues

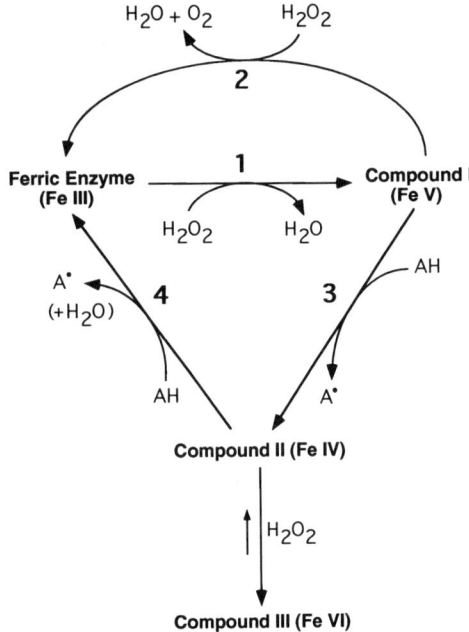

Figure 3. The reaction cycles of catalase. Formal oxidation states of the heme iron are shown in *parentheses*. The first step in the catalase cycle involves a 2e⁻ equivalent reduction of H_2O_2 to H_2O and the corresponding oxidation of the ferric enzyme (ferricatalase) to compound I (*step 1*). Compound I is converted back to ferricatalase by a 2e⁻ equivalent reduction and the corresponding oxidation of a second molecule of H_2O_2 to O_2 (*step 2*), thus completing the catalatic (a) cycle. In the peroxidative (b) cycle of catalase, compound I is converted to compound II by a 1e⁻ equivalent reduction (*step 3*). Compound II is inactive with respect to the catalatic cycle. Through a second 1e⁻ equivalent reduction, compound II can be converted back to ferricatalase (*step 4*). AH represents an electron donor (*e.g.* a phenolic compound), while A• denotes the resulting radical formed after donation of an electron. Compound III is an inactive form of catalase produced at high H_2O_2 levels, which is not readily converted back to compound II. (Reprinted by permission from *J. Biol. Chem.*).[58]

induced lipid peroxidation, while inactive analogues, which do not induce *PR* genes or inhibit catalase and APX, failed to induced lipid peroxidation. Moreover, exogenously supplied lipid peroxides activated *PR* genes while diethyldithiocarbamic acid, a compound that converts lipid peroxides into their hydroxyl derivatives, blocked this activation. Since lipid peroxidation is a self-perpetuating chain reaction, a small amount of SA free radical might result in the formation of an effective lipid peroxide signal, without a readily discernible inhibition of catalase or ascorbate peroxidase. This may be particularly relevant in uninfected systemic tissues where SA levels are too low to effectively inhibit these H_2O_2-scavenging enzymes, unless SA is concentrated in a subcellular compartment such as peroxisomes.

SA interacts with other proteins besides those involved in redox regulation. One of these is a ~25 kD, soluble protein termed SABP2, whose high affinity for SA (K_d = 90 nM) is ~150 times higher than that of catalase.[54] The ability of SABP2 to bind SA and its biologically active analogues tighter than inactive analogues suggests that SABP2 plays a role in SA-mediated disease resistance. Interestingly, SABP2 has a 15-fold higher affinity for the plant defense activating agent benzothiadiazole (BTH; Fig. 1)[68] than for SA. This is consistent with BTH's greater efficacy in inducing plant defense responses.

SA has also been shown to inhibit 1-aminocyclopropane-1-carboxylic acid oxidase and aconitase.[69,70] Inhibition of 1-aminocyclopropane-1-carboxylic acid oxidase, a key enzyme in ethylene biosynthesis, may reduce senescence and ethylene production, two processes known to be affected by SA.[69] Similarly, inhibition of aconitase is likely related to thermogenesis, the other process for which SA has been shown to play a signaling role.[71] During thermogenesis, the alternative pathway of respiration and its key enzyme and encoding gene, alternative oxidase (*Aox*), are activated by SA. Recently, Vanlerberghe and McIntosh[72] demonstrated that citrate, the substrate of aconitase, is an important signal metabolite regulating *Aox* expression. Thus, SA may activate the *Aox* gene and the alternative pathway of respiration, at least in part, by inhibiting aconitase and thereby elevating citrate levels. Since H_2O_2 also induces the *Aox* gene,[72,73] inhibition of the two H_2O_2-scavenging enzymes may also play a role in thermogenesis. Thus, it appears that SA interacts with several different proteins to exert its physiological effects. Future analyses will likely identify other SA-effector proteins that might play roles in disease resistance and/or other SA-mediated processes.

SA-MEDIATED GENE INDUCTION

Over the past two decades, SA has been shown to induce the expression of many defense-related genes.[41] SA also appears to potentiate expression of certain genes and defense responses. There is increasing evidence that pre- or co-treatment with SA can positively influence the magnitude (and kinetics) of several defense responses (*e.g.* H_2O_2 production, gene activation, and cell death) that are induced by a variety of stimuli. These stimuli include infection,[74–76] fungal cell wall fragments, chitosan, ergosterol, mastoparan,[77] wounding,[74,78] and H_2O_2.[76] These effects of SA have been termed potentiation or conditioning, and they can often be seen with low to moderate levels of SA (50–200 μM), which by themselves are insufficient to induce these defense responses.

Here, we focus on recent studies of genes whose expression is induced, rather than potentiated, by SA. These genes can be divided into two classes. Genes in the first class are often called immediate early genes, as their expression does not require protein synthesis. They include plant GST, *Agrobacterium tumefaciens* nopaline and octopine synthase, and the 35S promoter of cauliflower mosaic virus. Promoters of this class of genes contain copies of *as-1*-like *cis* elements, which mediate SA-induced expression. Several transcription factors belonging to the TGA family of bZIP proteins have been identified and shown to bind these elements.[79,80] An activity that binds the *as-1* element in the 35S promoter was shown to be induced by SA or cycloheximide treatment of tobacco leaves, while phosphatase treatment of nuclear extracts decreased its activity.[81] This result led to the suggestion that the *as-1* binding factor is sequestered by an inhibitory protein; this inhibition is relieved by SA-mediated phosphorylation of either the factor or the inhibitory protein. Interestingly, SA treatment of tobacco suspension cells induces a rapid and transient activation of a novel MAP kinase.[82] Whether this kinase is responsible for the SA-dependent phosphorylation of the above protein is presently unknown. The *Arabidopsis GST6* promoter also contains an *as-1*-like element which is bound by a factor termed OBF.[79] Interestingly, a second factor, termed OBP1 and identified by its interaction with OBF, binds next to the OBF-binding site and stimulates OBF binding to the *GST6* promoter. These results suggest that the interaction of these two factors may be important for *GST6* expression.

The second class of SA-inducible genes includes the acidic *PR* genes. Their induction by SA is sensitive to inhibitors of protein synthesis. Promoters of the tobacco *PR-1a* and *PR-2* genes have been analyzed in the most detail. No common SA-responsive element has yet been defined in these genes. A 10 bp TCA element that is common to the promoters of several tobacco *PR* genes, as well as several stress-induced genes, was shown to bind to a 40 kD nuclear protein in a SA-dependent manner.[83] However, this TCA element is neither sufficient nor necessary for SA-mediated induction of the tobacco *PR-2d* promoter *in vivo*.[84] Rather, *in vivo* analysis of this promoter has defined a 25 bp SA-responsive element which contains the sequence TTCGACC.[84,85] This sequence is related to the W boxes present in the promoters of several elicitor- and wound-induced genes,[86] whose expression can be potentiated by SA.[74,76,87] These results suggest the interesting possibility that related factors may be involved in the regulation of genes induced by SA and genes whose induction is potentiated by SA treatment. Klessig and colleagues (Shah, Durner, Zhang and Klessig, unpublished) have identified and cloned several proteins that bind the SA-responsive element or nearby sequences. One of these factors, SRRB2, is readily phosphorylated by the SA-activated MAP kinase,[82] which dramatically increases its DNA binding activity.

Current evidence suggests that several factors may also be involved in SA induction of the tobacco *PR-1a* promoter. *In vivo* analysis of this promoter argues that more than one region is involved in its SA-mediated activation.[88] A GT-1-like protein(s) has been shown to bind various fragments of this promoter *in vitro*.[89] The *PR-1a* promoter also contains several Myb consensus binding sites, some of which can be bound by recombinant Myb1 *in vitro*.[90] Tobacco *myb1* gene expression is induced by pathogen infection or SA treatment. Its rapid kinetics of induction by SA are similar to those of immediate early genes. Though it is tempting to speculate that the GT-1 and Myb1-like protein(s) might be involved in SA-dependent expression of the *PR-1a* gene, *in vivo* evidence is still lacking. Interestingly, Myb binding sites are also present in the promoters of PAL genes, whose expression is potentiated by SA.[87,91] This finding again raises the possibility that the same or similar factors may participate in the regulation of both SA-potentiated and SA-induced genes.

GENETIC ANALYSIS OF SA-MEDIATED SIGNAL TRANSDUCTION

The generation of *Arabidopsis* mutants has provided powerful tools with which to dissect the SA signal transduction pathway(s). Analysis of the various mutant phenotypes has further underscored the importance of SA in the activation of disease resistance. Two general classes of mutants have been identified: those in which resistance, including SAR, is compromised and those in which resistance to microbial pathogens is enhanced (Table 1). Of the former, only the allelic *npr1/sai1/nim1* group of mutants appears to affect SA signal transduction.[92–95] Mutations in this gene prevent both the development of SAR and the induction of *PR* genes by SA. These mutants also fail to respond to INA or BTH, which supports the biochemical evidence that both compounds induce defense responses via the SA signal transduction pathway and probably are functional analogues of SA.[49,54,55,59,68,96] The recessive nature of these mutations suggests that the wt protein acts as a positive regulator of the SA signal transduction pathway. *NPR1* was recently cloned and shown to encode a unique 60 kD protein containing ankyrin repeats.[97,98] Since ankyrin repeats facilitate protein–protein interactions and one of the *npr1* mutations is in these repeats, Npr1 probably functions in signal transduction by interacting with other proteins. In addition to ankyrin repeats, *NPR1/NIM1* shares other homology with IkB.[98] IkB binds to NF-kB and represses NF-kB's ability to activate genes involved in mammalian immune, acute phase, and inflammatory responses. Based on analogy with the mammalian system, it has been suggested that the transcription factor targeted by *NPR1/NIM1* serves as a repressor of defense gene expression, either by direct or indirect action.

Many mutants have been obtained in the past few years which exhibit increased resistance to pathogens. Some were identified by their enhanced resistance phenotype (*cim*),[8] others by their constitutive expression of *PR* genes or promoters (*cpr* and *cep*),[99,100] and still others based on their ability to spontaneously develop lesions (*lsd* and *acd2*).[101–103] All of these mutants have constitutively elevated levels of SA and *PR* gene expression. Since enhanced resistance in these mutants is suppressed in the presence of the *nahG* gene, the elevated levels of SA likely play a causal role in the development of enhanced resistance

Table 1. Arabidopsis mutants used to dissect the SA signal transduction pathway(s)

Mutant	Dominant/ recessive	SA levels	Pr gene expression	Comments
npr1, sai1, nim1	Recessive	Normal	SA/INA/BTH-non-inducible	Enhanced susceptibility to avirulent pathogens; normal HR
eds	Recessive	Normal	Normal	Enhanced susceptibility to pathogens[94]
ndr1	Recessive	Not determined	Normal	Enhanced susceptibility to pathogens; normal HR to most avirulent pathogens tested[108]
acd1	Recessive	Not determined	Normal	Develop spontaneous lesions and have enhanced susceptibility to pathogens[109]
acd2	Recessive	Elevated	Constitutively high	Develop spontaneous lesions and have enhanced resistance to pathogens
lsd1, lsd3, lsd5	Recessive	Elevated	Constitutively high	Develop spontaneous lesions and have enhanced resistance to pathogens
lsd2, lsd4, lsd6, lsd7	Dominant	Elevated	Constitutively high	Develop spontaneous lesions and have enhanced resistance to pathogens; all phenotypic properties suppressed by nahG in lsd6 and lsd7; lesion formation not suppressed by nahG in lsd2 and lsd4
cpr1	Recessive	Elevated	Constitutively high	Enhanced resistance to pathogens; all phenotypic properties suppressed by nahG
cep1	Recessive	Elevated	Constitutively high	Develop spontaneous lesions and have enhanced resistance to pathogens; constitutive PR expression suppressed by nahG; not allelic to cpr1
cim3	Dominant	Elevated	Constitutively high	Enhanced resistance to pathogens; all phenotypic properties suppressed by nahG

and SAR. For a detailed review of mutants affecting the SA signaling pathway, see Dangl et al.,[5] Ryals et al.,[8] and Delaney.[104]

CONCLUDING REMARKS

This review has focused on SA and its importance in plant disease resistance. However, an increasing body of evidence suggests that the SA-mediated signal transduction pathway(s) is only one aspect of the many responses activated by pathogen attack. The complexities of the defense responses, as well as SA's role in them, are emerging themes. For example, SA is present in multiple forms which are readily interconvertible and may play important roles in regulation, storage, and transmission of the SA signal. Similarly, multiple SA-effector proteins have been identified. Some of these will likely function in other SA-mediated processes such as alternative respiration, thermogenesis, ethylene biosynthesis, and senescence. Nonetheless, it is highly probable that several of these proteins with which SA interacts participate in the development of disease resistance. It is also likely that there are other SA-effector proteins yet to be identified. These findings argue that SA has multiple modes of action, not only for controlling different plant processes, but perhaps even in facilitating various defense responses.

There are multiple forms of SA-mediated gene expression. Expression of some genes is potentiated by SA, while induction of others, such as the *PR* genes, requires only SA and no other stimulus. The latter group can be further divided into those which require protein synthesis for SA-mediated induction and those which do not. In addition, analyses of *PR* gene promoters suggest that multiple DNA binding factors may be necessary to convey SA responsiveness.

Finally, a complex network of interacting signal transduction pathways likely exists in plants which allows them to respond in a coordinated manner to diverse external environmental stimuli and to internal physiological states. Evidence is emerging for substantial cross-talk between and among various signaling pathways. For instance, Seo *et al.*[105] found that inactivation of a MAP kinase associated with wounding in transgenic tobacco results in unusual responses of these plants to wounding. After wounding, plants fail to produce the wound-associated hormone, jasmonic acid, but rather activate disease resistance responses including SA and PR protein biosynthesis. In addition, a growing number of transgenic plants, which express a foreign gene or a mutated form of an endogenous gene, spontaneously develop lesions, constitutively produce SA and PR proteins, and exhibit enhanced resistance to pathogens.[16,19,106,107] Expression of these foreign or mutated genes may, in some cases, mimic part of the defense signaling pathway leading to SA synthesis. Alternatively, their expression may disrupt various metabolic processes (such as hexose partitioning or ubiquitin-dependent proteolysis) which then impinge on the SA signaling pathway. Some of the many mutants of *Arabidopsis* which exhibit enhanced disease resistance and constitutive production of SA and PR proteins may also indirectly affect SA signal transduction by altering other pathways or processes which then activate the SA signaling pathway via cross-talk. Defining the components of the SA-mediated defense signaling pathway, including the SA-effector proteins, as well as the points of connection with other pathways or processes that regulate responses to various external and internal stimuli, will prove both challenging and exciting.

ACKNOWLEDGMENTS

Studies carried out in the authors' lab were supported by grants numbers MCB 9310371 and MCB 9514239 from the National Science Foundation.

REFERENCES

1. BENT, A.F. 1996. Plant disease resistance genes: Function meets structure. Plant Cell 8:1757–1771.
2. DANGL, J.L. 1995. Piéce de résistance: Novel classes of plant disease resistance genes. Cell 80:363–366.
3. JONES, J.D.G. 1996. Plant disease resistance genes: Structure, function and evolution. Curr. Opin. Biotechnol. 7:155–160.
4. STASKAWICZ, B.J., AUSUBEL, F.M., BAKER, B.J., ELLIS, J.G., JONES, J.D.G. 1995. Molecular genetics of plant disease resistance. Science 268:661–667.
5. DANGL, J.L., DIETRICH, R.A., RICHBERG, M.H. 1996. Death don't have no mercy: Cell death programs in plant-microbe interactions. Plant Cell 8:1793–1807.
6. MITTLER, R., LAM, E. 1996. Sacrifice in the face of foes: Pathogen-induced programmed cell death in plants. Trends Microbiol. 4:10–15.
7. GOODMAN, R.N., NOVACKY, A. 1994. The Hypersensitive Reaction in Plants to Pathogens. A Resistance Phenomenon. American Phytopathological Society Press, St. Paul, Minnesota.
8. RYALS, J.A., NEUENSCHWANDER, U.H., WILLITS, M.G., MOLINA, A., STEINER, H.-Y., HUNT, M.D. 1996. Systemic acquired resistance. Plant Cell 8:1809–1819.
9. HAMMOND-KOSACK, K.E., JONES, J.D.G. 1996. Resistance gene-dependent plant defense responses. Plant Cell 8:1773–1791.
10. DEMPSEY, D.A., KLESSIG, D.F. 1995. Signals in plant disease resistance. Bull. Inst. Pasteur 93:167–186.
11. LINTHORST, H.J.M. 1991. Pathogenesis-related proteins in plants. Crit. Rev. Plant Sci. 10:123–150.
12. WOBBE, K.K., KLESSIG, D.F. 1996. Salicylic acid-an important signal in plants. In: Plant Gene Research (E.S. Dennis, B. Hohn, Th. Hohn, F. Meins, Jr., J. Schell, D.P.S. Verma, eds.). Springer-Verlag, Wien and New York, pp. 167–196.
13. SCOFIELD, S.R., TOBIAS, C.M., RATHJEN, J.P., CHANG, J.H., LAVELLE, D.T., MICHELMORE, R.W., STASKAWICZ, B.J. 1996. Molecular basis of gene-for-gene specificity in bacterial speck disease of tomato. Science 274:2063–2065.
14. TANG, X., FREDERICK, R.D., ZHOU, J., HALTERMAN, D.A., JIA, Y., MARTIN, G.B. 1996. Initiation of plant disease resistance by physical interaction of AvrPto and Pto kinase. Science 274:2060–2063.
15. UMEMOTO, N., KAKITANI, M., IWAMATSU, A., YOSHIKAWA, M., YAMAOKA, N., ISHIDA, I. 1997. The structure and function of a soybean β-glucan-elicitor-binding protein. Proc. Natl. Acad. Sci. USA 94:1029–1034.
16. BEFFA, R., SZELL, M., MEUWLY, P., PAY, A., VOELI-LANGE, R., MÉTRAUX, J.-P., NEUHAUS, G., MEINS, F. JR., NAGY, F. 1995. Cholera toxin elevates pathogen resistance and induces pathogenesis-related gene expression in tobacco. EMBO J. 14:5753–5761.
17. SANO, H., SEO, S., ORUDGEV, E., YOUSSEFIAN, S., ISHIZUKA, K., OHASHI, Y. 1994. Expression of the gene for a small GTP binding protein in transgenic tobacco elevates endogenous cytokinin levels, abnormally induces salicylic acid in response to wounding, and increases resistance to tobacco mosaic virus infection. Proc. Natl. Acad. Sci. USA 91:10556–10560.
18. HAHLBROCK, K., SCHEEL, D., LOGEMENN, E., NURNBERGER, T., PAPNISKE, M., REINOLD, S., SACKS, W.R., SCHMELZER, E. 1995. Oligopeptide-elicited defense gene activation in cultured parsley cells. Proc. Natl. Acad. Sci. USA 92:4150–4157.
19. MITTLER, R., SHULAEV, V., LAM, E. 1995. Coordinated activation of programmed cell death and defense mechanisms in transgenic tobacco plants expressing a bacterial proton pump. Plant Cell 7:29–42.
20. LEVINE, A., TENHAKEN, R., DIXON, R., LAMB, C. 1994. H_2O_2 from the oxidative burst orchestrates the plant hypersensitive disease resistance response. Cell 79:583–595.

21. MEHDY, M.C. 1994. Active oxygen species in plant defense against pathogens. Plant Physiol. 105:467–472.
22. SUZUKI, K., SHINSHI, H. 1995. Transient activation and tyrosine phosphorylation of a protein kinase in tobacco cells treated with a fungal elicitor. Plant Cell 7:639–647.
23. XING, T., HIGGINS, V.J., BLUMWARD, E. 1996. Regulation of plant defense response to fungal pathogens: Two types of protein kinases in the reversible phosphorylation of the host plasma membrane H^+-ATPase. Plant Cell 8:555–564.
24. ZHOU, J., LOH, Y.-T., BRESSAN, R.A., MARTIN, G.B. 1995. The tomato gene *pti* encodes a serine/threonine kinase that is phosphorylated by *Pto* and is involved in the hypersensitive response. Cell 83:925–935.
25. WHITE, R.F. 1979. Acetylsalicylic acid (aspirin) induces resistance to tobacco mosaic virus in tobacco. Virology 99:410–412.
26. ANTONIW, J.F., WHITE, R.F. 1980. The effects of aspirin and polyacrylic acid on soluble leaf proteins and resistance to virus infection in five cultivars of tobacco. Phytopathol. Z. 98:331–341.
27. MALAMY, J., CARR, J.P., KLESSIG, D.F., RASKIN, I. 1990. Salicylic acid: A likely endogenous signal in the resistance response of tobacco to viral infection. Science 250:1002–1004.
28. MALAMY, J., HENNIG, J., KLESSIG, D.F. 1992. Temperature-dependent induction of salicylic acid and its conjugates during the resistance response to tobacco mosaic virus infection. Plant Cell 4:359–365.
29. YALPANI, N., SILVERMAN, P., WILSON, T.M.A., KLEIER, D.A., RASKIN, I. 1991. Salicylic acid is a systemic signal and an inducer of pathogenesis-related proteins in virus-infected tobacco. Plant Cell 3:809–818.
30. MÉTRAUX, J.-P., SIGNER, H., RYALS, J.A., WARD, E., WYSS-BENZ, M., GAUDIN, J., RASCHDORF, K., SCHMID, E., BLUM, W., INVERARDI, B. 1990. Increase in salicylic acid at the onset of systemic acquired resistance in cucumber. Science 250:1004–1006.
31. WARD, E.R., UKNES, S.J., WILLIAMS, S.C., DINCHER, S.S., WIEDERHOLD, D.L., ALEXANDER, D.C., AHL-GOY, P., MÉTRAUX, J.-P., RYALS, J.A. 1991. Coordinate gene activity in response to agents that induce systemic acquired resistance. Plant Cell 3:1085–1094.
32. DELANEY T., UKNES, S., VERNOOIJ, B., FRIEDRICH, L., WEYMANN, K., NEGROTTO, D., GAFFNEY, T., GUT-RELLA, M., KESSMANN, H., WARD, E., RYALS, J. 1994. A central role of salicylic acid in plant disease resistance. Science 266:1247–1250.
33. GAFFNEY, T., FRIEDRICH, L., VERNOOIJ, B., NEGROTTO, D., NYE, G., UKNES, S., WARD, E., KESSMANN, H., RYALS, J. 1993. Requirement of salicylic acid for the induction of systemic acquired resistance. Science 261:754–756.
34. MAUCH-MANI, B., SLUSARENKO, A.J. 1996. Production of salicylic acid precursors is a major function of phenylalanine ammonia-lyase in the resistance of *Arabidopsis* to *Peronospora parasitica*. Plant Cell 8:203–212.
35. YALPANI, N., LEÓN, J., LAWTON, M.A., RASKIN, I. 1993. Pathway of salicylic acid biosynthesis in healthy and virus-inoculated tobacco. Plant Physiol. 103:315–321.
36. LEÓN, J., YALPANI, N., RASKIN, I., LAWTON, M.A. 1993. Induction of benzoic acid 2-hydroxylase in virus-inoculated tobacco. Plant Physiol. 103:323–328.
37. ENYEDI, A.J., YALPANI, N., SILVERMAN, P., RASKIN, I. 1992. Localization, conjugation and function of salicylic acid in tobacco during the hypersensitive reaction to tobacco mosaic virus. Proc. Natl. Acad. Sci. USA 89:2480–2484.
38. ENYEDI, A.J., RASKIN, I. 1993. Induction of UDP-glucose:salicylic acid glucosyltransferase activity in tobacco mosaic virus-inoculated tobacco (*Nicotiana tabacum*) leaves. Plant Physiol. 101:1375–1380.
39. TANAKA, S., HAYAKAWA, K., UMETANI, Y., TABATA, M. 1990. Glucosylation of isomeric hydroxybenzoic acids by cell suspension cultures of *Mallotus japonicus*. Phytochemistry 29:1555–1558.

40. YALPANI, N., SCHULZ, M., DAVIS, M.P., BALKE, N.E. 1992. Partial purification and properties of an inducible uridine 5'-disphosphate-glucose:salicylic acid glucosyltransferase from oat roots. Plant Physiol. 100:457–463.
41. KLESSIG, D.F., MALAMY, J. 1994. The salicylic acid signal in plants. Plant Mol. Biol. 26:1439–1458.
42. MEUWLY, P. MÖLDERS, W., SUMMERMATTER, K., STICHER, L., MÉTRAUX, J.-P. 1994. Salicylic acid and chitinase in infected cucumber plants. In: International Symposium on Natural Phenols in Plant Resistance, Vol. 1. Acta Horticulture, pp. 371–374.
43. SUMMERMATTER, K., MEUWLY, P., MÖLDERS, W., MÉTRAUX, J.-P. 1994. Salicylic acid levels in *Arabidopsis thaliana* after treatments with *Pseudomonas syringae* or synthetic inducers. In: International Symposium on Natural Phenols in Plant Resistance, Vol. 1. Acta Horticulture, pp. 367–370.
44. HENNIG, J., MALAMY, J., GRYNKIEWICZ, G., INDULSKI, J., KLESSIG, D.F. 1993. Interconversion of the salicylic acid signal and its glucoside in tobacco. Plant J. 4:593–600.
45. CHEN, Z., MALAMY, J., HENNIG, J., CONRATH, U., SÁNCHEZ-CASAS, P., SILVA, H., RICIGLIANO, J., KLESSIG, D.F. 1995. Induction, modification, and transduction of the salicylic acid signal in plant defense responses. Proc. Natl. Acad. Sci. USA. 92:4134–4137.
46. GIANINAZZI, S., AHL, P. 1983. The genetic and molecular basis of b-proteins in the genus *Nicotiana*. Neth. J. Plant Pathol. 89:275–281.
47. SHULAEV, V., LEÓN, J., RASKIN, I. 1995. Is salicylic acid a translocated signal of systemic acquired resistance in plants? Plant Cell 7:1691–1701.
48. MÖLDERS, W., BUCHALA, A., MÉTRAUX, J.-P. 1996. Transport of salicylic acid in tobacco necrosis virus-infected cucumber plants. Plant Physiol. 112:787–792.
49. MALAMY, J., SÁNCHEZ-CASAS, P., HENNIG, J., GUO, A., KLESSIG, D.F. 1996. Dissection of the salicylic acid signaling pathway in tobacco. Mol. Plant-Microbe Interact. 9:474–482.
50. RASMUSSEN, J.B., HAMMERSCHMIDT, R., ZOCK, M.N. 1991. Systemic induction of salicylic acid accumulation in cucumber after inoculation with *Pseudomonas syringae* pv. syringae. Plant Physiol. 97:1342–1347.
51. VERNOOIJ, B., FRIEDRICH, L., MORSE, A., REIST, R., KOLDITZ-JAWHAR, R., WARD, E., UKNES, S., KESSMANN, H., RYALS, J. 1994. Salicylic acid is not the translocated signal responsible for inducing systemic acquired resistance but is required in signal transduction. Plant Cell 6:959–969.
52. SHULAEV, V., SILVERMAN, P., RASKIN, I. 1997. Airborne signalling by methyl salicylate in plant pathogen resistance. Nature 385:718–721.
53. CHEN, Z., SILVA, H., KLESSIG, D.F. 1993. Active oxygen species in the induction of plant systemic acquired resistance by salicylic acid. Science 262:1883–1886.
54. DU, H., KLESSIG, D.F. 1997. Identification of a soluble, high affinity salicylic acid-binding protein from tobacco. Plant Physiol. 113:1319–1327.
55. DURNER, J., KLESSIG, D.F. 1995. Inhibition of ascorbate peroxidase by salicylic acid and 2,6-dichloroisonicotinic acid, two inducers of plant defense responses. Proc. Natl. Acad. Sci. USA 92:11312–11316.
56. CHEN, Z., KLESSIG, D.F. 1991. Identification of a soluble salicylic acid-binding protein that may function in signal transduction in the plant disease resistance response. Proc. Natl. Acad. Sci. USA 88:8179–8183.
57. CHEN, Z., RICIGLIANO, J., KLESSIG, D.F. 1993. Purification and characterization of a soluble salicylic acid binding protein from tobacco. Proc. Natl. Acad. Sci. USA. 90:9533–9537.
58. DURNER, J., KLESSIG, D.F. 1996. Salicylic acid is a modulator of tobacco and mammalian catalases. J. Biol. Chem. 271:28492–28501.
59. CONRATH, U., CHEN, Z., RICIGLIANO, J.R., KLESSIG, D.F. 1995. Two inducers of plant defense responses, 2,6-dichloroisonicotinic acid and salicylic acid, inhibit catalase activity in tobacco. Proc. Natl. Acad. Sci. USA 92:7143–7147.

60. BI, Y.-M., KENTON, P., MUR, L., DARBY, R., DRAPER, J. 1995. Hydrogen peroxide does not function downstream of salicylic acid in the induction of PR protein expression. Plant J. 8:235–245.
61. NEUENSCHWANDER, U., VERNOOIJ, B., FRIEDRICH, L., UKNES, S., KESSMANN, H., RYALS, J. 1995. Is hydrogen peroxide a second messenger of salicylic acid in systemic acquired resistance? Plant J. 8:227–233.
62. LEÓN, J., LAWTON, M.A., RASKIN, I. 1995. Hydrogen peroxide stimulates salicylic acid biosynthesis in tobacco. Plant Physiol. 108:1673–1678.
63. SUMMERMATTER, K., STICHER, L., MÉTRAUX, J.-P. 1995. Systemic responses in *Arabidopsis thaliana* infected and challenged with *Pseudomonas syringae* pv *syringae*. Plant Physiol. 108:1379–1385.
64. TAKAHASHI, H., CHEN, Z., DU. H., LIU, Y., KLESSIG, D.F. 1997. Development of necrosis and activation of disease resistance in transgenic tobacco plants with severely reduced catalase levels. Plant J. 11:993–1005.
65. CHAMNONGPOL, S., WILLEKENS, H., LANGEBARTELS, C., VAN MONTAGU, M., INZÉ, D., VAN CAMP, W. 1996. Transgenic tobacco with a reduced catalase activity develop necrotic lesions and induce pathogenesis-related gene expression under high light. Plant J. 10:491–503.
66. DU, H., KLESSIG, D.F. Role for salicylic acid in the activation of defense responses in catalase-deficient transgenic tobacco. Mol. Plant-Microbe Interact., in press.
67. ANDERSON, M., CHEN, Z., KLESSIG, D.F. 1997. Possible involvement of lipid peroxidation in SA-mediated induction of *PR-1* gene expression. Phytochemistry, in press.
68. GÖRLACH, J. VOLRATH, S., KNAUF-BEITER, G., HENGY, G., BECKHOVE, U., KOGEL, K.-H., OOSTENDORP, M., STAUB, T., WARD, E., KESSMANN, H., RYALS, J. 1996. Benzothiadiazole, a novel class of inducers of systemic acquired resistance, activates gene expression and disease resistance in wheat. Plant Cell 8:629–643.
69. LESLIE, C.A., ROMANI, R.J. 1988. Inhibition of ethylene biosynthesis by salicylic acid. Plant Physiol. 88:833–837.
70. RÜFFER, M., STEIPE, B., ZENK, M.H. 1995. Evidence against specific binding of salicylic acid to plant catalase. FEBS Lett. 377:175–180.
71. RASKIN, I., EHMANN, A., MELANDER W.R., MEEUSE, B.J.D. 1987. Salicylic acid - a natural inducer of heat production in Arum lilies. Science 237:1601–1602.
72. VANLERBERGHE, G.C., McINTOSH, L. 1996. Signals regulating the expression of the nuclear gene encoding alternative oxidase of plant mitochondria. Plant Physiol. 111:589–595.
73. WAGNER, A.M. 1995. A role for active oxygen species as second messengers in the induction of alternative oxidase gene expression in *Petunia hybrida* cells. FEBS Lett. 368:339–342.
74. MUR, L.A.J., NAYLOR, G., WARNER, S.A.J., SUGARS, J.M., WHITE, R.F., DRAPER, J. 1996. Salicylic acid potentiates defense gene expression in tissue exhibiting resistance to pathogen attack. Plant J. 9:559–571.
75. SIEGRIST, J., JEBLICK, W., KAUSS, H. 1994. Defense responses in infected and elicited cucumber (*Cucumis sativus* L.) hypocotyl segments exhibiting acquired resistance. Plant Physiol 105:1365–1374.
76. SHIRASU, K., NAKAJIMA, H., RAJASEKHAR, V.K., DIXON, R.A., LAMB, C. 1997. Salicylic acid potentiates an agonist-dependent gain control that amplifies pathogen signals in the activation of defense mechanisms. Plant Cell 9:261–270.
77. KAUSS, H., JEBLICK, W. 1996. Influence of salicylic acid on the induction of competence for H_2O_2 elicitation. Plant Physiol. 111:755–763.
78. FAUTH, M., MERTEN, A., HAHN, M.G., JEBLICK, W., KAUSS, H. 1996. Competence for elicitation of H_2O_2 in hypocotyls of cucumber is induced by breaching the cuticle and is enhanced by salicylic acid. Plant Physiol. 110:347–353.
79. CHEN, W., CHAO, G., SINGH, K.B. 1996. The promoter of a H_2O_2-inducible, *Arabidopsis* glutathione *S*-transferase gene contains closely linked OBF- and OBP1-binding sites. Plant J. 10:955–966.

80. ZHU, Q., DRÖGE-LASER, W., DIXON, R.A., LAMB, C. 1996. Transcriptional activation of plant defense genes. Curr. Opinion Gen. Dev. 6:624–630.
81. JUPIN, I., CHUA, N.-H. 1996. Activation of the CaMV *as-1 cis*-element by salicylic acid: Differential DNA-binding of a factor related to TGA1a. EMBO J. 15:5679–5689.
82. ZHANG, S., KLESSIG, D.F. 1997. Salicylic acid activates a 48-kD MAP kinase in tobacco. Plant Cell, 9:809–824.
83. GOLDSBROUGH, A.P., ALBRECHT, H., STRATFORD, R. 1993. Salicylic acid-inducible binding of a tobacco nuclear protein to a 10 bp sequence which is highly conserved amongst stress-inducible genes. Plant J. 3:563–571.
84. SHAH, J., KLESSIG, D.F. 1996. Identification of a salicylic acid-responsive element in the promoter of the tobacco pathogenesis-related β-1,3-glucanase gene, *PR-2d*. Plant J. 10:1089–1101.
85. HENNIG, J., DEWEY, R.E., CUTT, J.R., KLESSIG, D.F. 1993. Pathogen, salicylic acid and developmental dependent expression of a β-1,3-glucanase/GUS gene fusion in transgenic tobacco plant. Plant J. 4:481–493.
86. RUSHTON, P.J., TORRES, J.T., PARNISKE, M., WERNERT, P., HAHLBROCK, K., SOMSSICH, I.E. 1996. Interaction of elicitor-induced DNA-binding proteins with elicitor response elements in the promoter of parsley PR1 gene. EMBO J. 15:5690–5700.
87. KAUSS, H., THEISINGER-HINKEL, E., MINDERMANN,R., CONRATH, U. 1992. Dichloroisonicotinic and salicylic acid, inducers of systemic acquired resistance, enhance fungal elicitor responses in parsley. Plant J. 2:655–660.
88. VAN DE RHEE, M.D., BOL, J.F. 1993. Induction of the tobacco *PR-1a* gene by virus infection and salicylate treatment involves an interaction between multiple regulatory elements. Plant J. 3:71–82.
89. BUCHEL, A.S., MOLEMKAMP, R., BOL, J.F., LINTHORST, H.J.M. 1996. The *PR-1a* promoter contains a number of regulatory elements that bind GT-1-like factors with different affinity. Plant Mol. Biol. 30:493–504.
90. YANG, Y., KLESSIG, D.F. 1996. Isolation and characterization of a tobacco mosaic virus-inducible *myb* oncogene homolog from tobacco. Proc. Natl. Acad. Sci. USA 93:14972–14977.
91. SABLOWSKI, R.W.M., MOYANO, E., CULIANEZ-MACIA, F.A., SCHUCH, W., MARTIN, C., BEVAN, M.A. 1994. A flower-specific Myb protein activates transcription of phenylpropanoid biosynthetic genes. EMBO J. 13:128–137.
92. CAO, H., BOWLING, S.A., GORDON, A.S., DONG, X. 1994. Characterization of an *Arabidopsis* mutant that is nonresponsive to inducers of systemic acquired resistance. Plant Cell 6:1583–1592.
93. DELANEY, T.P., FRIEDRICH, L., RYALS, J.A. 1995. *Arabidopsis* signal transduction mutant defective in chemically and biologically induced disease resistance. Proc. Natl. Acad. Sci. USA 92:6602–6606.
94. GLAZEBROOK, J., ROGERS, E.E., AUSUBEL, F.M. 1996. Isolation of *Arabidopsis* mutants with enhanced disease susceptibility by direct screening. Genetics 143:973–982.
95. SHAH, J., TSUI, F., KLESSIG, D.F. 1997. Characterization of a salicylic acid-insensitive mutant (*sai*1) of *Arabidopsis thaliana*, identified in a selective screen utilizing the SA-inducible expression of the *tms*2 gene. Mol. Plant-Microbe Interact. 10:69–78.
96. VERNOOIJ, B., FRIEDRICH, L., AHL GOY, P., STAUB, T., KESSMANN, H., RYALS, J. 1995. 2,6-Dichloroisonicotinic acid-induced resistance to pathogens without the accumulation of salicylic acid. Mol. Plant Microbe Interact. 8:228–234.
97. CAO, H., GLAZEBROOK, J., CLARKE, J.D., VOLKO, S., DONG, X. 1997. The *Arabidopsis NPR1* gene that controls systemic acquired resistance encodes a novel protein containing ankyrin repeats. Cell 88:57–63.
98. RYALS, J., WEYMANN, K., LAWTON, K., FRIEDRICH, L., ELLIS, D., STEINER, H.-Y., JOHNSON, J., DELANEY, T.P., JESSE, T., VOS, P., UKNES, S. 1997. The Arabidopsis *NIM1*

protein shows homology to the mammalian transcription factor inhibitor kB. Plant Cell 9:425–439.
99. BOWLING, S.A., GUO, A., CAO, H., GORDON, A.S., KLESSIG, D.F., DONG, X. 1994. A mutation in *Arabidopsis* that leads to constitutive expression of systemic acquired resistance. Plant Cell 6:1845–1857.
100. SILVA, H., KLESSIG, D.F. 1997. Characterization of *cep1*, a new *Arabidopsis* mutant leading to systemic acquired resistance. Plant J., in review.
101. DIETRICH, R.A., DELANEY, T.P., UKNES, S.J., WARD, E.R., RYALS, J.A., DANGL, J.L. 1994. *Arabidopsis* mutants simulating disease resistance response. Cell 77:565–577.
102. GREENBERG, J.T., GUO, A., KLESSIG, D.F., AUSUBEL, F.M. 1994. Programmed cell death in plants: A pathogen-triggered response activated coordinately with multiple defense functions. Cell 77:551–563.
103. WEYMANN, K., HUNT, M., UKNES, S., NEUENSCHWANDER, U., LAWTON, K., STEINER, H.-Y., RYALS, J. 1995. Suppression and restoration of lesion formation in Arabidopsis *lsd* mutants. Plant Cell 7:2013–2022.
104. DELANEY, T.P. 1997. Genetic dissection of acquired resistance to disease. Plant Physiol. 113:5–12.
105. SEO, S., OKAMOTO, M., SETO, H., ISHIZUKA, K., SANO H., OHASHI, Y. 1995. Tobacco MAP kinase: A possible mediator in wound signal transduction pathways. Science 270:1988–1992.
106. BACHMAIR, A., POTUSCHAK, T., BECKER, F., NEJINSKAIA, V. 1994. Ubiquitin-dependent proteolysis in plants—a key metabolic pathway influencing plant-pathogen interaction. In: Advances in Molecular Genetics of Plant-Microbe Interactions, Vol. 3 (M.J. Daniels, J.A. Downie, A.E. Osbourn, eds.). Kluwer Academic Publishers, Dordrecht, The Netherlands, pp. 375–379.
107. HERBERS, K., MEUWLY, P., FROMMER, W.B., MÉTRAUX, J.-P., SONNEWALD, U. 1996. Systemic acquired resistance mediated by the ectopic expression of invertase: Possible hexose sensing in the secretory pathway. Plant Cell 8:793–803.
108. CENTURY, K.S., HOLUB, E.B., STASKAWICZ, B.J. 1995. *NDR1*, a locus of *Arabidopsis thaliana* that is required for disease resistance to both a bacterial and a fungal pathogen. Proc. Natl. Acad. Sci. USA 92:6597–6601.
109. GREENBERG, J.T., AUSUBEL, F.M. 1993. *Arabidopsis* mutants compromised for the control of cellular damage during pathogenesis and aging. Plant J. 4:327–341.

Chapter Eight

BIOSYNTHESIS OF RHIZOBIAL EXOPOLYSACCHARIDES AND THEIR ROLE IN THE ROOT NODULE SYMBIOSIS OF LEGUMINOUS PLANTS

Wilbert A. T. van Workum and Jan W. Kijne

Institute of Molecular Plant Sciences
Leiden University
Wassenaarseweg 64
2333 AL Leiden, The Netherlands

Introduction 139
Structural Features of Rhizobial Polysaccharides 141
Biosynthesis 144
    Succinoglycan 145
    Regulation of Succinoglycan 147
Symbiotic Importance 148
    Mutant Analysis 148
    EPS Function as Signaling Molecules? 154
Conclusion 157

INTRODUCTION

    Bacteria belonging to the genera *Rhizobium, Bradyrhizobium, Sinorhizobium, Mesorhizobium,* and *Azorhizobium,* commonly called rhizobia, are able to elicit the formation of nodules on the roots of leguminous plants which can be considered as new plant organs. Inside these organs, the rhizobia are present as dinitrogen-fixing organelles maintained by the plant in return for ammonia. Characteristic of this symbiosis is that rhizobia infect living plant cells by an endocytotical process. For a detailed survey of the *Rhizobium* infection process, we refer to a review by Kijne.[1] Briefly, the rhizobia attach to growing root hairs

Phytochemical Signals and Plant–Microbe Interactions, edited by Romeo *et al.*
Plenum Press, New York, 1998.

and provoke tight root hair curls in which they are trapped. Simultaneously, rhizobia induce cell divisions in the root cortex, resulting in the formation of a nodule primordium. Starting from the root hair curl, the rhizobia invade the plant through infection threads (tip-growing tubular structures containing rhizobia), are endocytosed by young nodule cells, and differentiate into dinitrogen-fixing bacteroids. This symbiotic association is host-plant-specific, *i.e.*, one rhizobial strain can infect only a limited number of different hosts. For example, *R. leguminosarum* bv. *viciae* can induce formation of dinitrogen-fixing nodules on *Pisum, Vicia, Lathyrus*, and *Lens*, whereas *R. leguminosarum* bv. *trifolii* nodulates *Trifolium* species (see Table 1).

A key factor in determination of specificity of nodulation is the structure of Nod factors or lipochitin oligosaccharides (LCOs), signal molecules produced by rhizobia after induction of their synthesis by plant flavonoids. Nod factors produced by all rhizobia consist of an oligosaccharide backbone of β-1,4-linked *N*-acetyl-D-glucosamine (chitin) to which a fatty acid is attached at the nitrogen of the nonreducing sugar residue. Nod factors show strain-dependent variations in length of the chitin backbone, structure of the fatty acyl moiety, and presence of specific substituents attached to one or more of the sugar residues. It has been shown that Nod factors are primarily responsible for host-specific processes such as root hair curling, formation of nodule primordia, and preparation for infection thread formation (formation of so-called preinfection threads); for a recent review, see Spaink.[2]

Besides Nod factors, rhizobial exopolysaccharides (EPS) are thought to play a role in determining the host plant specificity of nodulation.[3] In this chapter, examples of structures of rhizobial EPS will be given, followed by a detailed

Table 1. Host specificity of nodulation

Bacterial species	Host plants
R. leguminosarum bv. *viciae*	*Pisum, Vicia, Lathyrus, Lens*
R. leguminosarum bv. *trifolii*	*Trifolium*
R. leguminosarum bv. *phaseoli*	*Phaseolus*
R. etli	*Phaseolus*
R. tropici	broad host-range, *e.g.*, *Phaseolus, Leucaena*
R. lupini	*Lupinus*
M. loti	*Lotus*
S. meliloti	*Medicago, Melilotus, Trigonella*
S. fredii	broad host-range, *e.g.*, *Phaseolus, Glycine*
R. sp. GRH2	broad host-range, *e.g.*, *Phaseolus, Acacia*
R. sp. NGR234	broad host-range, over 75 plant genera, incl. the non-legume *Parasponia*
B. japonicum	*Glycine, Macroptilium*
B. elkanii	*Glycine, Macroptilium*
A. caulinodans	*Sesbania*

description of biosynthesis of succinoglycan of *Sinorhizobium meliloti* (previously called *Rhizobium meliloti*), which represents one of the best-studied examples of biosynthesis of bacterial (exo)polysaccharides. Then, the importance of rhizobial EPS in the interaction of rhizobia with leguminous plants will be discussed, with a focus on the role of EPS as "infochemicals" in cell-to-cell communication.

STRUCTURAL FEATURES OF RHIZOBIAL EXOPOLYSACCHARIDES

Numerous bacteria produce EPS, sugar polymers which show by definition little or no cell adhesion and thus can be found in the extracellular environment in the form of slime. Rhizobial EPS are heteropolysaccharides (*i.e.*, they contain different kinds of monosaccharides) formed from repeating unit structures. Although the rhizobia comprise a relatively small group of bacteria, a lot of variation can be found among the chemical structures of the EPS they produce. This variation can be due to sugar composition, condensation linkages, subunit size, and non-carbohydrate substitutions.

In general, EPS secreted by strains of *R. leguminosarum* bv. *viciae*, *R. leguminosarum* bv. *trifolii*, *R. leguminosarum* bv. *phaseoli*, and *R. etli* (previously included in *R. leguminosarum* bv. *phaseoli*) have the same conserved octasaccharide repeating unit, composed of glucose, glucuronic acid, and galactose in the ratio 5:2:1,[4–7] substituted with O-acetyl, pyruvyl, and hydroxybutanoyl groups (Fig. 1).[8–10] The distribution of O-acetyl and hydroxybutanoyl groups was found to be different for several *R. leguminosarum* strains, however, no correlation between substitution pattern and the ability to nodulate certain host plants was found.[11,12] EPS having the same octasaccharide repeating unit were found for the broad host-range *R.* sp. GRH2.[13] Certain strains of *R. leguminosarum* bv. *viciae*, *R. leguminosarum* bv. *trifolii*, and *R. leguminosarum* bv. *phaseoli* secrete EPS of which the repeating unit differs in sugar composition and side chain length.[14–18] However, the EPS secreted by all strains mentioned above are structurally related by virtue of possessing identical backbones and the same, β1,6-linked glucosyl residue starting the side chain (Fig. 1).

Strains of *S. meliloti* produce EPS containing glucose and galactose in a molar ratio of 7:1,[19,20] substituted with O-acetyl, pyruvyl, and succinyl groups. These molecules, usually called succinoglycan or EPS I, are also produced by several strains of *Alcaligenes* and *Agrobacterium*.[21–24] The structure of EPS produced by *R. tropici* strain CIAT899 (previously called *R. leguminosarum* bv. *phaseoli* CIAT899) is similar to that of succinoglycan, with the major difference being that in succinoglycan, the non-reducing terminal-sugar residue is glucose instead of galactose and only one pyruvyl substituent is present on the side chain (Fig. 2).[25] Almost the same EPS was found to be produced by

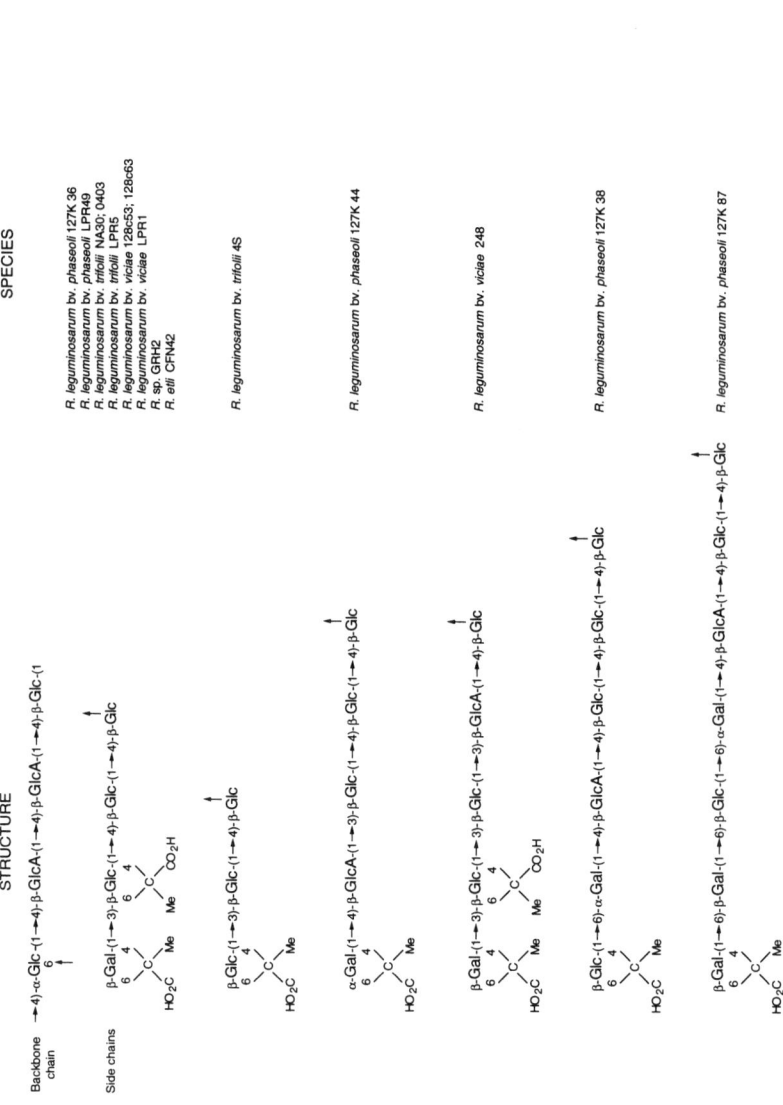

Figure 1. Chemical structure of repeating units of EPS of *R. etli* CFN42, *R.* sp. GRH2 and 13 different strains of *R. leguminosarum*. Non-stoichiometric substituents, namely acetate groups and 3-hydroxybutanoate have not been indicated. Abbreviations: Glc, glucose; GlcA, glucuronic acid; and Gal, galactose.

R. leguminosarum bv. *trifolii* AHU1134, *R. leguminosarum* bv. *phaseoli* AHU1133, and *R. lupini* KLU, only differing with respect to 4,6-pyruvylation of the side chain.[26] It should be noted that the latter three strains lost their nodulation ability during storage,[27] and it is not known whether this can be ascribed to the properties of the EPS produced or to loss of other functions

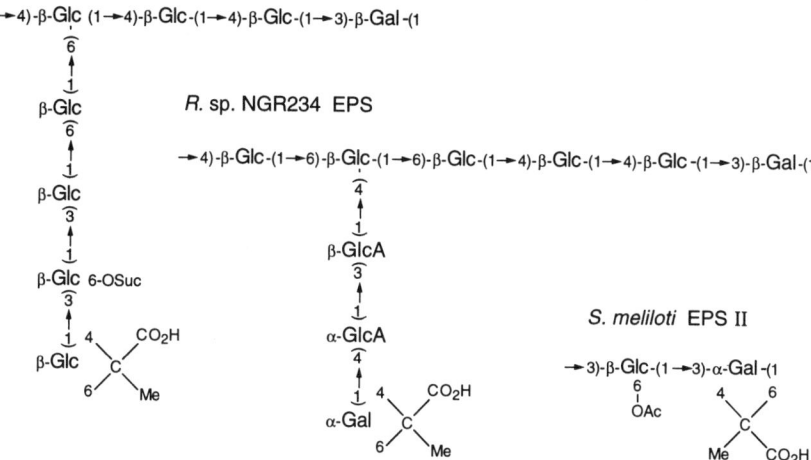

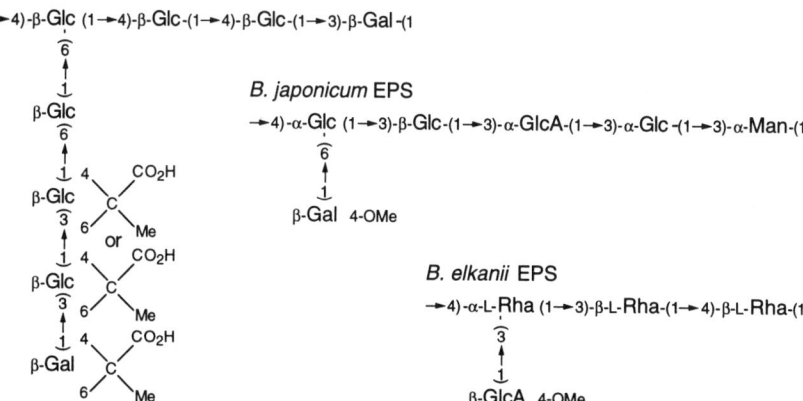

Figure 2. Chemical structure of repeating units of EPS of *S. meliloti*, *R.* sp. NGR234, *R. tropici* CIAT899, *B. elkanii*, and *B. japonicum*. Acetate substitutions have not been indicated. Abbreviations: Glc, glucose; GlcA, glucuronic acid; and Gal, galactose, GalA, galacturonic acid; Man, mannose; L-Rha, L-rhamnose; Me, methyl; Suc, succinate.

such as Nod factor production. The EPS structure of *R.* sp. NGR234 also resembles that of succinoglycan with respect to sugar composition and condensation linkages of the backbone, but otherwise is totally different in having a high concentration of acidic groups in the branches (Fig. 2).[28]

The EPS of *Bradyrhizobium* strains and other "slow-growing" rhizobia have been found to be diverse in composition, often varying from strain to strain,[29–31] unlike the situation for "fast-growing" rhizobia (*R. leguminosarum, S. meliloti, R. etli, R. tropici*) in which strains of a species generally appear to produce similar EPS. Several *B. japonicum* strains (*e.g.*, USDA5, USDA110, and USDA123) produce EPS composed of glucose, mannose, galactose, and galacturonic acid in the molar ratio of 2:1:1:1,[32,33] whereas *B. elkanii* (previously called type II *B. japonicum*) strains (*e.g.*, USDA31 and USDA39) produce EPS consisting of L-rhamnose and glucuronic acid in the ratio 5:1 (Fig. 2).[34] To our knowledge, detailed structural studies of EPS of other *Bradyrhizobium* strains or other "slow-growing" rhizobia have not yet been performed.

The examples of EPS structures discussed so far concern exopolysaccharides produced by rhizobial strains in a free-living state under normal (laboratory) culturing conditions. However, some rhizobial strains have the ability to produce another EPS under special circumstances. Some *S. meliloti* strains can produce a galactoglucan (also called EPS II)[35] when grown in phosphate-limited media[36] or when harboring a mutation in the *expR*[37] or *mucR*[38] genes. More recently, production of novel polysaccharides by *Bradyrhizobium* strains within soybean nodules was described.[39–41] These were called nodule polysaccharides (NPS). For *B. elkanii* strains, these NPS are identical to EPS, whereas *B. japonicum* strains produce NPS which are structurally totally different from EPS.[41]

BIOSYNTHESIS

Rhizobial EPS belong to the most heterogeneous group of bacterial EPS which are composed of those heteropolysaccharides formed from repeating unit structures.[42] The biosynthesis of this type of polysaccharide is presumably performed by a multi-enzyme complex at the inner side of the cytoplasmic membrane of the bacterium. Biosynthesis proceeds as follows: First, the repeating units are built on isoprenoid lipid carriers,[43] a process in which each glycosyl residue is transferred to the growing subunit by a specific glycosyl transferase. Non-carbohydrate substituents are usually added during synthesis of the repeating unit.[44–47] When the repeating units are ready, they are removed from the isoprenoid lipid carriers, polymerized, and exported to the cell surface. EPS biosynthesis is dependent on the availability of sugar nucleotides. These compounds, mainly nucleotide diphosphate monosaccharides (*e.g.*, UDP-glucose, GDP-mannose), represent an activated form of the monosaccharide and are the immediate precursors of the oligosaccharide repeating units.

Table 2. Function of gene products involved in EPS I biosynthesis of *S. meliloti*

Protein	Function	References
ExoC	phosphoglucomutase	60
ExoN	UDP-glucose pyrophosphorylase	49,52
ExoB	UDP-glucose-4-epimerase	51,56
ExoY	galactosyl transferase	54,47
ExoF	galactose transfer	54,47
ExoA	first glucosyl transferase	49,53,47
ExoL	second glucosyl transferase	49,53,47
ExoZ	acetyl transferase	51,57
ExoM	third glucosyl transferase	49,53,47
ExoO	fourth glucosyl transferase	49,53,47
ExoU	fifth glucosyl transferase	50,53,47
ExoW	sixth glucosyl transferase	50,53,47
ExoH	succinyl transferase	48,55
ExoV	pyruvyl transferase	50,52,47
ExoP	export and/or polymerization	49,52,47
ExoQ	export and/or polymerization	54,47
ExoT	export and/or polymerization	50,52,47
ExsA	export	63
ExoK	endo-β-1,3-1,4 glucanase	48
ExoX	negative regulator	54,74,70
ExoR	negative regulator	66,94,142
ExoS	negative regulator	66,94,143
ExsB	negative regulator	63
MucR	positive regulator	38,72
ExpR	positive regulator	37
ExoI	unknown function	50
ExoD	unknown function	73,74,142

Succinoglycan

Of the rhizobial EPS described above, EPS I biosynthesis by *S. meliloti* has been most studied. So far, 26 genes involved in this process have been cloned and sequenced,[48–54] and for many of the corresponding gene products the biochemical functions have been elucidated.[47,55–57] In the following, we will describe how *S. meliloti* synthesizes EPS I, as an example of how bacteria produce such heteropolysaccharides. For a description of EPS II biosynthetic genes, we refer to a recent paper of Becker et al.[58] EPS I biosynthesis by *S. meliloti* is directed by the so-called *exo* and *exs* genes (Table 2), most of which are located in a cluster on megaplasmid 2.[59]

ExoC and *exoN* were found to code for phosphoglucomutase and UDP-glucose pyrophosphorylase, respectively, two enzymes involved in the biosynthesis of UDP-glucose.[49,52,60] *ExoB* codes for UDP-glucose-4-epimerase,[51,56] which is responsible for the biosynthesis of UDP-galactose, the second sugar nucleotide necessary for EPS I biosynthesis. Mutations in *exoB* and *exoC* completely abolish EPS I production, whereas a mutation in *exoN* only has a minor effect, indicating that at least one other gene might substitute for *exoN*. Mutations in genes involved in the biosynthesis of sugar precursors are usually pleiotropic, *i.e.*, they also affect biosynthesis of other polysaccharides. Indeed, it was found that the enzymatic activity of ExoC is also required for the biosynthesis of EPS II, the O-antigen of LPS, and β(1-2)glucans,[38,60] whereas ExoB is indispensable for the biosynthesis of the O-antigen of LPS.[38,56]

The biosynthetic role of the gene products of 12 *S. meliloti exo* genes was assigned by isolation and characterization of lipid-linked succinoglycan biosynthetic intermediates from these *exo* mutant cells.[47] The *exoY* and *exoF* gene products were found to be required for the first step of EPS I biosynthesis, the transfer of galactose onto the lipid carrier.[47] The amino acid sequence of ExoY shows significant homology to a family of glycosyl transferases catalyzing the addition of either glucose or galactose to the lipid carrier in polysaccharide biosynthesis.[54,61] Therefore, it was proposed that ExoY is the galactose transferase in *S. meliloti*. The precise function of ExoF in the transfer of galactose is not known. The *exoA*, *exoL*, *exoM*, *exoO*, *exoU*, and *exoW* gene products were found to catalyze the addition of the consecutive sugars of the repeating unit, respectively.[47] However, the gene encoding the last glucosyl transferase has not yet been characterized. Addition of the non-carbohydrate substituents is catalyzed by the gene products of *exoZ*, *exoH*, and *exoV*. Mutations in *exoZ*[51] cause production of succinoglycan that lacks the acetyl modification,[57] whereas mutations in *exoH* cause production of succinoglycan that lacks the succinyl modification.[55] Strains harboring a mutation in *exoV* accumulate an octasaccharide subunit that lacks the pyruvate substituent,[47] which is consistent with the fact that the deduced amino acid sequence of ExoV shows homology to GumL, a pyruvyl transferase from *Xanthomonas campestris*.[52] Oligosaccharides from *exoP*, *exoQ*, or *exoT* mutants were indistinguishable from those isolated from wild-type intermediates, indicating that the ExoP, ExoQ, and ExoT proteins may all be involved in stages of succinoglycan biosynthesis after subunit assembly, such as the polymerization of the octasaccharide subunits or the export of the completed polymer.[47]

A mutant harboring a truncated *exoP* gene was shown to produce only low-molecular-weight (LMW) EPS I,[49] and therefore the N-terminal domain of ExoP was proposed to be a member of a polypeptide family involved in polysaccharide chain-length determination.[62] The amino acid sequence of ExsA displayed significant homology to ATP binding cassette (ABC) transporter proteins. Mutants in the *exsA* gene were found to produce a relatively low amount of high-molecular-weight (HMW) EPS I compared to the amount of LMW EPS I, indicating that ExsA is involved in (especially HMW) EPS I transport.[63] Tn5 insertions within the *exoK* gene cause a decrease in EPS production but no significant effect on nodulation.[59] In contrast, the *exoK* gene product does not show homology to polysaccharide biosynthesis proteins but to endo-β-1,3-1,4-glucanases of bacilli and *Clostridium thermocellum*.[48] The importance of ExoK for the efficiency of EPS I biosynthesis remains to be established.

Regulation of Succinoglycan Biosynthesis

Production of EPS I by *S. meliloti* is influenced by various environmental conditions, so a complex regulatory machinery is required. *S. meliloti* EPS I production is greatly increased if the cells are limited for nitrogen, phosphorus, or sulfur in the presence of a good carbon source.[64] Studies on the regulation of EPS I biosynthesis suggest that different osmotic conditions also affect EPS production in *S. meliloti*. Low osmotic pressure resulted in the predominant production of LMW EPS I, whereas the production of HMW EPS I was stimulated by an increased osmotic pressure.[65]

Genetic studies resulted in the identification of 6 regulators of EPS I biosynthesis in *S. meliloti*. Insertions in *exoS* and *exoR* resulted in an increase in EPS I production, the structure of which was indistinguishable from that produced by the parental strain Rm1021.[66] Synthesis of EPS I by strain Rm1021 was greatly stimulated by ammonia starvation. In contrast, the *exoR* mutant produced high levels of EPS I regardless of the presence or absence of ammonia in the medium. The *exoS* mutant produced elevated amounts of EPS I in the presence of ammonia, but much larger amounts were formed after ammonia starvation.[66] Analysis of *exo* mRNA levels demonstrated that ExoR acts as a negative regulator of the transcription of *exo* genes.[67] Furthermore, it was found that the level of expression of *exoA*-, *exoF*-, *exoP*-, and *exoQ-phoA* translational fusions was much higher in an *exoR* or *exoS* background, indicating that both *exoR* and *exoS* affect *exo* gene expression.[68] Another gene that negatively regulates EPS I biosynthesis is *exoX*. Mutants in *exoX* overproduce EPS I, whereas inhibition of EPS I production occurs when *exoX* is present at a gene copy number higher than that of *exoY*.[61,69,70] Since *exo* gene expression was unaffected by the gene dosage of *exoX* and *exoY*, it was suggested that ExoX functions as a posttranscriptional inhibitor of ExoY. Recently, sequence analysis has revealed another EPS I biosynthesis gene, *exsB*, in the *exo/exs* gene cluster, which might encode a negative regulator which posttranscriptionally affects the level of *exo* gene expression.[63] Knockout of the *exsB* gene resulted in a larger EPS I production by the mutant strain.

As has already been mentioned, some *S. meliloti* strains (*e.g.*, Rm2011) have the cryptic ability to synthesize the alternative exopolysaccharide EPS II or galactoglucan.[35,71] *S. meliloti* strains containing a mutation in the genes *expR*[37] or *mucR*[38] produce EPS II instead of EPS I. Therefore, these genes are most likely regulators of both EPS I and EPS II synthesis. The deduced amino acid sequence of MucR was highly homologous to the *Agrobacterium tumefaciens* Ros protein, a negative regulator of *vir* genes and necessary for succinoglycan production.

MucR, like Ros, contains a putative zinc finger sequence of the C_2H_2 type, and negatively regulates the transcription of EPS II biosynthesis genes. In contrast, a posttranscriptional regulation by MucR of EPS I biosynthesis genes has been postulated.[72]

Finally, one locus has been found, *exoD*, which plays a yet unknown role in EPS I biosynthesis. *ExoD* mutants produce a reduced amount of EPS I,[64] of which the distribution of HMW and LMW forms depends on the growth medium.[73] Furthermore, the *exoD* mutants are sensitive to alkaline conditions, and effective nodulation of alfalfa by these mutants can only occur in slightly acidic plant growth media.[74]

SYMBIOTIC IMPORTANCE

Mutant Analysis

Since many wild-type rhizobia produce copious amounts of EPS, it was postulated a few decades ago that EPS might play a role in the symbiosis with leguminous plants.[3] First evidence for such a role was obtained by testing the symbiotic behavior of EPS-deficient (*exo* mutants) mutants which were generated by transposon mutagenesis. It was shown that *exo* mutants of certain rhizobial species are unable to induce the formation of nitrogen-fixing nodules on certain hosts. This deficiency was demonstrated for *R. leguminosarum* bv. *viciae* and pea,[75] *R. leguminosarum* bv. *trifolii* and clover,[76] *S. meliloti* and alfalfa,[64] *R.* sp. NGR234 and *Leucaena*,[77] and *M. loti* and *Leucaena*.[78] In contrast, non-pleiotropic *exo* mutations did not interfere with the formation of nitrogen-fixing nodules by *R. leguminosarum* bv. *phaseoli* in bean,[75,79] *S. fredii* in soybean,[80] *M. loti* in *Lotus*,[78] *R. tropici* in bean,[81] or *B. japonicum* in soybean.[82]

In order to understand the discrepancy in EPS requirement for a certain rhizobial strain to nodulate its host plant, it is important to consider the differences in development of the root nodule in the various plant species studied. In certain plant species, nodule formation is initiated by the formation of primordia in the *inner* root cortex. The formation of a nodule primordium preceeds the formation of a persistent meristem that continues to grow during the processes of infection of the root nodule cells, resulting in a protracting structure. This type of nodulation is therefore commonly classified as indeterminate nodulation. In other plant species, nodule formation is initiated by the formation of a primordium in the *outer* root cortex. This nodule primordium gives rise to the formation of a round nodular structure that eventually does not contain meristematic activity, and therefore is called a determinate nodule. Furthermore, it is known that rhizobia induce broad, matrix-rich infection threads in indeterminate nodule-type legumes, whereas narrow infection threads can be found in determinate nodule-type legumes (Fig. 3).[1]

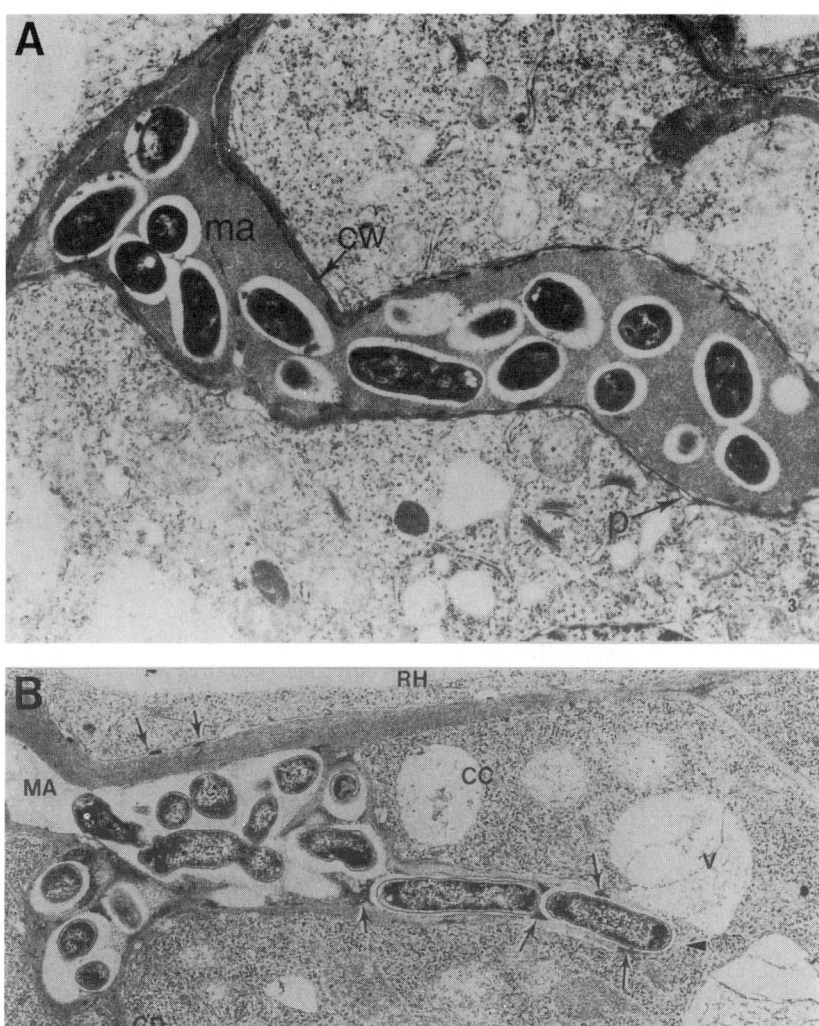

Figure 3. (A) Longitudinal section of a part of a broad infection thread in pea. Bacteria in the infection thread are embedded in an electrondense matrix. The electron transparent zone around the bacteria probably is due to shrinkage. MA: matrix; P: plasmalemma.[144] (B) Development of a narrow infection thread in soybean. RH: root hair cell base; CC: subadjacent cortical cell; MA: intercellular matrix; CD: young cell wall; V: vacuole; CW: cell wall. The arrows indicate unidentified dark material in the walls and between the bacteria.[145]

Apparently, a matrix-type of infection thread enables rhizobia to pass several host cells on the way to the inner cortex. Generally, EPS biosynthesis by rhizobia is required for the formation of nitrogen-fixing nodules on indeterminate nodule-type legumes (*e.g.*, *Leucaena*, *Medicago*, *Pisum*, *Trifolium*, and *Vicia* species). Determinate nodule-type legumes do not require rhizobial EPS biosynthesis for nodulation. This observation is supported by the fact that the same mutation that blocks EPS biosynthesis and normal nodulation of indeterminate nodule-type legumes does not abolish effective nodule development of the determinate nodule-type legumes, *Phaseolus* or *Lotus*.[75,78,79] EPS biosynthesis seems only to be involved in nodulation competitiveness or nodulation speed of *B. japonicum*, the symbiont of the determinate nodule-type legume soybean, albeit reports are conflicting.[82–84] It is tempting to speculate that this difference in EPS requirement between indeterminate nodule-type and determinate nodule-type hosts is due to the fact that indeterminate nodule-type hosts are nodulated via broad infection threads. EPS could either be part of a functional infection thread matrix or contribute to its synthesis.

Recently, it was reported that *exo* mutants of astragalus rhizobia form nitrogen-fixing nodules on *Astralagus sinicus*, an indeterminate nodule-type host.[85] This is the first report of an indeterminate nodule-type host not requiring rhizobial EPS for an effective symbiosis, without evidence for the existence of other polysaccharides substituting for EPS deficiency. Further study of this symbiosis is required to explain this exceptional case. However, it should be noted that the differences between indeterminate nodulation and determinate nodulation are not strict. For instance, in *Sesbania rostrata*, the formation of determinate root nodules is preceded by the formation of nodule primordia in the inner root cortex (at the basis of emerging lateral roots), instead of in the outer cortex.[86] Therefore, the general assumption that effective nodulation of indeterminate nodule-type legumes requires EPS production by the microsymbiont might turn out to be a rule with a notable number of exceptions.

Besides the differences in EPS requirement of indeterminate and determinate nodule-type legumes, there are also striking differences in symbiotic behavior of *exo* mutants on different indeterminately nodulating host plants. For instance, *exo* mutants of *R. leguminosarum* bv. *viciae* normally fail to nodulate pea,[75,79] whereas *exo* mutants of *R. leguminosarum* bv. *trifolii* and *S. meliloti* induce formation of small white bumps on clover and empty ineffective nodules on *Medicago*, respectively.[64,79,87,88] The differences are most likely related to the response of the host plant to Nod factors secreted by the *Rhizobium* bacteria rather than to EPS activity. In most plant species, the nodule primordia induced by Nod factors are only small bumps which are the result of a few cell divisions in the cortex. In contrast, the formation of full-size nodular structures, including fully completed vascular systems, has been shown for *Medicago* plants after induction by Nod factors.

Despite the differences in symbiotic behavior of *exo* mutants on different indeterminately nodulating host plants, all mutants fail to colonize the interior of the host plant (for an example, see Fig. 4). *Exo* mutants of *R. leguminosarum* bv. *trifolii* were slower to induce root hair curling and were disturbed in infection thread formation,[87] whereas delayed root hair curling and abortive infection thread was reported for alfalfa seedlings inoculated with *exo* mutants of *S. meliloti*.[89] Due to defective infection thread formation, *exo* mutants of *R. leguminosarum* bv.

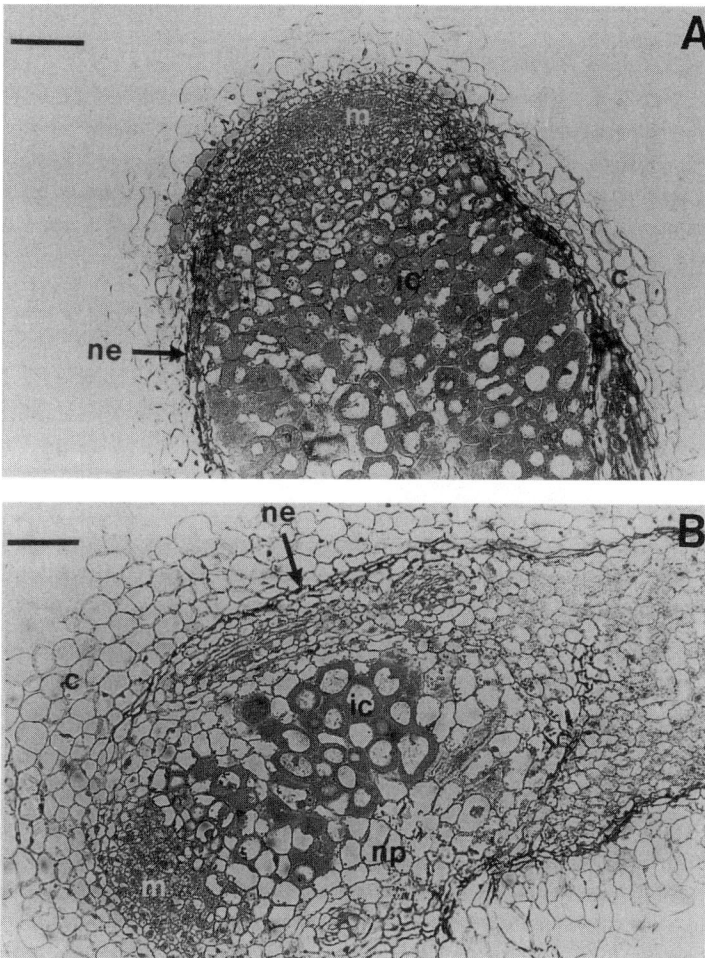

Figure 4. Light microscopy of 7 μm sections of *Vicia sativa* ssp. *nigra* (vetch) nodules. (A) Eighteen-day old vetch nodule induced by wild-type *R. leguminosarum* bv. *viciae*. (B) Twenty four-day old vetch nodule induced by an *exo* mutant of *R. leguminosarum* bv. *viciae*. Bar indicates 0.1 mm, ne = nodule endodermis, ic = infected cells, m = nodule meristem, c = nodule cortex, np = nodule parenchyma. Starch granules in non-infected cells are indicated by arrowheads. Adapted from van Workum et al. (1995).[91]

viciae, *R. leguminosarum* bv. *trifolii*, and *S. meliloti* induce delayed formation of non-infected or partially infected nodules on their respective host plants, vetch, clover, and alfalfa,[87–91] with the extent of the symbiotic defects depending on the plant growth conditions used.[87,91]

During a successful *Rhizobium* infection, the bacteria grow biotrophically in the plant, without triggering an adverse plant response. The host may respond to invading microorganisms in several ways; for example, by producing antimicrobial substances such as phytoalexins (phytoalexins are antimicrobial compounds produced by plants following microbial infection and therefore important elements of the plant defense response), by the hypersensitive reaction, or by modification of the plant cell wall.[92] EPS molecules might function by masking potential elicitors such as lipopolysaccharides on the bacterial surface, thus preventing recognition and elicitation of responses of the host defense system.[93] If EPS molecules have a masking role, one may not expect specific structural requirements for these macromolecules to function. Indeed, it was shown that the acetyl substituent of EPS I is not required for induction of nitrogen-fixing nodules by *S. meliloti*.[57] However, *exoH* mutants of *S. meliloti* that fail to succinylate EPS I are affected in infection thread formation on alfalfa and thus induce formation of empty nodules that do not fix nitrogen.[55] As *exoH* mutants produce only a trace of LMW EPS I, either the lack of succinylation or the lack of LMW EPS I could explain the nodule entry defect.[73] Some studies have shown that regulation of EPS synthesis by *S. meliloti* is essential not only for nodule cell invasion but also for successful development of the symbiotic process after nodule cell invasion.[66,94] Similarly, it has been shown that regulation of EPS synthesis by *R. leguminosarum* bv. *phaseoli* is required for formation of nitrogen-fixing nodules on bean.[95] Obviously, the development of a nitrogen-fixing bacteroid is disturbed by abnormal production of EPS. However, nothing is known about the molecular working mechanism.

It has been shown in *S. meliloti* that two polysaccharides other than EPS I can have the same function in the plant-bacterium interaction. *S. meliloti* strains containing a mutation in the genes *expR*[37] or *mucR*[38] produce predominantly EPS II (also called galactoglucan)[35,71] and minor amounts of EPS I. Both mutants were found to form nitrogen-fixing nodules on alfalfa, suggesting that EPS II can replace EPS I in the nodulation process.[37,38,72] In order to exclude the possibility that the minor amount of EPS I is responsible for effective nodulation, double mutants were constructed which failed to produce any succinoglycan and were only producing EPS II. It was found that an *expR exoA* double mutant of *S. meliloti* did not produce EPS I, but was able to enter alfalfa nodules and fix nitrogen.[37] However, synthesis of EPS II did not suppress the symbiotic defects of EPS I-deficient strains on *Medicago caerulea*, *Medicago truncatula*, *Melilotus alba*, and fenugreek, plants that are normal hosts for *S. meliloti*.[37] Apparently, exopolysaccharides produced by *S. meliloti* play a role in determination of host range. In contrast, it was reported that synthesis of EPS II by a *mucR* mutant did not suppress the symbiotic effect of EPS I-deficiency on alfalfa.[72] NMR analyses of the EPS II produced by the *mucR*⁻ strains failed to show any obvious structural differences when compared with the EPS II produced by an *expR exoA* double mutant. Interestingly, it turned out that *mucR*⁻ EPS II-producing strains do not make an LMW fraction of the EPS.[96] Apparently, the molecular weight of EPS II is important for its symbiotic function.

Evidence for the existence of another polysaccharide that might substitute for EPS I in symbiosis was obtained in studies performed with derivatives of *S. meliloti* Rm41. Rm41 is a separate isolate of *S. meliloti* that is able to produce both EPS I and EPS II, and exhibits a normal host range.[97,98] *S. meliloti* AK631, a derivative of Rm41, carries a mutation in *exoB* and is therefore deficient in synthesis of UDP-galactose. As a consequence of this mutation, AK631 can not produce EPS I or EPS II, yet it is fully infective on alfalfa.[99] Subsequent studies showed that mutations in either the *lpsX*, *lpsY*, or *lpsZ* genes or the *fix-23* gene region of strain AK631 resulted in the loss of ability of this strain to induce formation of nitrogen-fixing nodules on the *Medicago*, *Melilotus*, and *Trigonella* species tested.[98–100] Analysis of crude polysaccharide extracts of these mutants by gel filtration chromatography and sugar analysis showed changes in elution characteristics and KDO content (KDO = 3-deoxy-D-manno-2-octulosonic acid) compared to the parental strain.[98–100] Furthermore, it was found that *lpsX*-, *lpsY*-, *lpsZ*-, and *fix-23*-mutants were differently interacting with certain bacteriophages, pointing to changes on the cell surface.[98–100]

Recently, it was found that *S. meliloti* AK631 produces acidic KDO-rich polysaccharides (KPS) that are analogous to the group II K-antigens of *Escherichia coli*,[101] and that the *lpsXYZ* genes and *fix-23* genes are involved in the synthesis of these polysaccharides.[102,103] K-antigens are produced by many well studied bacteria, such as *E. coli*, *Haemophilus influenzae*, *Neisseria* ssp., and *Klebsiella* spp.. In contrast to EPS, the K-antigens are tightly associated with the bacterial cells. In some cases, it has been shown that they help to protect pathogenic bacteria against host defense responses.[104] Sequence analysis of the *fix-23* region revealed six open reading frames, which show striking homology to genes encoding different fatty acid synthases (FAS) and polyketide synthase (PKS) proteins, which are most likely involved in synthesis of a lipid carrier needed for the biosynthesis and export of the capsule.[102]

Mutations in the *fix-23* region affect the synthesis of the HMW form of the KPS and the ability to react with polyclonal antibodies.[102] The predicted amino acid sequence of *lpsZ* is homologous to KpsC, which is involved in the polymerization and export of the group II K-antigens of *E. coli*.[105] A mutation in *lpsZ* results in a distinct expression of the KPS, particularly in the molecular weight distribution, and loss of infectivity of strain AK631.[103] Probably, the size range of the *S. meliloti* KPS is important for its ability to substitute for the EPS in the nodulation process. Detailed studies of more than twenty strains of *S. fredii*, *S. meliloti*, and *R.* sp. NGR234 have revealed that KPS are common products of rhizobia.[106] It was found that these capsular polysaccharides are highly variable in sugar composition, linkages, non-carbohydrate substitutions, or size range. Thus, in contrast to EPS, the structure of which is mostly conserved within species, the KPS are strain-specific antigens.[106]

It can be concluded that in the case of the *S. meliloti*–alfalfa symbiosis, three different rhizobial polysaccharides, EPS I, EPS II, and KPS, can function in the infection process. Given the structural differences between these molecules, it is conceivable that these three polysaccharides act as signals which promote infection by different pathways.[106] Alternatively, these molecules may have certain physicochemical properties in common which are required for functioning in the infection process. The size of these polysaccharides is a determining factor for activity.

EPS Function as Signaling Molecules?

The establishment of the *Rhizobium*-legume symbiosis requires specific chemical signaling between the symbionts. EPS have long been suspected to be signaling molecules because they are located on the outside of the rhizobial cell and because of the diversity of EPS structures among rhizobia. First evidence for a role of LMW EPS as host-specific signaling molecules was obtained by Djordjevic et al.[107] They found that purified EPS fractions of *R. leguminosarum* bv. *trifolii* could restore nodulation of *exo* mutants of this strain, whereas purified EPS I fractions of *S. meliloti* could not. This might be due to the fact that the octasaccharide subunit of EPS I of *S. meliloti* is not recognized by clover plants. Likewise, it was shown that the symbiotic deficiencies of *exo* mutants of *S. meliloti* could be rescued by the addition of LMW EPS I at the time of inoculation.[108,109] Of this LMW EPS I, only the most charged tetramer of the repeating unit was active.[109] Furthermore, nonsuccinylated LMW EPS I and HMW EPS I did not restore the nodulation deficiency of an *exo* mutant, nor did LMW EPS from *R. leguminosarum* bv. *trifolii* or *Rhizobium* sp. strain NGR234.[109] Further indications for a specific role of LMW EPS as signaling molecules was obtained recently by Gonzalez and coworkers.[96] They showed that a LMW EPS II fraction consisting of 15–20 EPS II disaccharide subunits isolated from a *S. meliloti expR* mutant efficiently restored nodule invasion by noninfective strains when present in amounts as low as 7 pmol per plant, demonstrating that also LMW EPS II may act as a symbiotic signal during infection.

Microscopic and biochemical analysis of noninfected pseudonodules, induced by an EPS I-deficient *S. meliloti* mutant on *M. sativa*, revealed strong evidence for the induction of a plant defense response by the mutated microsymbiont. The cortical cell walls of these pseudonodules were abnormally thick and encrusted with an autofluorescent material, while parts of these cell walls and wall appositions contained callose.[110] The authors proposed that rhizobial EPS might act as a suppressor of the alfalfa plant defense system, enabling *S. meliloti*

BIOSYNTHESIS OF RHIZOBIAL EXOPOLYSACCHARIDES   155

to infect the plant.[110] Further support for this hypothesis was obtained by testing purified EPS fractions from *S. meliloti* and *Xanthomonas campestris* for putative suppressor activity in elicitor-responsive cell cultures of alfalfa and, as controls, the non-leguminous plants, tobacco and tomato.[111] Apart from other defense-related reactions, each cell culture reacted to the addition of small amounts of a non-specific yeast elicitor with a strong transient alkalinization of the culture medium. In alfalfa cell cultures, the elicitor-induced alkalinization could be suppressed by the simultaneous application of LMW EPS I of *S. meliloti*, whereas neither HMW EPS I, HMW EPS II, nor the heterologous EPS xanthan from *X. campestris* yielded a reduction of the elicitor response.[111]

LMW EPS I might block receptors that are part of the elicitation process. Indications for such a mechanism were found for the *Bradyrhizobium*-soybean symbiosis. In soybean, 1,3-1,6-β-glucans derived from the fungal pathogen, *Phytophthora sojae*, usually elicit synthesis of the phytoalexin glyceollin. In contrast, it was found that cyclic 1,3-1,6-β-glucans from *B. japonicum* USDA110 are inactive in stimulating phytoalexin synthesis. Interestingly, the bacterial β-glucans were shown to be ligands for the β-glucan-binding sites, the putative receptors for the fungal elicitor in soybean roots, and thus were proposed to be suppressors of defense responses induced by the *P. sojae* β-glucans.[112]

On the other hand, plant defense reactions seem to be a normal phenomenon in symbiotic interactions. Leguminous plants have developed a way of regulating the number of nitrogen-fixing nodules on their roots, a process which is called autoregulation. For alfalfa, it was shown that the plant reacts to infection by its symbiont *S. meliloti* by eliciting a defense mechanism similar to the hypersensitive reaction observed in incompatible plant-pathogen interactions.[113] After the first nodule primordia have been induced, an increasing number of infection threads abort in a single or a few root cortical cells in which both symbionts simultaneously undergo necrosis.[113] Therefore, a successful *Rhizobium*-legume symbiosis can be regarded as the balance between the infective ability of the *Rhizobium* bacterium and the defense reaction of the plant.

Based on observations that inhibitors of ethylene biosynthesis, such as aminoethoxyvinyl glycine (AVG), cause an increase in persistent rhizobial infections, it was postulated that ethylene might be an endogenous signal for regulation of rhizobial infection.[114,115] Furthermore, it was reported that nodulation of *Vicia sativa* ssp. *nigra* by EPS-deficient mutants of *R. leguminosarum* bv. *viciae* could be partially restored if ethylene production by the host plant root, resulting from rhizobial inoculation, was minimized.[91] Genetic evidence that ethylene is indeed a component of the signaling pathway controlling autoregulation of nodulation was obtained by studying the infection and nodulation phenotypes of an ethylene-insensitive mutant of *Medicago truncatula*.[116]

Other factors possibly controlling rhizobial infection of legumes are CHS (chalcone synthase), CHI (chalcone isomerase), and IFR (isoflavone reducetase), enzymes of the phenylpropanoid pathway or "defense" pathway in plants. The compounds produced through this pathway, e.g., flavonoids and isoflavonoids, are known to act as signals in plant-microbe interactions. Flavonoids and isoflavonoids were found to activate[117–120] or inhibit[117,121] Nod factor biosynthesis by several *Rhizobium* species and to induce changes in polysaccharide synthesis of *S. fredii* and *R. etli*.[93,122,123] Furthermore, it was reported that flavonoids might be endogenous auxin transport inhibitors in plants,[124] and, therefore, it was hypothesized that local induction of flavonoid synthesis coordinates nodule formation by modulating auxin distribution.[125,126] Experimental support for this hypothesis was obtained by Mathesius et al., who showed that microtargeting of flavonoids into white clover roots results in a local increase of auxin concentration.[127] Flavonoids might also participate in the host defense reaction in response to *Rhizobium*. Isoflavonoids include precursors of phytoalexins, the synthesis of which is part of the hypersensitive response to incompatible pathogens.[128]

Analogous to plant-pathogen interactions, induction of CHS expression and phytoalexin accumulation was also observed in several ineffective symbioses, suggesting that plants use the phenylpropanoid pathway to control ineffective rhizobia.[126,129–131] Interestingly, it was shown by Northern blot analyses that the cognate Nod factor of *S. meliloti* induced high transcript levels of the CHS, CHR, and IFR genes as well as phytoalexin production when present at a concentration of 10^{-6} M, indicating that the Nod factor itself could be the elicitor of plant defense responses.[132] However, it should be noted that at a concentration of 10^{-9} M, the expression of these genes was not stimulated, despite the fact that this concentration is probably sufficient for triggering nodule organogenesis. On the other hand, it was found that alfalfa expresses CHS in root hairs and the root epidermis following inoculation with wild-type *S. meliloti*, but not after inoculation with an *exoB* mutant of the same strain.[133] Furthermore, *nod* gene-inducing flavonoids were not detected in exudates prepared from alfalfa roots inoculated with the *S. meliloti exoB* mutant, in contrast to the situation with exudates of wild-type inoculated roots.[133] Infiltration of alfalfa leaves with purified KPS, isolated from *S. meliloti* strain AK631, led to a rapid induction of CHS and CHR genes, whereas neither EPS I nor EPS II were biologically active in this system.[134] In many plant species, CHS is encoded by multigene families,[135–138] and it has been shown for soybean that CHS genes induced early in symbiosis are different from homologs induced in a host defense or stress response during the later stages of an ineffective symbiosis.[139]

Considering all available data, it can be speculated that LMW EPS (and other rhizobial polysaccharides such as KPS) control the balance between rhizobial infection and host defense by regulating the expression of genes related to (iso)flavonoid synthesis and host defense. However, the mechanism by which this might occur remains to be studied. In addition, EPS may have other roles in the symbiotic process, analogous to functions proposed for EPS of phytopathogenic bacteria.[140] For instance, EPS might benefit the *Rhizobium* bacteria by reducing contact with toxic molecules,[141] or by minimizing interactions with plant cells so as to reduce host responses and promote colonization.[140] Alternatively, EPS might induce changes in the plant cell wall by inducing the synthesis of hydrolytic enzymes during formation of infection threads. Ljunggren and Fåhraeus[3] found that (crude) rhizobial EPS was able to induce polygalacturonase activity in plant roots, which would facilitate the entrance of the bacteria into the host plant. However, attempts to verify this finding have not yet been successful.

CONCLUSION

Detailed studies performed on the symbiotic interaction between rhizobia and indeterminate nodule-type legumes (*Leucaena*, *Medicago*, *Pisum*, *Trifolium*, and *Vicia* species) have revealed that EPS synthesis by the *Rhizobium* bacterium is essential for normal infection thread formation and, therefore, for the formation of nitrogen-fixing nodules on these host plants. It is now clear that EPS of a certain size can serve as signal molecules that influence plant root metabolism in such a way that plants allow the *Rhizobium* bacterium to infect their roots. In the case of the *S. meliloti-Medicago sativa* symbiosis, it has been shown that at least three molecules, EPS I, EPS II, and KPS, can have the same function in symbiosis,[37,64,99] whereas synthesis of EPS II does not suppress the symbiotic defects of EPS I-deficient strains on other host plants.[37] Apparently, exopolysaccharides produced by *S. meliloti* play a role in determining host range. It remains to be studied whether other indeterminate nodule-type host plants, require certain structural elements of EPS for successful interactions with their microsymbionts.

With regard to its function, LMW EPS may control the balance between infection and defense. Especially in indeterminate nodule-type plants, in which rhizobia need a longer time to reach the dividing nodule primordium, in comparison to the situation with determinate nodule-type plants, where the risk of interference by stress compounds such as ethylene is larger, such a function might be appropriate. At present, it is unknown how the plant senses the presence of EPS with a specific structure and a certain size. An important element in future studies will be development of a bioassay in which activity of purified LMW EPS in legume roots can be tested.

ACKNOWLEDGMENTS

W.A.T. van Workum is supported by the Foundation for Life Sciences (SLW), which is subsidized by the Netherlands Organization of Scientific Research (NWO). The authors are indebted to Peter Hock for the graphical work.

REFERENCES

1. KIJNE, J.W. 1992. The Rhizobium infection process. In: Biological Nitrogen Fixation (G. Stacey, R.H. Burris, H.J. Evans, eds.), Chapman and Hall, New York, London, pp. 349–398.
2. SPAINK, H.P. 1996. Regulation of plant morphogenesis by lipo-chitin oligosaccharides. Crit. Rev. Plant Sci. 15: 559–582.
3. LJUNGGREN, H., FÅHRAEUS, G. 1959. Effect of *Rhizobium* polysaccharide on the formation of polygalacturonase in lucerne and clover. Nature (London) 184: 1578–1579.
4. GIL-SERRANO, A., GONZÁLEZ-JIMÉNEZ, I., TEJERO-MATEO, P., SÁNCHEZ DEL JUNCO, A. 1992. Structure of the acidic exopolysaccharide secreted by *Rhizobium leguminosarum* biovar. *phaseoli* CFN42. Carbohydr. Res. 225: 169–174.
5. ROBERTSON, B.K., ÅMAN, P., DARVILL, A.G., McNEIL, M., ALBERSHEIM, P. 1981. Host-symbiont interactions. V. The structure of acidic extracellular polysaccharides secreted by *Rhizobium leguminosarum* and *Rhizobium trifolii*. Plant Physiol. 67: 389–400.
6. McNEIL, M., DARVILL, J., DARVILL, A., ALBERSHEIM, P., VAN VEEN, R., HOOYKAAS, P., SCHILPEROORT, R., DELL, A. 1986. The discernible structural features of the acidic exopolysaccharides secreted by different *Rhizobium* species are the same. Carbohydr. Res. 146: 307–326.
7. DUDMAN, W.F., FRANZÉN, L.-E., DARVILL, J.E., McNEIL, M., DARVILL, A.G., ALBERSHEIM, P. 1983. The structure of the acidic polysaccharide secreted by *Rhizobium phaseoli* strain 127 K36. Carbohydr. Res. 117: 141–156.
8. PHILIP-HOLLINGSWORTH, S., HOLLINGSWORTH, R.I., DAZZO, F.B., DJORDJEVIC, M.A., ROLFE, B.G. 1989. The effect of interspecies transfer of *Rhizobium* host-specific nodulation genes on acidic polysaccharide structure and *in situ* binding by host lectin. J. Biol. Chem. 264: 5710–5714.
9. KUO, M.-S.K., MORT, A.J. 1986. Location and identity of acyl substituents on the extracellular polysaccharides of *R. trifolii* and *R. leguminosarum*. Carbohydr. Res. 145: 247–265.
10. PHILIP-HOLLINGSWORTH, S., HOLLINGSWORTH, R.I., DAZZO, F.B. 1989. Host-range related structural features of the acidic extracellular polysaccharides of *Rhizobium trifolii* and *Rhizobium leguminosarum*. J. Biol. Chem. 264: 1461–1466.
11. CANTER CREMERS, H.C.J., BATLEY, M., REDMOND, J.W., WIJFJES, A.H.M., LUGTENBERG, B.J.J., WIJFFELMAN, C.A. 1991. Distribution of O-acetyl groups in the exopolysaccharide synthesized by *Rhizobium leguminosarum* strains is not determined by the Sym plasmid. J. Biol. Chem. 266: 9556–9564.
12. ORGAMBIDE, G., PHILIP-HOLLINGSWORTH, S., CARGILL, L., DAZZO, F. 1992. Evaluation of acidic heteropolysaccharide structures in *Rhizobium leguminosarum* biovars altered in nodulation genes and host range. Mol. Plant-Microbe Interact. 5: 484–488.
13. LOPEZ-LARA, I.M., ORGAMBIDE, G., DAZZO, F.B., OLIVARES, J., TORO, N. 1993. Characterization and symbiotic importance of acidic extracellular polysaccharides of *Rhizobium* sp. strain GRH2 isolated from acacia nodules. J. Bacteriol. 175: 2826–2832.
14. FRANZÉN, L.-E., DUDMAN, W.F., McNEIL, M., DARVILL, A.G., ALBERSHEIM, P. 1983. The structure of the acidic polysaccharide secreted by *Rhizobium phaseoli* strain 127 K44. Carbohydr. Res. 117: 157–167.
15. AMEMURA, A., HARADA, T., ABE, M., HIGASHI, S. 1983. Structural studies on the extracellular acidic polysaccharide from *Rhizobium trifolii* 4S. Carbohydr. Res. 115: 165–174.
16. DUDMAN, W.F., FRANZÉN, L.-E., McNEIL, M., DARVILL, A.G., ALBERSHEIM, P. 1983. The structure of the acidic polysaccharide secreted by *Rhizobium phaseoli* strain 127K87. Carbohydr. Res. 117: 169–183.
17. ÅMAN, P., FRANZÉN, L.-E., DARVILL, J.E., McNEIL, M., DARVILL, A.G., ALBERSHEIM, P. 1982. The structure of the acidic polysaccharide secreted by *Rhizobium phaseoli* strain 127 K38. Carbohydr. Res. 103: 77–100.

18. CANTER CREMERS, H.C.J., STEVENS, K., LUGTENBERG, B.J.J., WIJFFELMAN, C.A., BATLEY, M., REDMOND, J.W., BREEDVELD, M.W., ZEVENHUIZEN, L.P.T.M. 1991. Unusual structure of the exopolysaccharide of *Rhizobium leguminosarum* bv. *viciae* strain 248. Carbohydr. Res. 218: 185–200.
19. AMAN, P., McNEIL, M., FRANZEN, L.-E., DARVILL, A.G., ALBERSHEIM, P. 1981. Structural elucidation, using H.P.L.C.-M.S. and G.L.C.-M.S., of the acidic polysaccharide secreted by *Rhizobium meliloti* strain 1021. Carbohydr. Res. 95: 263–282.
20. JANSSON, P.-E., KENNE, L., LINDBERG, B., LJUNGGREN, H., LONNGREN, J., RUDEN, U., SVENSSON, S. 1977. Demonstration of an octasaccharide repeating unit in the extracellular polysaccharide of *Rhizobium meliloti* by sequential degradation. J. Am. Chem. Soc. 99: 3812–3815.
21. HARADA, T., AMEMURA, A., JANSSON, P.-E., LINDBERG, B. 1979. Comparative studies of polysaccharides elaborated by *Rhizobium, Alcaligenes* and *Agrobacterium*. Carbohydr. Res. 77: 285–288.
22. HISAMATSU, M., SANO, K., AMEMURA, A., HARADA, T. 1978. Acidic polysaccharides containing succinic acid in various strains of *Agrobacterium*. Carbohydr. Res. 61: 89–96.
23. HISAMATSU, M., ABE, J., AMEMURA, A., HARADA, T. 1980. Structural elucidation of succinoglycan and related polysaccharides from *Agrobacterium* and *Rhizobium* by fragmentation with two special β-D-glycanases and methylation analysis. Agric. Biol. Chem. 44: 1049–1055.
24. ZEVENHUIZEN, L.P.T.M. 1973. Methylation analysis of acidic exopolysaccharides of *Rhizobium* and *Agrobacterium*. Carbohydr. Res. 26: 409–419.
25. GIL-SERRANO, A., SANCHEZ DEL JUNCO, A., TEJERO-MATEO, P. 1990. Structure of the extracellular polysaccharide secreted by *Rhizobium leguminosarum* var. *phaseoli* CIAT 899. Carbohydr. Res. 204: 103–107.
26. AMEMURA, A., HARADA, T. 1983. Structural studies on extracellular acidic polysaccharides secreted by three non-nodulating rhizobia. Carbohydr. Res. 112: 85–93.
27. HIGASHI, S., ABE, M. 1978. Phage induced depolymerase for exopolysaccharide of rhizobiaceae. J. Gen. Appl. Microbiol. 24: 143–153.
28. DJORDJEVIC, S.P., BATLEY, M., REDMOND, J.R., ROLFE, B.G. 1986. The structure of the exopolysaccharide from *Rhizobium* sp. strain ANU280 (NGR234). Carbohydr. Res. 148: 87–99.
29. DUDMAN, W.F. 1976. The extracellular polysaccharides of *Rhizobium japonicum*: compositional studies. Carbohydr. Res. 46: 97–110.
30. KENNEDY, L.D., BAILEY, R.W. 1976. Monomethyl sugars in extracellular polysaccharides from slow-growing *Rhizobia*. Carbohydr. Res. 49: 451–454.
31. KENNEDY, L.D. 1976. Isolation of 3-O-methyl-D-ribose from a *Rhizobium* polysaccharide. Carbohydr. Res. 52: 259–261.
32. MORT, A.J., BAUER, W.D. 1982. Application of two new methods for cleavage of polysaccharides into specific oligosaccharide fragments. J. Biol. Chem. 257: 1870–1875.
33. PUVANESARAJAH, V., SCHELL, F.M., GERHOLD, D., STACEY, G. 1987. Cell surface polysaccharide from *Bradyrhizobium japonicum* and a non-nodulating mutant. J. Bacteriol. 169: 137–141.
34. DUDMAN, W.F. 1978. Structural studies of the extracellular polysaccharides of *Rhizobium japonicum* strains 71A, CC708 and CB1795. Carbohydr. Res. 66: 9–23.
35. HER, G.-R., GLAZEBROOK, J., WALKER, G.C., REINHOLD, V.N. 1990. Structural studies of a novel exopolysaccharide produced by a mutant of *Rhizobium meliloti* strain Rm1021. Carbohydr. Res. 198: 305–312.
36. ZHAN, H., LEE, C.C., LEIGH, J.A. 1991. Induction of the second exopolysaccharide (EPSb) in *Rhizobium meliloti* SU47 by low phosphate concentrations. J. Bacteriol. 173: 7391–7394.
37. GLAZEBROOK, J., WALKER, G.C. 1989. A novel exopolysaccharide can function in place of the calcofluor-binding exopolysaccharide in nodulation of alfalfa by *Rhizobium meliloti*. Cell 56: 661–672.

38. ZHAN, H., LEVERY, S.B., LEE, C.C., LEIGH, J.A. 1989. A second exopolysaccharide of *Rhizobium meliloti* strain SU47 that can function in root nodule invasion. Proc. Natl. Acad. Sci. USA 86: 3055–3059.
39. STREETER, J.G., SALMINEN, S.O., WHITMOYER, R.E., CARLSON, R.W. 1992. Formation of novel polysaccharides by *Bradyrhizobium japonicum* bacteroids in soybean nodules. Appl. Environ. Microbiol. 58: 607–613.
40. STREETER, J.G., SALMINEN, S.O., BEUERLEIN, J.E., SCHMIDT, W.H. 1994. Factors influencing the synthesis of polysaccharide by *Bradyrhizobium japonicum* bacteroids in field-grown soybean nodules. Appl. Environ. Microbiol. 60: 2939–2943.
41. AN, J.H., CARLSON, R.W., GLUSHKA, J., STREETER, J.G. 1995. The structure of a novel polysaccharide produced by *Bradyrhizobium* species within soybean nodules. Carbohydr. Res. 269: 303–317.
42. SUTHERLAND, I.W. 1982. Biosynthesis of microbial exopolysaccharides. Adv. Microbiol. Physiol. 23: 79–150.
43. TROY, F.A., FRERMAN, F.E., HEATH, E.C. 1971. The biosynthesis of capsular polysaccharide in *Aerobacter aerogenes*. J. Biol. Chem. 246: 118–133.
44. VANDERSLICE, R.W., DOHERTY, D.H., CAPAGE, M.A., BETLACH, M.R., HASSLER, R.A., HENDERSON, N.M., RYAN-GRANIERO, J., TECKLENBURG, M. 1989. Genetic engineering of polysaccharide structure in *Xanthomonas campestris*. In: Biomedical and Biotechnological Advances in Industrial Polysaccharides (V. Crescenzi, I.C.M. Dea, S. Paoletti, S.S. Stivala, I.W. Sutherland, eds.), Gordon and Breach Science Publishers, New York, pp. 145–156.
45. CAPAGE, M.A., DOHERTY, D.H., BETLACH, M.R., VANDERSLICE, R.W. 1987. Recombinant-DNA mediated production of xanthan gum. International patent W087/05938.
46. IELPI, L., COUSO, R.O., DANKERT, M.A. 1993. Sequential assembly and polymerization of the polyprenol-linked pentasaccharide repeating unit of the xanthan polysaccharide in *Xanthomonas campestris*. J. Bacteriol. 175: 2490–2500.
47. REUBER, T.L., WALKER, G.C. 1993. Biosynthesis of succinoglucan, a symbiotically important exopolysaccharide of *Rhizobium meliloti*. Cell 74: 269–280.
48. BECKER, A., KLEICKMANN, A., ARNOLD, W., PÜHLER, A. 1993. Analysis of the *Rhizobium meliloti exoH/exoK/exoL* fragment: ExoK shows homology to excreted endo-β-1,3–1,4-glucanases and ExoH resembles membrane proteins. Mol. Gen. Genet. 238: 145–154.
49. BECKER, A., KLEICKMANN, A., KELLER, M., ARNOLD, W., PÜHLER, A. 1993. Identification and analysis of the *Rhizobium meliloti exoAMONP* genes involved in exopolysaccharide biosynthesis and mapping of promoters located on the *exoHKLAMONP* fragment. Mol. Gen. Genet. 241: 367–379.
50. BECKER, A., KLEICKMANN, A., KÜSTER, H., KELLER, M., ARNOLD, W., PÜHLER, A. 1993. Analysis of the *Rhizobium meliloti* genes *exoU*, *exoV*, *exoW*, *exoT* and *exoI* involved in exopolysaccharide biosynthesis and nodule invasion: *exoU* and *exoW* probably encode glucosyltransferases. Mol. Plant-Microbe Interact. 6: 735–744.
51. BUENDIA, A.M., ENENKEL, B., KÖPLIN, R., NIEHAUS, K., ARNOLD, W., PÜHLER, A. 1991. The *Rhizobium meliloti exoZ/exoB* fragment of megaplasmid2: ExoB functions as a UDP-glucose 4-epimerase and ExoZ shows homology to NodX of *Rhizobium leguminosarum* biovar *viciae* strain TOM. Mol. Microbiol. 5: 1519–1530.
52. GLUCKSMANN, M.A., REUBER, T.L., WALKER, G.C. 1993. Genes needed for the modification, polymerization, export, and processing of succinoglycan by *Rhizobium meliloti*: A model for succinoglycan biosynthesis. J. Bacteriol. 175: 7045–7055.
53. GLUCKSMANN, M.A., REUBER, T.L., WALKER, G.C. 1993. Family of glycosyl transferases needed for the synthesis of succinoglycan by *Rhizobium meliloti*. J. Bacteriol. 175: 7033–7044.
54. MÜLLER, P., KELLER, M., WENG, W.M., QUANDT, J., ARNOLD, W., PÜHLER, A. 1993. Genetic analysis of the *Rhizobium meliloti exoYFQ* operon: ExoY is homologous to sugar

transferases and ExoQ represents a transmembrane protein. Mol. Plant-Microbe Interact. 6: 55–65.
55. LEIGH, J.A., REED, J.W., HANKS, J.F., HIRSCH, A.M., WALKER, G.C. 1987. *Rhizobium meliloti* mutants that fail to succinylate their calcofluor-binding exopolysaccharide are defective in nodule invasion. Cell 51: 579–587.
56. CANTER CREMERS, H.C.J., BATLEY, M., REDMOND, J.W., EYSDEMS, L., BREEDVELD, M.W., ZEVENHUIZEN, L.P.T.M., PEES, E., WIJFFELMAN, C.A., LUGTENBERG, B.J.J. 1990. *Rhizobium leguminosarum exoB* mutants are deficient in the synthesis of UDP-glucose 4'-epimerase. J. Biol. Chem. 265: 21122–21127.
57. REUBER, T.L., WALKER, G.C. 1993. The acetyl substituent of succinoglycan is not neccessary for alfalfa invasion by *Rhizobium meliloti* Rm1021. J. Bacteriol. 175: 3653–3655.
58. BECKER, A., RÜBERG, S., KÜSTER, H., ROXLAU, A.A., KELLER, M., IVASHINA, T., CHENG, H.-P., WALKER, G.C., PÜHLER, A. 1997. The 32-kilobase *exp* gene cluster of *Rhizobium meliloti* directing the biosynthesis of galactoglucan: Genetic organization and properties of the encoded gene products. J. Bacteriol. 179: 1375–1384.
59. LONG, S., REED, J.W., HIMAWAN, J., WALKER, G.C. 1988. Genetic analysis of a cluster of genes required for synthesis of the calcofluor-binding exopolysaccharide of *Rhizobium meliloti*. J. Bacteriol. 170: 4239–4248.
60. UTTARO, A.D., CANGELOSI, G.A., GEREMIA, R.A., NESTER, E.W., UGALDE, R.A. 1990. Biochemical characterization of avirulent *exoC* mutants of *Agrobacterium tumefaciens*. J. Bacteriol. 171: 1640–1646.
61. REED, J.W., CAPAGE, M., WALKER, G.C. 1991. *Rhizobium meliloti exoG* and *exoJ* mutations affect the ExoX- ExoY system for modulation and exopolysaccharide production. J. Bacteriol. 173: 3776–3788.
62. BECKER, A., NIEHAUS, K., PÜHLER, A. 1995. Low-molecular-weight succinoglycan is predominantly produced by *Rhizobium meliloti* strains carrying a mutated ExoP protein characterized by a periplasmic N-terminal domain and a missing C-terminal domain. Mol. Microbiol. 16: 191–203.
63. BECKER, A., KÜSTER, H., NIEHAUS, K., PÜHLER, A. 1995. Extension of the rhizobium meliloti succinoglycan biosynthesis gene cluster: Identification of the *exsA* gene encoding an ABC transporter protein, and the *exsB* gene which probably codes for a regulator of succinoglycan biosynthesis. Mol. Gen. Genet. 249: 487–497.
64. LEIGH, J.A., SIGNER, E.R., WALKER, G.C. 1985. Exopolysaccharide deficient mutants of *Rhizobium meliloti* that form ineffective nodules. Proc. Natl. Acad. Sci. USA 82: 6231–6235.
65. BREEDVELD, M.W., ZEVENHUIZEN, L.P.T.M., ZEHNDER, A.J.B. 1990. Osmotically induced oligo- and polysaccharide synthesis by *Rhizobium meliloti* SU-47. J. Gen. Microbiol. 136: 2511–2519.
66. DOHERTY, D., LEIGH, J.A., GLAZEBROOK, J., WALKER, G.C. 1988. *Rhizobium meliloti* mutants that overproduce the *R. meliloti* acidic calcofluor-binding exopolysaccharide. J. Bacteriol. 170: 4249–4256.
67. REED, J., GLAZEBROOK, J., WALKER, G.C. 1991. The *exoR* gene of *Rhizobium meliloti* affects RNA levels of other *exo* genes but lacks homology to known transcriptional regulators. J. Bacteriol. 173: 3789–3794.
68. REUBER, T.L., LONG, S., WALKER, G.C. 1991. Regulation of *Rhizobium meliloti exo* genes in free-living cells and in planta examined using Tn*phoA* fusions. J. Bacteriol. 173: 426–434.
69. GRAY, J.X., DJORDJEVIC, M.A., ROLFE, B.G. 1990. Two genes that regulate exopolysaccharide production in *Rhizobium* sp. strain NGR234: DNA sequences and resultant phenotypes. J. Bacteriol. 172: 193–203.
70. ZHAN, H., LEIGH, J.A. 1990. Two genes that regulate exopolysaccharide production in *Rhizobium meliloti*. J. Bacteriol. 172: 5254–5259.

71. LEVERY, S.B., ZHAN, H., LEE, C.C., LEIGH, J.A., HAKOMORI, S. 1991. Structural analysis of a second acidic exopolysaccharide of *Rhizobium meliloti* that can function in alfalfa root nodule invasion. Carbohydr. Res. 210: 339–347.
72. KELLER, M., ROXLAU, A., WENG, W.M., SCHMIDT, M., QUANDT, J., NIEHAUS, K., JORDING, D., ARNOLD, W., PÜHLER, A. 1995. Molecular analysis of the *Rhizobium meliloti mucR* gene regulating the biosynthesis of the exopolysaccharides succinoglycan and galactoglucan. Mol. Plant-Microbe Interact. 8: 267–277.
73. LEIGH, J.A., LEE, C.C. 1988. Characterization of polysaccharides of *Rhizobium meliloti exo* mutants that form ineffective nodules. J. Bacteriol. 170: 3327–3332.
74. REED, J.W., WALKER, G.C. 1991. Acidic conditions permit effective nodulation of alfalfa by invasion-deficient *Rhizobium meliloti exoD* mutants. Genes & Development 5: 2274–2287.
75. BORTHAKUR, D., BARBER, C.E., LAMB, J.W., DANIELS, M.J., DOWNIE, J.A., JOHNSTON, A.W.B. 1986. A mutation that blocks exopolysaccharide synthesis prevents nodulation of peas by *Rhizobium leguminosarum* but not of beans by *R. phaseoli* and is corrected by cloned DNA from *Rhizobium* or the phytopathogen *Xanthomonas*. Mol. Gen. Genet. 203: 320–323.
76. CHAKRAVORTY, A.K., ZURKOWSKI, W., SHINE, J., ROLFE, B.G. 1982. Symbiotic nitrogen fixation: Molecular cloning of *Rhizobium* genes involved in exopolysaccharide synthesis and effective nodulation. J. Mol. Appl. Genet. 1: 585–596.
77. CHEN, H., BATLEY, M., REDMOND, J., ROLFE, B.G. 1985. Alteration of the effective nodulation properties of fast growing broad host range *Rhizobium* due to changes in exopolysaccharide synthesis. J. Plant Physiol. 120: 331–349.
78. HOTTER, G.S., SCOTT, D.B. 1991. Exopolysaccharide mutants of *Rhizobium loti* are fully effective on a determinate nodulating host but are ineffective on an indeterminate nodulating host. J. Bacteriol. 173: 851–859.
79. DIEBOLD, R., NOEL, K.D. 1989. *Rhizobium leguminosarum* exopolysaccharide mutants: Biochemical and genetic analyses and symbiotic behavior on three hosts. J. Bacteriol. 171: 4821–4830.
80. KIM, C.-H., TULLY, R.E., KEISTER, D.L. 1989. Exopolysaccharide-deficient mutants of *Rhizobium fredii* HH303 which are symbiotically effective. Appl. Environ. Microbiol. 55: 1852–1854.
81. MILNER, J.L., ARAUJO, R.S., HANDELSMAN, J. 1992. Molecular and symbiotic characterization of exopolysaccharide-deficient mutants of *Rhizobium tropici* strain CIAT899. Mol. Microbiol. 6: 3137–3147.
82. PARNISKE, M., KOSCH, K., WERNER, D., MÜLLER, P. 1993. ExoB mutants of *Bradyrhizobium japonicum* with reduced competitiveness for nodulation of *Glycine max*. Mol. Plant-Microbe Interact. 6: 99–106.
83. ZDOR, H., PUEPPKE, S.G. 1991. Nodulation competitiveness of Tn5- induced mutants of *Rhizobium fredii* USDA208 that are altered in motility and extracellular polysaccharide production. Can. J. Microbiol. 37: 52–58.
84. EGGLESTON, G., HUBER, M.C., LIANG, R.T., KARR, A.L., EMERICH, D.W. 1996. *Bradyrhizobium japonicum* mutants deficient in exo and capsular polysaccharides cause delayed infection and nodule initiation. Mol. Plant-Microbe Interact. 9: 419–423.
85. CHEN, H.C., LONG, B.G., SONG, H.Y. 1996. Exopolysaccharide deficient mutants of astragali rhizobia are symbiotically effective on A*stragalus sinicus*, an indeterminate nodulating host. Plant and Soil 179: 217–221.
86. NDOYE, I., DEBILLY, S.F., VASSE, J., DREYFUS, B., TRUCHET, G. 1994. Root nodulation of *Sesbania rostrata*. J. Bacteriol. 176: 1060–1068.
87. ROLFE, B.G., CARLSON, R.W., RIDGE, R.W., DAZZO, F.B., MATEOS, P.F., PANKHURST, C.E. 1996. Defective infection and nodulation of clovers by exopolysaccharide mutants of *Rhizobium leguminosarum* bv. *trifolii*. Austr. J. Plant Physiol. 23: 285–303.

88. SKORUPSKA, A., BIALEK, U., URBANIK-SYPNIEWSKA, T., VAN LAMMEREN, A. 1995. Two types of nodules induced on *Trifolium pratense* by mutants of *Rhizobium leguminosarum* bv. *trifolii* deficient in exopolysaccharide production. J. Plant Physiol. 147: 93–100.
89. YANG, C., SIGNER, E.R., HIRSCH, A.M. 1992. Nodules initiated by *Rhizobium meliloti* exopolysaccharide mutants lack a discrete, persistent nodule meristem. Plant Physiol. 96: 143–151.
90. FINAN, T.M., HIRSCH, A.M., LEIGH, J.A., JOHANSEN, E., KULDAU, G.A., DEEGAN, S., WALKER, G.C., SIGNER, E.R. 1985. Symbiotic mutants of *Rhizobium meliloti* that uncouple plant from bacterial differentiation. Cell 40: 869–877.
91. VAN WORKUM, W.A.T., VAN BRUSSEL, A.A.N., TAK, T., WIJFFELMAN, C.A., KIJNE, J.W. 1995. Ethylene prevents nodulation of *Vicia sativa* ssp. *nigra* by exopolysaccharide-deficient mutants of *Rhizobium leguminosarum* bv. *viciae*. Mol. Plant-Microbe Interact. 8: 278–285.
92. DARVILL, A.G., ALBERSHEIM, P. 1984. Phytoalexins and their elicitors - a defense against microbial infection in plants. Ann. Rev. Plant Physiol. 35: 243–275.
93. NOEL, K.D. 1992. Rhizobial polysaccharides required in symbioses with legumes. In: Molecular Signals in Plant-Microbe Interactions (D.P. Verma, ed.), CRC Press, Inc, Boca Rato, FL, pp. 341–357.
94. OZGA, D.A., LARA, J.C., LEIGH, J.A. 1994. The regulation of exopolysaccharide production is important at two levels of nodule development in *Rhizobium meliloti*. Mol. Plant-Microbe Interact. 7: 758–765.
95. BORTHAKUR, D., DOWNIE, J.A., JOHNSTON, A.W.B., LAMB, J.W. 1985. Psi, a plasmid-linked *Rhizobium phaseoli* gene that inhibits exopolysaccharide production and which is required for symbiotic nitrogen fixation. Mol. Gen. Genet. 200: 278–282.
96. GONZALEZ, J.E., REUHS, B.L., WALKER, G.C. 1996. Low molecular weight EPS II of *Rhizobium meliloti* allows nodule invasion in *Medicago sativa*. Proc. Natl. Acad. Sci. USA 93: 8636–8641.
97. PUTNOKY, P., GROSSKOPF, E., CAM HA, D.T., KISS, G.B., KONDOROSI, A. 1988. *Rhizobium fix* genes mediate at least two communication steps in symbiotic nodule development. J. Cell. Biol. 106: 597–607.
98. WILLIAMS, M.N.V., HOLLINGSWORTH, R.I., KLEIN, S., SIGNER, E.R. 1990. The symbiotic defect of *Rhizobium meliloti* exopolysaccharide mutants is suppressed by *lpsZ+*, a gene involved in lipopolysaccharide biosynthesis. J. Bacteriol. 172: 2622–2632.
99. PUTNOKY, P., PETROVICS, G., KERESZT, A., GROSSKOPF, E., CAM HA, D., BANFALVI, Z., KONDOROSI, A. 1990. *Rhizobium meliloti* lipopolysaccharide and exopolysaccharide can have the same function in the plant-bacterium interaction. J. Bacteriol. 172: 5450–5458.
100. WILLIAMS, M.N.V., HOLLINGWORTH, R.I., BRZOSKA, P.M., SIGNER, E.R. 1990. *Rhizobium meliloti* chromosomal loci required for suppression of exopolysaccharide mutations by lipopolysaccharide. J. Bacteriol. 172: 6596–6598.
101. REUHS, B.R., CARLSON, R.W., KIM, J.S. 1993. *Rhizobium fredii* and *Rhizobium meliloti* produce 3-deoxy-D-manno-2-octulosonic acid-containing polysaccharides that are structurally analogous to group II K antigens (capsular polysaccharides) found in *Escherichia coli*. J. Bacteriol. 8: 3570–3580.
102. PETROVICS, G., PUTNOKY, P., REUHS, B., KIM, J., THORP, T.A., NOEL, K.D., CARLSON, R.W., KONDOROSI, A. 1993. The presence of a novel type of surface polysaccharide in *Rhizobium meliloti* requires a fatty acid synthase-like gene cluster involved in symbiotic nodule development. Mol. Microbiol. 8: 1083–1094.
103. REUHS, B.L., WILLIAMS, M.N.V., KIM, J.S., CARLSON, R.W., CÔTÉ, F. 1995. Suppression of the fix⁻ phenotype of *Rhizobium meliloti exoB* mutants by *lpsZ* is correlated to a modified expression of the K polysaccharide. J. Bacteriol. 177: 4289–4296.

104. JANN, B., JANN, K. 1990. Structure and biosynthesis of the capsular antigens of *Escherichia coli*. Curr. Top. Microbiol. Immunol. 150: 19–42.
105. PAZZANI, C., ROSENOW, C., BOULNOIS, G.J., BRONNER, D., JANN, K., ROBERTS, I.S. 1994. Molecular analysis of Region 1 of the *Escherichia coli* K5 antigen gene cluster: A region encoding proteins involved in cell surface expression of capsular polysaccharide. J. Bacteriol. 175: 5978–5983.
106. REUHS, B.L. 1996. Acidic capsular polysaccharides (K antigens) of *Rhizobium*. In: Biology of Plant-Microbe Interactions (G. Stacey, B. Mullin, P.M. Gresshoff, eds.), Int. Soc. Mol. Plant-Microbe Interact. St. Paul, pp. 331–336.
107. DJORDJEVIC, S.P., CHEN, H., BATLEY, M., REDMOND, J.W., ROLFE, B.G. 1987. Nitrogen fixation ability of exopolysaccharide synthesis mutants of *Rhizobium* sp NGR234 and *Rhizobium trifolii* is restored by the addition of homologous exopolysaccharides. J. Bacteriol. 169: 53–60.
108. URZAINQUI, A., WALKER, G.C. 1992. Exogenous suppression of the symbiotic deficiencies of *Rhizobium meliloti exo* mutants. J. Bacteriol. 174: 3403–3406.
109. BATTISTI, L., LARA, J.C., LEIGH, J.A. 1992. Specific oligosaccharide form of the *Rhizobium meliloti* exopolysaccharide promotes nodule invasion in alfalfa. Proc. Natl. Acad. Sci. USA 89: 5625–5629.
110. NIEHAUS, K., KAPP, D., PÜHLER, A. 1993. Plant defense and delayed infection of alfalfa pseudonodules induced by an exopolysaccharide (EPSI)-deficient *Rhizobium meliloti* mutant. Planta 190: 415–425.
111. NIEHAUS, K., BAIER, R., BECKER, A., PÜHLER, A. 1996. Symbiotic suppression of the *Medicago sativa* defense system - the key of *Rhizobium meliloti* to enter the host plant? In: Biology of Plant-Microbe Interactions (G. Stacey, B. Mullin, P.M. Gresshoff, eds.), Int. Soc. for Mol. Plant-Microbe Interact. St. Paul, pp. 349–352.
112. MITHÖFER, A., BHAGWAT, A.A., FEGER, M., EBEL, J. 1996. Suppression of fungal β-glucan-induced plant defense in soybean (*Glycine max* L.) by cyclic 1,3–1,6-β-glucans from the symbiont *Bradyrhizobium japonicum*. Planta 199: 270–275.
113. VASSE, J., DE BILLY, F., TRUCHET, G. 1993. Abortion of infection during the *Rhizobium meliloti*-alfalfa symbiotic interaction is accompanied by a hypersensitive reaction. Plant Journal 4: 555–566.
114. ZAAT, S.A.J., VAN BRUSSEL, A.A.N., TAK, T., LUGTENBERG, B.J.J., KIJNE, J.W. 1989. The ethylene-inhibitor aminoethoxyvinylglycine restores normal nodulation by *Rhizobium leguminosarum* biovar. *viciae* on *Vicia sativa* subsp. *nigra* by suppressing the 'Thick and short roots' phenotype. Planta 177: 141–150.
115. PETERS, N.K., CRIST-ESTES, D.K. 1989. Nodule formation is stimulated by the ethylene inhibitor aminoethoxyvinylglycine. Plant Physiol. 91: 690–693.
116. PENMETSA, R.V., COOK, D.R. 1997. A legume ethylene-insensitive mutant hyperinfected by its rhizobial symbiont. Science 275: 527–530.
117. PETERS, N.K., LONG, S.R. 1988. Alfalfa root exudates and compounds which promote or inhibit induction of *Rhizobium meliloti* nodulation genes. Plant Physiol. 88: 396–400.
118. RECOURT, K., SCHRIPSEMA, J., KIJNE, J.W., VAN BRUSSEL, A.A.N., LUGTENBERG, B.J.J. 1991. Inoculation of *Vicia sativa* subsp. *nigra* roots with *Rhizobium leguminosarum* biovar *viciae* results in release of *nod* gene activating flavanones and chalcones. Plant Mol. Biol. 16: 841–852.
119. RECOURT, K., VAN TUNEN, A.J., MUR, L.A., VAN BRUSSEL, A.A.N., LUGTENBERG, B.J.J., KIJNE, J.W. 1992. Activation of flavonoid biosynthesis in roots of *Vicia sativa* ssp. *nigra* plants by inoculation with *Rhizobium leguminosarum* biovar *viciae*. Plant Mol. Biol. 19: 411–420.
120. VAN BRUSSEL, A.A.N., RECOURT, K., PEES, E., SPAINK, H.P., TAK, T., WIJFFELMAN, C.A., KIJNE, J.W., LUGTENBERG, B.J.J. 1990. A biovar-specific signal of *Rhizobium*

leguminosarum bv. *viciae* induces increased nodulation gene-inducing activity in root exudate of *Vicia sativa* subsp. *nigra*. J. Bacteriol. 172: 5394–5401.
121. FIRMIN, J.L., WILSON, K.E., ROSSEN, L., JOHNSTON, A.W.B. 1986. Flavonoid activation of nodulation genes in *Rhizobium* reversed by other compounds present in plants. Nature (London) 324: 90–92.
122. DUNN, M.F., PUEPPKE, S.G., KRISHNAN, H.B. 1992. The *nod* gene inducer genistein alters the composition and molecular mass distribution of extracellular polysaccharides produced by *Rhizobium fredii*. FEMS Microbiol. Lett. 97: 107–112.
123. REUHS, B.L., KIM, J.S., BADGETT, A., CARLSON, R.W. 1994. Production of cell-associated polysaccharides of *Rhizobium fredii* USDA205 is modulated by apigenin and host root extract. Mol. Plant-Microbe Interact. 7: 240–247.
124. JACOBS, M., RUBERY, P.H. 1988. Naturally occurring auxin transport regulators. Science 241: 346–349.
125. HIRSCH, A.M. 1992. Tansley Review No. 40. Developmental biology of legume nodulation. New Phytol. 122: 211–237.
126. YANG, W.-C., CANTER CREMERS, H.C.J., HOGENDIJK, P., KATINAKIS, P., WIJFFELMAN, C.A., FRANSSEN, H., VAN KAMMEN, A., BISSELING, T. 1992. In-situ localization of chalcone synthase mRNA in pea root nodule development. Plant Journal 2: 143–151.
127. MATHESIUS, U., SCHLAMAN, H.R.M., MEIJER, D., LUGTENBERG, B.J.J., SPAINK, H.P., WEINMAN, J.J., RODDAM, L.F., SAUTTER, C., ROLFE, B.G., DJORDJEVIC, M.A. 1996. New tools for investigating nodule initiation and ontogeny: Spot inoculation and microtargeting of transgenic white clover roots shows auxin involvement and suggests a role for flavonoids. In: Biology of Plant-Microbe Interactions (G. Stacey, B. Mullin, P.M. Gresshoff, eds.), Int. Soc. Mol. Plant-Microbe Interact. St. Paul, pp. 353–358.
128. DIXON, R.A., LAMB, C.J. 1990. Molecular communication in interactions between plants and microbial pathogens. Ann Rev Plant Physiol. Plant Mol. Biol. 41: 339–367.
129. PARNISKE, M., FISCHER, H.-M., HENNECKE, H., WERNER, D. 1991. Accumulation of the phytoalexin glyceollin I in soybean nodules infected by a *Bradyrhizobium japonicum nifA* mutant. Z. Naturforsch. 46c: 318–320.
130. PARNISKE, M., SCHMIDT, P.E., KOSCH, K., MÜLLER, P. 1994. Plant defense responses of host plants with determinate nodules induced by EPS-defective *exoB* mutants of *Bradyrhizobium japonicum*. Mol. Plant-Microbe Interact. 7: 631–638.
131. WERNER, D., MELLOR, R.B., HAHN, M.G., GRISEBACH, H. 1985. Soybean root response to symbiotic infection: Glyceollin I accumulation in an ineffective type of soybean nodules with an early loss of the peribacteroid membrane. Z. Naturforsch. 40c: 179–181.
132. SAVOURÉ, A., SALLAUD, C., EL-TURK, J., ZUANAZZI, J., RATET, P., SCHULTZE, M., KONDOROSI, A., ESNAULT, R., KONDOROSI, E. 1997. Distinct response of *Medicago* suspension cultures and roots to Nod factors and chitin oligomers in the elicitation of defense-related responses. Plant Journal 11: 277–287.
133. McKHANN, H.I., PAIVA, N.L., DIXON, R.A., HIRSCH, A.M. 1997. Chalcone synthase transcripts are detected in alfalfa root hairs following inoculation with wild-type *Rhizobium meliloti*. Mol. Plant-Microbe Interact. 10: 50–58.
134. BECQUART-DE KOZAK, I., REUHS, B.L., BUFFARD, D., BREDA, C., KIM, J.S., ESNAULT, R., KONDOROSI, A. 1997. Role of the K-antigen subgroup of capsular polysaccharides in the early recognition process between *Rhizobium meliloti* and alfalfa leaves. Mol. Plant-Microbe Interact. 10: 114–123.
135. KOES, R.E., SPELT, C.E., MOL, J.N.M., GERATS, A.G.M. 1987. The chalcone synthase multigene family of *Petunia hybrida* (V30): Sequence homology, chromosomal localization and evolutionary aspects. Plant Mol. Biol. 10: 375–385.
136. NIESBACH-KLOSGEN, U., BARZEN, E., BERNHARDT, J., ROHDE, W., SCHWARZ-SOMMER, Z., REIF, H.J., WIENAND, U., SAEDLER, H. 1987. Chalcone synthase genes in plants: A tool to study evolutionary relationships. J. Mol. Evol. 26: 213–225.

137. RYDER, T.B., HEDRICK, S.A., BELL, J.N., LIANG, X., CLOUSE, S.D., LAMB, C.J. 1987. Organization and differential activation of a gene family encoding the plant defense enzyme chalcone synthase in *Phaseolus vulgaris*. Mol. Gen. Genet. 210: 219–233.
138. WINGENDER, R., ROHRIG, H., HORICKE, C., WING, D., SCHELL, J. 1989. Differential regulation of soybean chalcone synthase genes in plant defense, symbiosis and upon environmental stimuli. Mol. Gen. Genet. 218: 315–322.
139. ESTABROOK, E.M., SENGUPTA-GOPALAN, C. 1991. Differential expression of phenylalanine ammonia-lyase and chalcone synthase during soybean nodule development. Plant Cell 3: 299–308.
140. DENNY, T.P. 1995. Involvement of bacterial polysaccharides in plant pathogenesis. In: Annual Review of Phytopathology, 33, (R.K. Webster, G.A. Zentmyer, G. Shaner, eds.), Annual Reviews Inc. Palo Alto, pp. 173–197.
141. EISENSCHENK, L., DIEBOLD, R., PEREZ-LESHER, J., PETERSON, A.C., PETERS, N.K., NOEL, K.D. 1994. Inhibition of *Rhizobium etli* polysaccharide mutants by *Phaseolus vulgaris* root compounds. Appl. Environ. Microbiol. 60: 3315–3322.
142. REED, J.W., WALKER, G.C. 1991. The *exoD* gene of *Rhizobium meliloti* encodes a novel function needed for alfalfa nodule invasion. J. Bacteriol. 173: 664–677.
143. OSTERÅS, M., STANLEY, J., FINAN, T.M. 1995. Identification of Rhizobium-specific intergenic mosaic elements within an essential two-component regulatory system of *Rhizobium* species. J. Bacteriol. 177: 5485–5494.
144. KIJNE, J.W. 1975. The fine structure of pea root nodules. 1. Vacuolar changes after endocytotic host cell infection by *Rhizobium leguminosarum*. Phys. Plant Pathol. 5: 75–79.
145. TURGEON, B.G., BAUER, W.D. 1985. Ultrastructure of infection-thread development during infection of soybean by *Rhizobium japonicum*. Planta 163: 328–349.

Chapter Nine

FLAVONOIDS AS REGULATORS OF PLANT DEVELOPMENT
New Insights from Studies of Plant–Rhizobia Interactions

Herman P. Spaink

Leiden University
Institute of Molecular Plant Sciences
Wassenaarseweg 64
2333 AL Leiden, The Netherlands

Introduction 167
Flavonoid Biosynthesis as a Marker for Plant Morphogenetic Processes .. 169
Flavonoids as Regulators of Auxin Transport 170
A Model for the Evolutionary Basis of Flavonoids as Signal Molecules
    in Plant–Rhizobia Symbiosis 171
Conclusion 173

INTRODUCTION

In the study of plant–microbe interactions, flavonoids have attracted major attention as regulators of gene expression in symbiotic bacteria belonging to the family *Rhizobiaceae* (called rhizobia).[1,2] They are possibly also involved in the regulation of interactions of plants with other symbiotic microbes such as mycorrhizal fungi.[3,4] In legume-rhizobia interactions, several classes of flavonoids play a role in a signal transduction cascade resulting in the formation of root nodules (Fig. 1A). These legume nodules contain nitrogen-fixing differentiated forms of the rhizobia, called bacteroids.

Flavonoids play a role in the induction of rhizobial genes called *nod* (for nodulation) genes, which subsequently are involved in the production of signal molecules which govern the root infection and nodulation processes. A characteristic feature of this gene induction process is its host-specific nature. Rhizobial

Phytochemical Signals and Plant–Microbe Interactions, edited by Romeo *et al.*
Plenum Press, New York, 1998.

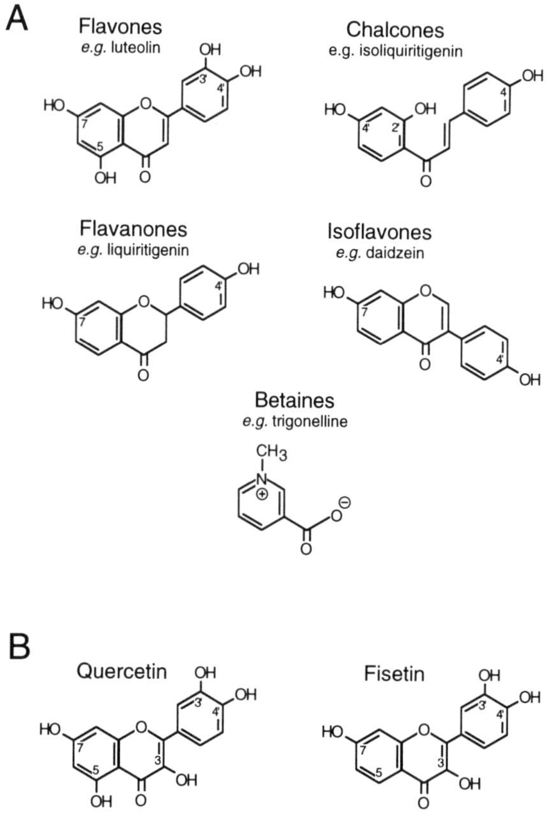

Figure 1. A. Classes and some examples of inducers of the rhizobial nodulation genes. For further details see ref.[57] **B.** Flavonoids that have been shown to be inhibitors of auxin transport in white clover roots.[37]

strains are adapted to characteristic mixtures of flavonoids secreted by the host plants. For each type of rhizobial strain only a limited number of host plants have evolved as partners for nitrogen-fixing symbiosis. Thus, flavonoids play an important role in the determination of host specificity. The specificity of this induction process is mediated by the rhizobial protein NodD, which, presumably, after binding to host-specific flavonoids, is able to activate the transcription of other *nod* genes. These other *nod* genes are involved in the production of mitogenic signal molecules, called lipo-chitin oligosaccharides (LCOs), which can induce the formation of root nodules.[5–7] An important application of these findings is that by changing the original *nodD* gene by a mutant form of *nodD* which is able to activate transcription in the absence of flavonoids, one can manipulate the host range of the rhizobia. This can, for example, be demonstrated with the strain *Rhizobium leguminosarum* biovar *trifolii*, which normally is restricted to associations with plants of the genus *Trifolium*. After introduction of a flavonoid-independent *nodD*, this strain is able to induce root nodules in various tropical leguminous plant genera.[8]

Important questions which remain are: (i) Why were flavonoids chosen during evolution as signal molecules for plant-rhizobial symbioses? (ii) How has the process of host-specificity developed? The latter question is instigated by the fact that flavonoids seem rather unlikely candidates to mediate specificity. Most plants have been shown to produce complex mixtures of flavonoids, and there seems to be no common denominator of the inducers produced by particular host plant genera.[9-11] In this review, some recent advances in the study of the role of flavonoids in plant-rhizobia interactions will be presented that suggest possible answers to these questions. These new insights will be discussed in the context of older literature, which pointed to a possible role of flavonoids as regulators of plant developmental processes.

FLAVONOID BIOSYNTHESIS AS A MARKER FOR PLANT MORPHOGENETIC PROCESSES

In many plants, including the legumes, several genes involved in the biosynthesis of flavonoids are encoded by multigene families.[12] The best studied examples are the genes encoding the enzymes phenylalanine ammonia-lyase (PAL), the enzyme which performs the first dedicated step in the phenylpropanoid pathway, and chalcone synthase (CHS), a β-ketoacylsynthase which elongates the coumaryl-CoA starter unit resulting from the phenylpropanoid pathway.[13,14] For example, CHS is encoded by at least 9 members in *Trifolium subterraneum*.[15,16] Gene expression of the PAL and CHS genes has been shown to be induced by inoculation with rhizobia bacteria in several leguminous plant species, such as *Vicia sativa*,[17] *Pisum sativum*,[18] *Trifolium subterraneum*,[16,19] *Medicago sativa*,[20] *Glycine soja*,[21] and *Vigna unguiculata*.[22] Gene induction appears mainly to be localized in cells of the developing root nodules or root hairs, whereas constitutive expression can be found in the main or lateral root meristems. Induction of CHS gene expression also can be detected after treatment of the roots with the LCO signal molecules.[22,23]

In a recent study,[23] it was show that different members of the CHS family in *T. subterraneum* are differently expressed in the region of the developing nodule primordia. The CHS1 is expressed in the epidermis and outer cortex, whereas CHS3 is expressed in the inner cortex beneath the site of inoculation.[23] The induction of CHS and PAL gene expression is concomitant with effects on the production and secretion of flavonoids. This was first shown by van Brussel *et al.*,[24] who showed that a biovar-specific signal of *Rhizobium leguminosarum* (later shown to be an LCO[25]) induces nodulation gene-inducing activity in root exudate of *Vicia sativa*. The nod-gene inducing factors were identified as 6 flavanones (including liquiritigenin, naringenin, and hesperitin) and 2 chalcones (including isoliquiritigenin), which are not produced in the absence of bacteria or LCOs.[26,27] Independent of inoculation with rhizobia, the roots of *V. sativa* contain four non-inducing 3-O-glycosides of the flavonol kaempferol. Later, an induction of the production of flavonoids also was demonstrated in other plant species.[20,22,28–30]

A major question emerging from these data is: What is the function of the flavonoids which are produced by roots? In other parts of the plant, flavonoids have been shown to function as precursors for various pigments, such as the anthocyanins occurring in flowers.[9] Flavonoids also have been shown to play a role in pollen tube growth.[13,31] They have been suggested as playing a role as defense compounds, similar to phytoalexins such as glyceolin, which is induced after inoculation with rhizobia.[32] In support of this hypothesis, incompatible legume-rhizobia interactions have been shown to lead to higher levels of flavonoid production than compatible interactions.[18,22,33] Furthermore, for the induction of flavonoid production in *Vicia sativa*, the hormone ethylene (a negative factor in root infection) plays a major role.[24,34–36] In *Glycine soja*, it was shown that the expression pattern of CHS genes is different during nodule development and defense responses. These results, together with the complex differential expression of the CHS family members,[23] indicate that a role in plant defense response (hitherto not proven) is probably not the only function of flavonoids in roots.

Alternatively, leguminous plants may have developed a mechanism to regulate the gene expression of the rhizobial symbionts. Although this is suggested by the fact that the rhizobia induce the production of *nod* gene inducers (see above), this explanation seems unlikely from an evolutionary point of view. Intuitively, it would seem difficult for higher organisms, such as plants, to adapt to the rapidly changing community of microbes. The same argument can be used against a species-specific role of flavonoids as defense compounds against particular pathogenic microbes.

Still another possible explanation is that flavonoids play a role in root development. Using microspectrofluorometric techniques, Mathesius *et al*, in a recently submitted manuscript,[37] have shown that flavonoid accumulation is common to all meristematic cells in the root. Furthermore, cells in different primordial zones (*e.g.* root nodule or lateral roots) or different stages of development (*e.g.* before or after cell division) are distinguished by the production of different types of flavonoids.[37] These results suggest that flavonoids play a structural role in development in the plant root.

FLAVONOIDS AS REGULATORS OF AUXIN TRANSPORT

Jacobs and Rubery[38] have shown that certain flavonoids (*e.g.* quercetin, tamarixetin, apigenin, kaempferol) and chalcones (*e.g.* butein) can act as auxin transport regulators in *Cucurbita pepo* hypocotyls. Strong support for a function of flavonoids in the regulation of auxin levels comes from the recent work of Mathesius et al.[39] These workers used transgenic *Trifolium repens* plants, which contained a fusion of the promoter of the GH3 gene with the β-glucuronidase (GUS) gene, in order to analyze the effects of local application of putative signal molecules on auxin levels.[40] The soybean GH3 gene responds rapidly and specifically to natural and artificial auxins.[41] The promoter of the soybean GH3 gene appears to be regulated in a similar way when introduced into heterologous backgrounds, such as *Nicotiana tabacum*[42] and *Trifolium repens*.[40] Mathesius et al.[39] showed that spot inoculation of various flavonoids on the roots of *Trifolium repens* leads to repression of the GH3 promoter in an area below the position of inoculation. The flavones, quercetin and fisetin (Fig. 1B), appeared to be the most active of a set of 10 flavonoids tested. Glycosidic flavonoids appeared to be inactive in this assay. The same results were obtained when the flavonoids were introduced inside the root tissue using the novel method of ballistic microtargeting,[43,44] suggesting that the difference in activity of various flavonoids is not due to a difference in penetration capacity into the plant tissue. The results indicate that specific flavonoids in the root are involved in the regulation of auxin transport through the vascular system and perhaps also into the root cortex.

Spot inoculation experiments indicate that auxin transport inhibition plays a role in root nodule development.[45] The results of Mathesius et al.,[39] who showed that rhizobial LCOs are equally as effective as flavonoids in repressing the GH3 promoter, also support a role of auxin transport inhibition in root nodule formation. In contrast, a general role for flavonoids in root development is not supported by the results of Burbulis et al.,[46] who reported that a null mutation in the single CHS gene of *Arabidopsis* does not have obvious effects on root development or reproduction. However, it is possible that other, as yet unidentified, CHS-like genes could compensate for the loss of function in flavonoid biosynthesis in the reported *Arabidopsis* mutant. It is also possible that compounds that are structurally completely different from flavonoids could perform the same function and therefore compensate for the mutation. This would be analogous to the induction of the rhizobial *nod* genes by flavonoids, since different compounds such as betaines (Fig. 1A) are also able to induce *nod* genes in *Rhizobium meliloti*.[47] In future research, the reported mutant in *Arabidopsis* could be a useful tool to analyze the role of other secondary metabolites of plant roots (*e.g.* glycosidic triterpenoids and various other phenolics) involved in plant morphogenesis.

A MODEL FOR THE EVOLUTIONARY BASIS OF FLAVONOIDS AS SIGNAL MOLECULES IN PLANT–RHIZOBIA SYMBIOSIS

In *Trifolium*, the flavonoid inducers of the *nod* genes have been shown to be exuded just behind the root tip in the zone where the root hairs emerge.[48,49] This is consistent with the hypothesis that particular flavonoids play a role in the initiation of lateral root formation, which also is confined mainly to the region of young developing root hairs. As mentioned above, flavonoids also have been suggested as playing a role in root nodule formation. The role of LCO signal molecules in root nodulation could be explained by their effect upon the production of flavonoids. The production of new types or higher levels of flavonoids could result in a specific effect on the local level of auxin in the root cortex, subsequently leading to the formation of root nodule primordia. As is the case with lateral root formation, the formation of root nodule primordia is confined to the region of the roots where young root hairs emerge. There is strict confinement of the root nodulation process to this part of the root, since root nodules can not be formed in the higher regions of the roots, not even after spot inoculation or microtargeting of the LCO nodulation factors.[44] It has been speculated that radially localized development is regulated by the local concentrations of compounds such as uridine or ethylene.[36,44,50,51] Local auxin concentration seems a good candidate for determining longitudinal positioning of root nodule formation.

Together, these data, which are summarized in the model outlined in Figure 2, suggest an answer to the question: Why have flavonoids been chosen during evolution as inducers of the rhizobial *nod* genes? It is not fruitful for rhizobia to secrete LCO nodulation factors at positions high in the root. Therefore, the flavonoids, as markers for the region of the root which is capable of forming new primordia, might have been chosen as key regulators for the *nod* genes which produce the LCOs. Flavonoids also have been indicated as chemo-attractants for rhizobia.[52–54] Rhizobia which are present at high regions of the roots might, therefore, in the first instance, move to lower regions of the roots and subsequently the process of *nod* gene induction would be initiated. However, a

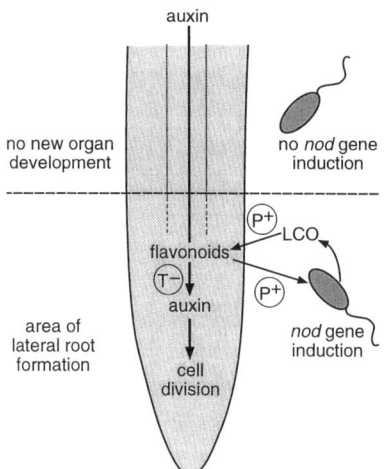

Figure 2. A model that postulates that flavonoids are used by rhizobia as indicators for the region of the root that is capable of forming root nodule primordia. The model explains why flavonoids could have been chosen a key regulators for the *nod* genes that produce the LCOs. Since it would not be fruitful for rhizobia to secrete LCO nodulation factors at positions high in the root (where nodules cannot be formed), the *nod* genes are switched on only in the lower regions of the root. Flavonoids are secreted in the lower region of the root because they are involved in the down regulation of vascular auxin transport during the process of lateral root formation. Symbols: p+, positive effect on production; t–, negative effect on transport.

positive feedback mechanism of LCO and flavonoid production (Fig. 2) leading to very high levels of both signal molecules would seem to have adverse effects at later stages of the symbiosis. Indeed, it has been shown that *nod* gene transcription is down regulated during later stages of the symbiosis in a flavonoid-independent way.[55–57]

An unexplained problem is the host-specific nature of the *nod* gene induction process. A possible evolutionary basis for this phenomenon could be the fact that secretion of flavonoids by the roots of other plant species in the nearby environment could have negative effects on root development. Flavonoids, being rather stable compounds which are active at low concentrations, could, therefore, play an important role in competition with other plant species. In order to be independent of flavonoid secretion by other plant species, it would have been advantageous to evolve towards the use of flavonoid signal molecules which are different from competing species. In other words, various types of flavonoids have developed during evolutionary specialization in order to avoid interference from competing plant species. The associating microbes, with their faster capacity of evolutionary adaptation, would have co-evolved with their preferred symbiotic host plants. Co-incidently, the great evolutionary success of the nitrogen-fixing symbiosis could have led to a greater diversity of flavonoids secreted by leguminous plants in order to influence growth of competing plant species which would likely profit from the surplus of fixed nitrogen.

Although this model is speculative, its strength lies in the fact that it can be tested. A thorough survey of the types of flavonoids of a large variety of plants could be correlated with: (i) the competitiveness of different species in one biotope. In this model, the most competitive species would be expected to have little overlap in their profile of secreted flavonoids; (ii) whether or not plants species occur in the same biotopes. Plants which are never present in each other's biotopes would be expected to have random overlap in the range of flavonoids produced.

CONCLUSION

Although flavonoids have been studied for many decades, mainly for their effects on animal development after ingestion and their possible use in human health care, their role in plant development has only become apparent in the last 10 years. Recently, the study of their role in the plant-rhizobia symbiosis has led to new insights into the possible function of flavonoids in root nodule development. These results support earlier suggestions for a role of flavonoids in root development. However, at the molecular level, nothing is yet known about the mechanism by which flavonoids regulate plant development. The most urgent objectives of future research on the underlying signal transduction pathway are to identify receptors for the flavonoids and to analyze how direct is their effect on auxin transport. Since research on most other plant signal molecules (including auxin itself and the LCOs) faces major bottlenecks towards similar objectives, such studies are not expected to be easy. However, the above mentioned results imply that a thorough understanding of root development can not be obtained without further studies on the function of flavonoids. Hopefully, such research will continue to profit from both the strong current drive to understand the process of root nodule formation as a model for organogenesis and the tools resulting from these studies.

ACKNOWLEDGMENTS

I thank Drs. U. Mathesius, M.A. Djordjevic (Australian National University, Canberra, Australia), and H.R.M. Schlaman (Leiden University) for communicating results prior to publication. The author is supported by the Netherlands Organization for Scientific Research (NWO).

REFERENCES

1. SPAINK, H.P. 1994. The molecular basis of the host specificity of the Rhizobium bacteria. Anton. Leeuwenhoek Int. J. Gen. M. 65: 81–98.
2. SPAINK, H.P. 1995. The molecular basis of infection and nodulation by rhizobia: The ins and outs of sympathogenesis. Ann. Rev. Phytopathol. 33: 345–368.
3. HARRISON, M. J. , DIXON, R.A. 1993. Isoflavonoid accumulation and expression of defense gene transcripts during the establishment of vesicular-arbuscular mycorrhizal associations in roots of *Medicago truncatula*. Mol. Plant-Microbe Int. 6: 643–654.
4. HARRISON, M.J., DIXON, R.A. 1994. Spatial patterns of expression of flavonoid/isoflavonoid pathway genes during interactions between roots of *Medicago truncatula* and the mycorrhizal fungus *Glomus versiforme*. Plant J. 6: 9–20.
5. SPAINK, H.P. 1996. Regulation of plant morphogenesis by lipo-chitin oligosaccharides. Crit. Reviews Plant Sciences 15: 559–582.
6. DÉNARIÉ, J., DEBELLÉ, F., PROMÉ, J.C. 1996. *Rhizobium* lipo-chitooligosaccharide nodulation factors: Signaling molecules mediating recognition and morphogenesis. Ann. Rev. Biochem. 65: 503–535.
7. MERGAERT, P., VAN MONTAGU, M., HOLSTERS, M. 1997. Molecular mechanisms of Nod factor diversity. Mol. Microbiol. 25: 811–817.
8. SPAINK, H.P., OKKER, R.J.H., WIJFFELMAN, C.A., TAK, T., GOOSEN-DEROO, L., PEES, E., VAN BRUSSEL, A.A.N., LUGTENBERG, B.J.J. 1989. Symbiotic properties of rhizobia containing a flavonoid- independent hybrid *nodD* product. J. Bacteriol. 171: 4045–4053.
9. Harborne, J.B. 1971. Distribution of flavonoids in the leguminosae. In: Chemotaxonomy of the Leguminosae, (J.B. Harborne, D. Boulter, B.L. Turner, eds.), Academic Press, New York, pp. 31–71.
10. ZAAT, S.A.J., WIJFFELMAN, C.A., MULDERS, I.H.M., VAN BRUSSEL, A.A.N., LUGTENBERG, B.J.J. 1988. Root exudates of various host plants of *Rhizobium leguminosarum* contain different sets of inducers of *Rhizobium* nodulation genes. Plant Physiol. 86: 1298–1303.

11. ZAAT, S.A.J., SCHRIPSEMA, J., WIJFFELMAN, C.A., VAN BRUSSEL, A.A N., LUGTENBERG, B.J.J. 1989. Analysis of the major inducers of the *Rhizobium nodA* promoter from *Vicia sativa* root exudate and their activity with different *nodD* genes. Plant Mol. Biol. 13: 175–188.
12. HOWLES, P.A., ARIOLI, T., WEINMAN, J. 1994. Characterization of phenylalanine ammonia-lyase multigene family in *Trifolium subterraneum*. Gene 138: 87–92.
13. KOES, E.R., QUATTROCCHIO, F., MOL, J.N.M. 1994. The flavonoid biosynthetic pathway in plants: Function and evolution. Bioessays 16: 123–132.
14. PHILLIPS, D.A., KAPULNIK, Y. 1995. Plant isoflavonoids, pathogens and symbionts. Trends Microbiol. 3: 58–64.
15. ARIOLI, T., HOWLES, P.A., WEINMAN, J.J., ROLFE, B.G. 1994. In *Trifolium subterraneum*, chalcone synthase is encoded by a multigene family. Gene 138: 79–86.
16. LAWSON, C.G.R., DJORDJEVIC, M.A., WEINMAN, J.J., ROLFE, B.G. 1994. *Rhizobium* inoculation and physical wounding result in the rapid induction of the same chalcone synthase copy in *Trifolium subterraneum*. Mol. Plant-Microbe Int. 7: 498–507.
17. RECOURT, K., VAN TUNEN, A.J., MUR, L.A., VAN BRUSSEL, A.A.N., LUGTENBERG, B.J.J., KIJNE, J.W. 1992. Activation of flavonoid biosynthesis in roots of *Vicia sativa* subsp *nigra* plants by inoculation with *Rhizobium leguminosarum* biovar *viciae*. Plant Mol. Biol. 19: 411–420.
18. YANG, W. -C., CANTER CREMERS, H.C.J., HOGENDIJK, P., KATINAKIS, P., WIJFFELMAN, C.A., FRANSSEN, H., VAN KAMMEN, A., BISSELING, T. 1992. *In-situ* localization of chalcone synthase mRNA in pea root nodule development. The Plant Journal 2: 143–151.
19. DJORDJEVIC, M.A., MATHESIUS, U., ARIOLI, T., WEINMAN, J., GÄRTNER, E. 1997. Chalcone synthase gene expression in transgenic subterranean clover correlates with localized accumulation of flavonoids. Aust. J. Plant Physiol. 24: 119–132.
20. McKAHN, H., PAIVA, N.L., DIXON, R.A., HIRSCH, A.M. 1997. Chalcone synthase transcripts are detected in alfalfa root hairs following inoculation with wild-type *Rhizobium meliloti*. Mol. Plant-Microbe Int. 10: 50–58.
21. ESTABROOK, E.M., SENGUPTA-GOPALAN, C. 1991. Differential expression of phenylalanine ammonia-lyase and chalcone synthase during soybean nodule development. Plant Cell 3: 299–308.
22. KRAUSE, A., VO, T.T.L., BROUGHTON, W.J. 1997. Induction of chalcone synthase expression by rhizobia and Nod factors in root hairs and roots. Mol. Plant-Microbe Int. 10: 388–393.
23. MATHESIUS, U., SCHLAMAN, W.R.M., MEIJER, D., LUGTENBERG, B.J.J., SPAINK, H.P., WEINMAN, J.J., RODDAM, L.F., SAUTTER, C., ROLFE, B.G., DJORDJEVIC, M.A. 1996. New tools for investigating nodule initiation and ontogeny: Spot inoculation and microtargetin of transgenic white clover roots shows auxin involvement and suggests a role for flavonoids. In: Biology of Plant-Microbe Interactions, (G. Stacey, B. Mullin, P.M. Gresshoff, eds.), Int. Society Mol. Plant-Microbe Interact., St. Paul, Minnesota, pp. 353–358.
24. VAN BRUSSEL, A.A.N., RECOURT, K., PEES, E., SPAINK, H.P., TAK, T. , WIJFFELMAN, C.A., KIJNE, J.W., LUGTENBERG, B.J.J. 1990. A biovar-specific signal of *Rhizobium leguminosarum* bv. *viciae* induces increased nodulation gene-inducing activity in root exudate of *Vicia sativa* subsp. *nigra*. J. Bacteriol. 172: 5394–5401.
25. SPAINK, H.P., SHEELEY, D.M., VAN BRUSSEL, A.A.N., GLUSHKA, J., YORK, W. S., TAK, T., GEIGER, O., KENNEDY, E.P., REINHOLD, V.N., LUGTENBERG, B.J. J. 1991. A novel highly unsaturated fatty acid moiety of lipo- oligosaccharide signals determines host specificity of *Rhizobium*. Nature (London) 354: 125–130.
26. RECOURT, K., SCHRIPSEMA, J., KIJNE, J.W., VAN BRUSSEL, A.A.N., LUGTENBERG, B.J.J. 1991. Inoculation of *Vicia sativa* subsp *nigra* roots with *R. leguminosarum* biovar *viciae* results in release of nod gene activating flavones and chalcones. Plant Mol. Biol. 16: 841–852.
27. RECOURT, K., VERKERKE, M., SCHRIPSEMA, J., VAN BRUSSEL, A.A.N., LUGTENBERG, B.J.J., KIJNE, J.W. 1992. Major flavonoids in uninoculated and inoculated roots of

Vicia sativa subsp. *nigra* are four conjugates of the nodulation gene- inhibitor kaempferol. Plant Mol. Biol. 18: 503–513.
28. BOLAÑOS-VÁSQUEZ, M.C., WERNER, D. 1997. Effects of *Rhizobium tropicii*, *R. etli*, and *R. leguminosarum* bv. *phaseoli* on *nod* gene-inducing flavonoids in root exudates of *Phaseolus vulgaris*. Mol. Plant-Microbe Int. 3: 339–346.
29. LAWSON, C.G.R., ROLFE, B.G., DJORDJEVIC, M.A. 1996. *Rhizobium* inoculation induces condition-dependent changes in the flavonoid composition of root exudates from *Trifolium subterraneum*. Aust. J. Plant Physiol. 23: 93–101.
30. SCHMIDT, P.E., BROUGHTON, W.J., WERNER, D. 1994. Nod factors of *Bradyrhizobium japonicum* and *Rhizobium* sp NGR234 induce flavonoid accumulation in soybean root exudate. Mol. Plant-Microbe Int. 7: 384–390.
31. YLSTRA, B., BUSSCHER, J., FRANKEN, J., HOLLMANN, P.C.H., MOL, J.N.M., VAN TUNEN, A.J. 1994. Flavonols and fertilization in *Petunia hybrida*: Localization and mode of action during pollen tube growth. Plant J. 6: 201–212.
32. SCHMIDT, P.E., PARNISKE, M., WERNER, D. 1991. Production of the phytoalexin glyceollin I by soybean roots in response to symbiotic and pathogenic infection. Botanica Acta 105: 18–25.
33. GROSSKOPF, E., HA, D.T.C., WINGENDER, R., RÖHRIG, H., SZECSI, J., KONDOROSI, E., SCHELL, J., KONDOROSI, A. 1993. Enhanced levels of chalcone synthase in alfalfa nodules induced by a Fix- mutant of Rhizobium meliloti. Mol. Plant-Microbe Int. 6: 173–181.
34. ZAAT, S.A.J., VAN BRUSSEL, A.A.N., TAK, T., LUGTENBERG, B.J.J., KIJNE, J.W. 1989. The ethylene-inhibitor aminoethoxyvinylglycine restores normal nodulation by *Rhizobium leguminosarum* biovar. *viciae* on *Vicia sativa* subsp. *nigra* by suppressing the 'Thick and short roots' phenotype. Planta 177: 141–150.
35. VANWORKUM, W.A.T., VANBRUSSEL, A.A.N., TAK, T., WIJFFELMAN, C.A., KIJNE, J.W. 1995. Ethylene prevents nodulation of *Vicia sativa* ssp *nigra* by exopolysaccharide-deficient mutants of *Rhizobium leguminosarum* bv *viciae*. Mol. Plant-Microbe Int. 8: 278–285.
36. SPAINK, H.P. 1997. Ethylene as a regulator of *Rhizobium* infection. Trends Plant Sci. 2: 203–204.
37. MATHESIUS, U., DJORDJEVIC, M.A., WEINMAN, J., ROLFE, B.G., SCHLAMAN, H.R.M., SPAINK, H.P., BAYLISS, C., McCULLY, M.E. 1998. Flavonoid accumulation is a specific marker during root morphogenesis and nodulation in white clover. submitted
38. JACOBS, M., RUBERY, P.H. 1988. Naturally occurring auxin transport regulators. Science 241: 346–349.
39. MATHESIUS, U., SCHLAMAN, H.R.M., SPAINK, H.P., SAUTTER, C., ROLFE, B. G., DJORDJEVIC, M.A. 1997. Auxin transport inhibition precedes root nodule formation in white clover roots and is regulated by flavonoids and derivatives of chitin oligosaccharides. Plant J. in press.
40. LARKIN, P.J., GIBSON, J.M., MATHESIUS, U., WEINMAN, J., GARTNER, E., HALL, E., TANNER, G.J., ROLFE, B.G., DJORDJEVIC, M.A. 1996. Transgenic white clover. Studies with the auxin-responsive promoter, GH3, in root gravitropism and lateral root development. Transgen. Res. 5: 325–335.
41. GUILFOYLE, T.J., HAGEN, G., LI, Y., ULMASOV, T., LIU, Z., GEE, M. 1993. Auxin-regulated transcription. Aust. J. Plant Physiol. 20: 489–502.
42. HAGEN, G., MARTIN, G., LI, Y., GUILFOYLE, T.J. 1991. Auxin-induced expression of the soybean GH3 promoter in transgenic tobacco plants. Plant Mol. Biol. 17: 567–579.
43. SAUTTER, C., WALDNER, H., NEUHAUS-URL, G., GALLI, A., NEUHAUS, G., POTRYKUS, I. 1991. Micro-targeting: High efficiency gene transfer using a novel approach for the acceleration of micro-projectiles. Biotechnology 9: 1080–1085.
44. SCHLAMAN, W.R.M., GISEL, A.A., QUAEDVLIEG, N.E.M., BLOEMBERG, G.V., LUGTENBERG, B.J.J., KIJNE, J.W., POTRYKUS, I., SPAINK, H.P., SAUTTER, C. 1997.

Chitin oligosaccharides can induce cortical cell division in roots of *Vicia sativa* when delivered by ballistic microtargeting. Development 124: 4887–4895.
45. HIRSCH, A.M., BHUVANESWARI, T.V., TORREY, J.G., BISSELING, T. 1989. Early nodulin genes are induced in alfalfa root outgrowths elicited by auxin transport inhibitors. Proc. Natl. Acad. Sci. USA 86: 1244–1248.
46. BURBULIS, I.E., IACOBUCCI, M., SHIRLEY, B.W. 1996. A null mutation in the first enzyme of flavonoid biosynthesis does not affect male fertility in Arabidopsis. Plant Cell 8: 1013–1025.
47. PHILLIPS, D.A., JOSEPH, C.M., MAXWELL, C.A. 1992. Trigonelline and stachydrine released from alfalfa seeds activate NodD2 protein in *Rhizobium meliloti*. Plant Physiol. 99: 1526–1531.
48. DJORDJEVIC, M.A., REDMOND, J.W., BATLEY, M., ROLFE, B.G. 1987. Clovers secrete specific phenolic compounds which either stimulate or repress *nod* gene expression in *Rhizobium trifolii*. EMBO J. 6: 1173–1179.
49. ROLFE, B.G. 1988. Flavones and isoflavones as inducing substances of legume nodulation. BioFactors 1: 3–10.
50. SMIT, G., DE KOSTER, C.C., SCHRIPSEMA, J., SPAINK, H.P., VAN BRUSSEL, A.A.N., KIJNE, J.W. 1995. Uridine, a cell division factor in pea roots. Plant Mol. Biol. 29: 869–873.
51. HEIDSTRA, R., YANG, W.-C., YALCIN, Y., PECK, S., EMONS, A.M., VAN KAMMEN, A., BISSELING, T. 1997. Ethylene provides positional information on cortical cell division but is not involved in Nod factor induced in tip growth. Development 124: 1781–1787.
52. AGUILAR, J.M.M., ASHBY, A.M., RICHARDS, A.J.M., LOAKE, G.J., WATSON, M.D., SHAW, C.H. 1988. Chemotaxis of *Rhizobium leguminosarum* biovar *phaseoli* towards flavonoid inducers of the symbiotic nodulation genes. J. Gen. Microbiol. 134: 2741–2746.
53. CAETANO-ANNOLLES, G., CHRIST-ESTES, K., BAUER, W.D. 1988. Chemotaxis of *Rhizobium meliloti* to the plant flavone luteolin requires functional nodulation genes. J. Bacteriol. 170: 3164–3169.
54. DHARMATILAKE, A.J., BAUER, W.D. 1992. Chemotaxis of *Rhizobium meliloti* towards nodulation gene- inducing compounds from alfalfa roots. Appl. Environ. Microbiol. 58: 1153–1158.
55. SHARMA, S.B., SIGNER, E.R. 1990. Temporal and spatial regulation of the symbiotic genes of *Rhizobium meliloti* in planta revealed by transposon Tn 5-*gusA*. Genes and Development 4: 344–356.
56. SCHLAMAN, H.R.M., HORVATH, B., VIJGENBOOM, E., OKKER, R.J.H., LUGTENBERG, B.J.J. 1991. Suppression of nodulation gene expression in bacteroids of *Rhizobium leguminosarum* biovar *viciae*. J. Bacteriol. 173: 4277–4287.
57. SCHLAMAN, H.R.M., PHILLIPS, D.A., KONDOROSI, E. 1998. Genetic organization and transcriptional regulation of rhizobial nodulation genes. In: The *Rhizobiaceae*, Molecular Biology of Model Plant-Associated Bacteria, (H.P. Spaink, A. Kondorosi, P.J.J. Hooykaas, eds.) Kluwer Academic Pub., Dordrecht, in press.

Chapter Ten

FATTY ACID-DERIVED SIGNALING MOLECULES IN THE INTERACTION OF PLANTS WITH THEIR ENVIRONMENT*

Elmar W. Weiler, Dietmar Laudert, Florian Schaller, Boguslava Stelmach, and Peter Hennig

Lehrstuhl für Pflanzenphysiologie
Ruhr-Universität
D-44780 Bochum, Germany

Occurrence and Structural Diversity of Octadecanoids 180
Biosynthesis and Metabolism of Octadecanoids 181
    Metabolites 181
    Enzymes 182
        Lipoxygenase 182
        Allene Oxide Synthase 183
        Allene Oxide Cyclase 184
        OPDA-10,11-Reductase 185
        β-Oxidation 185
    Regulation and Compartmentation 185
    Alternative Pathways 188
    Active Compounds 189
Molecular Physiology of Octadecanoid Action 191
    Plant-Herbivore Interactions 191
    Pathogen Defense 193
    Mechanotransduction 194
    Senescence 195
    Evolutionary Aspects 196
Conclusion 197

---

* This review is dedicated to Prof. M.H. Zenk on the occasion of his 65th birthday.

Phytochemical Signals and Plant–Microbe Interactions, edited by Romeo *et al.*
Plenum Press, New York, 1998.

OCCURRENCE AND STRUCTURAL DIVERSITY OF OCTADECANOIDS

It has become clear during the last few years that jasmonic acid (JA) (**1**) (Fig. 1), the first cyclic fatty acid metabolite with regulatory properties that became known from higher plants, is only one member of a large and growing family of compounds which regulate a broad spectrum of physiological responses, including plant defense reactions, against biotic and abiotic factors.[1] JA is probably of general occurrence among higher plants,[2] and has also been found in some fungi.[3,4] Its methyl ester is a key component of jasmine scent.[5] Many related structures have meanwhile been identified (see ref. 1 for review), but, in most cases, their physiological activity remains unknown. Dihydro-JA (**2**) may arise by reduction of JA or (*vide infra*) from linoleic acid and shows activity in some, but not all, JA bioassays.[6] Tuberonic acid (12-hydroxy-JA) (**3**) has been identified as a tuberization factor of potato,[7] while cucurbic acid (**4**) is regarded as an inactivation product of (**1**). Both compounds **3** and **4** may occur as glucosyl esters,[8,9] and this may help to sequester or translocate the compounds. Amino acid conjugates of JA, such as the L-isoleucine derivative (**5**), have been detected in many plant tissues,[10,11] and some exhibit significant biological activity which may reflect release of JA in the tissue and/or activity *per se* (*vide infra*). Likewise, biosynthetic precursors of JA, such as 12-oxo-10,15(Z)-phytodienoic acid (OPDA) (**9**), are biologically active. Again, this may reflect metabolic conversion to JA or an activity *per se*. The latest addition to the family of lipid-derived signaling compounds is volicitin (N-17-hydroxylinoleyl)-L-glutamine (**8**), an

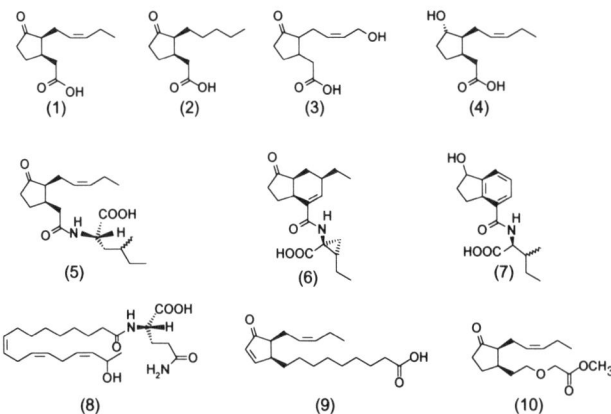

Figure 1. Structures of octadecanoids and related compounds. (**1**) jasmonic acid, (**2**) dihydrojasmonic acid, (**3**) 12-hydroxyjasmonic acid, (**4**) cucurbic acid, (**5**) jasmonoyl-L-isoleucine, (**6**) coronatine, (**7**) 1-hydroxyindanoyl-L-isoleucine; (**8**) volicitin, (**9**) 12-oxo-phytodienoic acid, (**10**) methyltrihomojasmonate.

inducer of plant alarm volatile emission from the oral secretions of beet armyworm caterpillars.[12]

It is almost certain that our present knowledge is far from complete and that further research will unveil yet other lipid-derived metabolites which add to a highly diverse family of regulatory molecules. We are only beginning to uncover the full biological significance of this class of compounds, for which the names jasmonates or octadecanoids are used, and to understand better the ways they act and interact. We are still far from elucidating the molecular basis of octadecanoid perception and signal transduction, while jasmonate response elements have meanwhile been identified in the promoters of a few genes.[13,14]

BIOSYNTHESIS AND METABOLISM OF OCTADECANOIDS

Metabolites

By 1984, through the work of Vick and Zimmerman,[15] it had been established that jasmonic acid is synthesized from α-linolenic acid by the pathway shown in Figure 2. The Vick-Zimmerman pathway starts with free α-linolenic acid (LA), presumably released from membrane lipids through the action of lipase(s). This, however, has not yet been shown unequivocally. LA is then converted by a

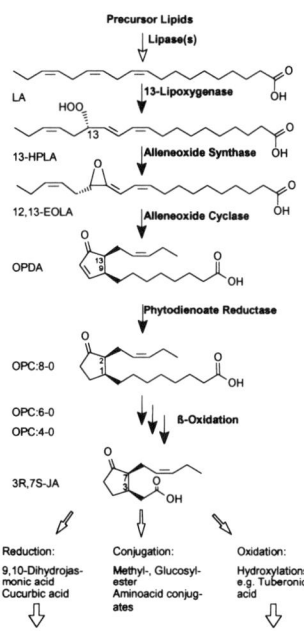

Figure 2. Biosynthesis of jasmonic acid (3R,7S-JA) from α-linolenic acid (LA) and metabolism of jasmonic acid. 13-HPLA, 13(S)-hydroperoxylinolenic acid; 12,13-EOLA, 12,13-epoxylinolenic acid; OPDA, 12-oxo-phytodienoic acid; OPC-8:0 (6:0, 4:0) 3-oxo-2(2′(Z)-pentenyl)-cyclopentane-1-octanoic (hexanoic, butyric) acid.

13-lipoxygenase and molecular oxygen to the 13(S)-hydroperoxide, 13(S)-hydroperoxylinolenic acid. This reaction cannot be regarded as specific for the jasmonate biosynthetic pathway, since 13(S)-hydroperoxylinolenic acid is a substrate for hydroperoxide lyases, glutathione peroxidases, and other enzymes.[16,17] The first specific reaction in the biosynthesis of jasmonate is the conversion of 13(S)-hydroperoxylinolenic acid to OPDA, via an unstable allene oxide, 12,13-epoxylinolenic acid, by the concerted action of allene oxide synthase/allene oxide cyclase. Reduction of the 10,11-double bond by a NADPH-dependent reductase then yields 3-oxo-2(2'(Z)-pentenyl)-cyclopentane-1-octanoic acid (OPC-8:0) which is metabolized further to JA. Intermediates in the conversion of OPC-8:0 to JA with shorter side chains, OPC-6:0 and OPC-4:0, were detected in radiotracer experiments.[18] From these findings, it has been deduced that JA is produced from OPC-8:0 by three cycles of β-oxidation. JA is initially biosynthesized as the (3R,7S)-enantiomer.[15] Enolization, however, quickly leads to formation of the energetically more favorable *trans*-isomer, (3R, 7R)-JA, which, in equilibrium, is the predominant form (90%).[19] JA extracted from tissue usually is in this equilibrium, unless in situations of induced *de novo* biosynthesis.

While, thus, some JA seems to be sequestered in pools of relatively slow turnover, some is metabolized further (Fig. 2). Metabolism involves reduction of the pentenyl side chain to dihydro-JA (**2**), of the carboxyl group to yield cucurbic acid (**4**), and hydroxylations of the pentenyl side chain to 11- and/or 12-hydroxylated JA-metabolites, which may then be conjugated to glucose and are as such found predominantly in tissues.[8,9] JA or dihydro-JA may also be conjugated to a range of amino acids, such as L-isoleucine,[10,11] and at least some of these amides are considered physiologically active.[20–22] An example is compound **5** (Fig. 1). Jasmine flowers[5] and other plants such as *Artemisia* species[23] release large quantities of methyl jasmonate into the environment as components of scent. Plants fed JA also have been shown to release some methyl jasmonate.[24,25] However, most plants normally produce very little, if any, of this volatile derivative. It is currently unknown whether methyl jasmonate has regulatory functions *in vivo*. Neither has the enzymology of its production been yet worked out. In contrast to OPC-8:0 which occurs only in traces, OPDA may accumulate to substantial amounts in a challenged tissue,[26] and the compound has been found in air-dried leaves of the composite, *Inulanthera calva*, in amounts of up to 3% of dry mass.[27,28] It is thus conceivable that OPDA may be metabolized to some further, as yet unknown, compounds other than OPC-8:0.

Enzymes

Lipoxygenase. In most plants, several isoenzymes with, among other things, different regioselectivity may occur, most commonly 9-lipoxygenases and 13-lipoxygenases.[29] Their abundance may vary between organs or tissues. In *Arabidopsis thaliana*, two isoforms, LOX1 and LOX2, have been cloned.[30,31]

Co-suppression resulting in specific elimination of the chloroplastidic isoform, LOX2, eliminated wound-induced accumulation of JA in leaves, but not wilt-induced accumulation of ABA,[31] suggesting functional (and/or spatial) separation of lipoxygenase isoforms. The enzymatic characteristics of LOX2 have not yet been determined. The co-suppressed plants lacking wound-induced accumulation of JA, however, exhibit normal JA levels in unstressed leaves.[31] Thus, an as yet unidentified pathway could be operative in leaves, or import from tissues other than leaves (*e.g.*, roots) could provide this basal level of JA.

Allene Oxide Synthase. This enzyme, originally named hydroperoxide isomerase,[32] later hydroperoxide dehydrase,[33] is now commonly addressed as allene oxide synthase, since it was shown by Hamberg that the enzyme produces an unstable allene oxide, 12,13(S)-epoxy-9(Z), 11-octadecadienoid acid (12,13-epoxylinoleic acid) from 13(S)-hydroperoxylinoleic acid which spontaneously hydrolyzes into the α-ketol, 12-keto-13-hydroxy-9(Z)-octadecenoic acid.[34] In an analogous manner, 13(S)-hydroperoxylinolenic acid is converted to the corresponding allene oxide 12,13(S)-epoxy-9(Z),11,15(Z)-octadecatrienoic acid (12,13-epoxylinolenic acid) (Fig. 2).[33] The allene oxides are extremely unstable in aqueous environment ($t_{1/2}$ in the order of 30 seconds), but they may be stabilized by polypeptides such as albumin.[35] They spontaneously hydrolyze to the corresponding α- and γ-ketols. In the case of 12,13-epoxylinolenic acid, considerable amounts of the cyclic metabolite, OPDA, are also formed spontaneously, while only traces of the analogous compound, 3-oxo-2-pentyl-cyclopent-4-ene-1-octanoic acid, are formed from 12,13-epoxylinoleic acid.[36]

Allene oxide synthase activity has been localized in the chloroplast in spinach leaves,[33] and the cDNAs encoding allene oxide synthase from flax[37] and *Arabidopsis thaliana*[36] encode transit peptides characteristic for chloroplasts. A single gene encoding allene oxide synthase, a polypeptide of 58.7 kDa, is present in *A. thaliana*,[36] and the enzyme is active as a monomer. The enzyme of *A. thaliana* has been expressed in functional form in *E. coli* and shown to yield, besides the ketols, racemic *cis*-OPDA,[36] while OPDA extracted from plant tissues is the *cis*-(+)-enantiomer.[38] OPDA accumulating as a result of wounding or mechanical stressing of a tissue also is exclusively the *cis*-(+)-enantiomer.[38] It is thus clear that OPDA produced *in vivo* does not stem from spontaneous reactions of an allene oxide released by allene oxide synthase, but originates in a stereospecifically controlled manner. Allene oxide synthases are cytochrome P 450 enzymes belonging to the CYP74 subfamily,[36,37,39] and are characterized by their lack of transmembrane segments and low affinities for CO.[40,41] They do not depend on NADPH as a reductant and do not bind oxygen, but rather an oxygenated substrate, *e.g.*, a fatty acid hydroperoxide. Consequently, allene oxide synthases carry characteristic amino acid substitutions in the heme-binding and oxygen-binding pockets (Fig. 3), which discriminate this subfamily of cytochrome P 450s from classical members. Interestingly, thromboxane synthase

		O$_2$-binding domain	heme-binding domain
P 450$_{cam}$	*Pseudomonas putida*	G G L D **T**	**F** G **H** G S H L C L **G**
CYP 71A1	*Persea americana*	G G T D **T**	**F** G A **G** R R G C P **G**
CYP 72	*Catharanthus roseus*	A G Q E **T**	**F** S W **G** P R V C L **G**
CYP 73A1	*Helianthus annus*	A A I E **T**	**F** G V **G** R R S C P **G**
CYP 75A2	*Solanum melongena*	A G T D **T**	**F** G A **G** R R I C A **G**
CYP 77A1	*Solanum melongena*	G G T D **T**	**F** G V **G** R R I C P **G**
CYP 78A1	*Zea mays*	R G T D **T**	**F** G A **G** R R V C P **G**
CYP 80	*Berberis stolonifera*	P G S D **T**	**F** G S **G** R R I C P **G**
CYP 83A1	*Arabidopsis thaliana*	A G T G **T**	**F** G S **G** R R M C P **G**
CYP 90	*Arabidopsis thaliana*	A G Y E **T**	**F** G G **G** P R L C P **G**
CYP 5A1	human	A G Y E I	**F** G A **G** P R S C L **G**
CYP 74 (AOS)	*Linum usitatissimum*	G G F K I	P S V A N K Q C A **G**
CYP 74 (AOS)	*Arabidopsis thaliana*	G G M K I	P T V G N K Q C A **G**
CYP 74 (AOS)	*Parthenium argentatum*	G G V K I	P T V E N K Q C A **G**

Figure 3. Sequence conservation among cytochrome P450 enzymes around the O$_2$-binding domain and the heme binding domain. Invariant, characteristic amino acids are shown in boldface. Allene oxide synthases are grouped in the CYP74 sub-family. CYP5A1, thromboxane synthase. GenBank accession numbers:[43] P450$_{cam}$ (M12546), CYP71A1 (M32885), CYP72 (L10081), CYP73A1 (Z17369), CYP75 (X70824), CYP77 (X71656), CYP78 (L23209), CYP80 (U09610), CYP83 (U18929), CYP90 (X97367), CYP5A1 (M80646), CYP74 *L. usitatissimum* (U00428), CYP74 *A. thaliana* (T20864), CYP74 *P. argentatum* (X78166).

(Fig. 3, CYP 5A1), which has a prostaglandin endoperoxide as substrate and also does not require molecular oxygen,[42] shares the invariant amino acids of the oxygen binding pocket with the allene oxide synthases and deviates from the O$_2$-requiring cytochrome P 450s (see ref. 43 for review).

Allene Oxide Cyclase. The enzyme responsible for cyclization of 12,13-epoxylinolenic acid is allene oxide cyclase,[44,45] recently purified from maize kernels.[46] Addition of potato allene oxide cyclase preparations to recombinant *A. thaliana* allene oxide synthase expressed in *E. coli* results in the production, *in vitro*, of highly asymmetrical *cis*-OPDA consisting, predominantly or exclusively, of the (+)-enantiomer.[38] This is also the isomer isolated from plant tissues.[26,38] Thus, the enzymes from heterologous species are able to interact highly efficiently, and stereocontrol of OPDA production is exerted by the cyclase. The maize enzyme is a dimer of 47 kDa apparent molecular mass which does accept 12,13(S)-epoxy-9(Z),11,15(Z)-octadecatrienoic acid, but not 12,13(S)-epoxy-9(Z),11-octadecadienoic acid, as a substrate. In contrast to this, allene oxide synthase produces both allene oxides from their respective substrates, 13(S)-hydroperoxylinolenic acid or 13(S)-hydroperoxylinoleic acid. Given the short half-life of the allene oxides and the optical purity of natural OPDA, it is to be expected that the actions of allene oxide synthase and allene oxide cyclase are closely coordinated. The two enzymes might, in fact, form an

OPDA-synthase complex in the cell, thus avoiding escape of the short-lived allene oxide from the pathway. In accordance with this is the absence of the ketols in tissues *in vivo* (B. Vick, personal communication). They thus represent side reactions originating *in vitro* from partially uncoupled synthase-cyclase complexes as a result of processing artifacts.

OPDA-10,11-Reductase. The enzyme converts OPDA to OPC-8:0 by reduction of the C=C double bond of the conjugated enone moiety and has a *cis:trans* preference of 6:1.[47] It is active as a monomer of 41 kDa apparent molecular mass and depends on NADPH as a reductant. Molecular cloning of the reductase from *A. thaliana*[47a] has shown it to be closely related to Warburg's Old Yellow Enzyme, a flavoprotein enone reductase, and morphinone reductase of *Pseudomonas putida*, an enzyme which reduces morphinone to hydromorphone or codeinone to hydrocodone[48] but which prefers NADH as reductant. Both the enzyme purified from *Corydalis sempervirens* cell cultures[47] and recombinant *A. thaliana* reductase are highly selective but not absolutely specific for their substrates in that they reduce the C=C double bond of 2-cyclohexenone (as does Old Yellow Enzyme). However, neither reacts with cyclic enones in which the hydrogen atom at the β-carbon has been replaced by a methyl or methylene, *e.g.* as in the plant hormone abscisic acid or steroids such as progesterone and testosterone.[47]

β-Oxidation. The terminal reaction in biosynthesis of JA from OPC-8:0 is clipping of the octanoic side chain to an acetyl side chain. Since labeled OPC-6:0 and OPC-4:0 have been isolated in small quantities from tissue fed labeled OPC-8:0, it has been proposed that the mechanism by which the side chain is removed involves three successive eliminations of C2-moieties, and thus most likely occurs *via* β-oxidation.[18] While this remains a plausible and even likely possibility, it has not been proven unequivocally and in a cell-free system. Alternatives should still be considered. For example, β-oxidation of unsaturated fatty acids, such as oleic acid, terminates after only two rounds, leaving a $\Delta^{5,6}$-enoyl-CoA which cannot be β-oxidized further unless the double bound is reduced.[49] It thus remains to be proven whether or not the plant's β-oxidation system would be capable of β-oxidizing an octanoic acid side chain attached to a cyclopentanone ring all the way to the acetyl derivative.

Regulation and Compartmentation

The enzymes of octadecanoid biosynthesis occur in low amounts in control tissues. Both OPDA and JA can be extracted, sometimes in substantial amounts,[26] from unchallenged tissues. Thus, a basal supply of these metabolites is provided by the cell in the absence of inducers of their accumulation. It is, however, unknown if JA or OPDA-found in unchallenged tissue reflects a more or less continuous synthesis at a basal rate, or material stored from active synthesis during an earlier stage of development, or represents leftovers from previous challenges. Principally, it should be possible to discriminate "new" from "old" JA or OPDA by analysis of their isomeric state (biosynthesis affords the *cis*-isomer which isomerizes *via* enolization to the energetically more favorable *trans*-isomer). In thermodynamic equilibrium, the *trans:cis* ratio is approximately 90:10. Unfortunately, acid or base catalyzed isomerization occurs during work-up of tissues and extracts, as well as thermally,[50] e.g., in the injection port of gas chromatographs, making it difficult to come up with estimates of *trans:cis* ratios in tissues. It is our experience that isomerization of *cis*-OPDA during processing and GC-MS analysis[26] amounts to 5–10% of the initial value, while the fraction may be higher for JA (B. Stelmach, personal communication). Against this background, OPDA from control tissue is largely present as *cis*-isomer, while JA is extracted as a *cis:trans* equilibrium mixture. This finding would be compatible with the notion that JA in control tissue is "ancient" or that the JA pool in control tissue turns over slowly. For OPDA, the situation could be different. Alternatively, rates of isomerization may be lower for *cis*-OPDA than for *cis*-JA.

Tissue wounding, elicitor challenge, or mechanical treatment lead to strong, and sometimes drastic, increases in the levels of octadecanoids.[6,26,51–53] This could reflect (a) *de novo* biosynthesis and/or (b) release from storage forms. While there is no proof for the latter process, an enzyme specifically hydrolyzing the amide bond of JA-amino acid conjugates has recently been found.[54] The enzyme, however, is specific for (−)-JA amides, while (+)-JA amides are much weaker substrates. Thus, this amidohydrolase is unlikely to be involved in signal induced release of JA since it is predominantly the *cis*-(+)-isomer, (3R,7S)-JA, that is accumulating under conditions of induction.[55]

Stimulus-induced accumulation of JA and/or OPDA is mostly, if not exclusively, due to *de novo* synthesis. This can be achieved through (a) regulated provision of the substrate, LA, and/or (b) regulation at the level of the biosynthetic enzymes or their genes. Evidence is now being obtained that both processes are of biological relevance and occur in the cell.

The release of α-linolenic acid from membrane lipids has been postulated as a trigger step of octadecanoid accumulation.[56] This is in accord with the fact that the level of free fatty acids is usually low in the cell and agrees with the finding that mechanostimulation of tendrils, which induces OPDA accumulation, results in substantial decrease in the amounts of monogalactosyldiglycerides,[6] a major membrane lipid class containing α-linolenic acid in plastids (*vide infra*). Upon infection of *Petroselinum crispum* with the fungus *Phytophthora sojae*, rapid, transient, and highly localized accumulation of a plastidial ω-3 fatty acid desaturase mRNA around the infection sites has been detected.[57] This enzyme was proposed as responsible for replenishing or raising the levels of LA in elicited cells in order to feed octadecanoid biosynthesis.[57] In fact, small and transient increases in the levels of free LA have been detected in tomato leaf tissue during herbivore attack[58] and in cell cultures of *Eschscholtzia californica* treated with yeast cell wall elicitor.[53] Phospholipases of the A_2-type should be involved in release of LA from membrane lipids, but theses enzymes are not well characterized from plants. It has, however, been shown recently that A-type phospholipase is activated by plant defense elicitors.[59-61] The supply of substrate for octadecanoid production and the processes regulating it constitute a major area of our ignorance, and these areas clearly need more attention.

At variance with these results are findings by Harms et al.,[62] who have overexpressed flax allene oxide synthase in potato plants. The transgenic plants had substantially higher levels of JA compared to controls but, unexpectedly, they showed no induction of several JA-inducible genes. Overexpression of the enzyme from *A. thaliana* in tobacco, however, clearly had no effect on JA levels under control conditions (Laudert and Weiler, unpublished). Evidently, in tobacco, output of the octadecanoid pathway is limited by the availability of substrate. The situation in potato needs to be clarified with respect to why jasmonte-response genes did not respond to the apparently elevated levels of JA in the tissue.

The recent demonstration that levels of allene oxide synthase mRNA increase within minutes following wounding of leaves of *A. thaliana* constitutes the first piece of evidence for a regulation of octadecanoid biosynthesis at the level of transcription and/or mRNA-stability.[36] Allene oxide synthase is the rate-limiting enzyme of octadecanoid biosynthesis (Laudert and Weiler, unpublished) and an ideal target for control of pathway capacity. Recent work in our laboratory has strengthened the notion that this enzyme is a central control point in biosynthesis of octadecanoid signaling compounds. A second key checkpoint may be OPDA-reductase, the enzyme which controls substrate flow between the cyclic C18- and C12-metabolites in the pathway (Fig. 2). OPDA levels relative to JA levels may vary considerably, and, in mechanostimulated tendrils, OPDA levels rise without concomitant increases in levels of JA.[26] In contrast to this, elicitation of *Rauvolfia serpentina* cell cultures results in JA accumulation while the OPDA level remains low.[53] OPDA reductase has recently been cloned,[47a] facilitating molecular studies on the regulation of this enzyme as well as studies on the relative importance of C18- vs. C12-metabolites of the octadecanoid pathway.

The compartmentation of the pathway of JA biosynthesis has recently become much clearer. In *A. thaliana*, the lipoxygenase required for the production of wound-inducible JA, LOX2, has been localized in the chloroplast.[31] Allene oxide synthase-encoding full length cDNAs from flax[37] and *A. thaliana*[36] bear characteristic, non-composite chloroplast transit peptides, and the enzyme activity has been localized in chloroplasts,[33] specifically the chloroplast envelope.[16] The enzyme lacks transmembrane segments but is extracted in membrane associated form, and its release requires detergents. In contrast to this, allene oxide cyclase is prepared as a soluble enzyme.[44,46] For reasons discussed above, it must be assumed to form a complex with the synthase and should consequently be localized in the chloroplast also, but this remains to be shown directly. OPDA produced in chloroplasts has to enter the cytosol where it is reduced to OPC-8:0 which, in turn, would be expected to enter the peroxisomes for β-oxidation. JA then would have to leave the peroxisome. As discussed below, both OPDA and JA have to be regarded as biologically active, and they may have different profiles of biological activities. In an inductive situation, two signals (OPDA and JA) may be released from two different compartments, not necessarily from the same cells. Their further fate is unknown. Both may act intracellularly, but they may also leave the cell to move to distant sites of action. JA has been shown to translocate from tobacco leaves, where it is produced upon herbivore attack, to the roots where it stimulates nicotine biosynthesis.[63]

Large amounts of enzymatically active allene oxide synthase and allene oxide cyclase can be obtained from non-green tissue, such as flax achenes[40] and maize caryopses.[46] JA biosynthesis occurs in many non-green plant cell cultures,[52,53,64] and co-suppression of the chloroplastidic LOX2 in *A. thaliana* did not change basal levels of JA in unwounded tissue.[31] Furthermore, the *def1* mutant of tomato, which can be normalized by OPDA application, is strongly impaired in the systemic accumulation of proteinase inhibitors but less impaired in their accumulation around the wound sites.[65] These results clearly show that JA can be synthesized in the absence of differentiated or intact chloroplasts. Thus, the first part of the biosynthetic sequence (LA to OPDA) seems to occur more generally in plastids and the third part (OPC-8:0 to JA) in microbodies (peroxisomes or glyoxysomes) (Fig. 4). However, fungi lack plastids and may synthesize JA.[4] It is therefore necessary to consider additional sites of octadecanoid biosynthesis in plant cells.[66] On the other hand, allene oxide synthase of *A. thaliana* is encoded by a single gene and as a precursor with a plastidic transit peptide,[36] making a separate, second pathway unlikely in this species. It nevertheless remains a possibility that enzymes of divergent primary structure exist which cannot be detected by the hybridization probes available currently.

Figure 4 summarizes our current, and still partly hypothetical, picture of the compartmentation of octadecanoid biosynthesis in higher plants. Open questions relate to the intracellular location in the pathway of enzymes that have not yet been cloned, metabolite transport between compartments, cells, and tissues, the signals initiating the release of LA, the nature of phase III (OPC-8:0 to JA) of the biosynthetic sequence, and most aspects of regulation of the pathway at the enzyme and gene level. It is expected that within the next few years, many of these pertinent questions will be answered.

Alternative Pathways

It is clear that OPDA and JA are synthesized from LA.[15] It is expected, however, that release of fatty acids from membrane lipids will not liberate exclusively LA, but also related, unsaturated fatty acids whose conversion may give rise to related cyclic metabolites. For example, allene oxide synthase will

FATTY ACID-DERIVED SIGNALING MOLECULES

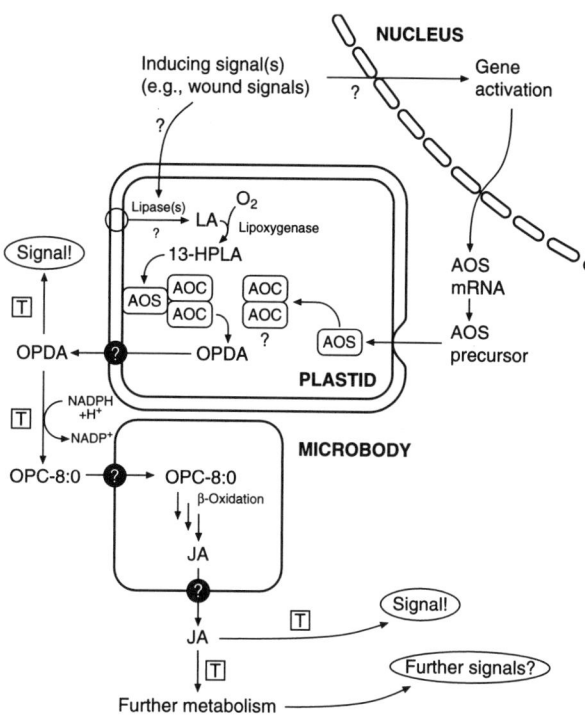

Figure 4. Compartmentation and regulation of the octadecanoid biosynthetic pathway. Question marks denote hypothetical steps. The boxed T's stand for possible transport processes between cells or tissues. AOS, allene oxide synthase; AOC, allene oxide cyclase. All other abbreviations as in the text.

accept the 13(S)-hydroperoxide of linoleic acid as well. From there, a formal pathway leading to dihydro-JA, a metabolite known to occur in plants, may be constructed.[64] However, spontaneous cyclization of the allene oxide, 12,13-epoxylinoleic acid, yields very little cyclization product,[36] and allene oxide cyclase from maize kernels has recently been shown to accept 12,13-epoxylinolenic acid but not 12,13-epoxylinoleic acid,[46] making it highly unlikely that linoleic acid is a substrate for octadecanoids of the dihydro-JA series *in vivo*.

Active Compounds

The question of active compounds other than JA in the octadecanoid pathway is now partially answered. The view that JA is the sole active octadecanoid has to be abandoned based on several findings:

1. It was first noted by Weiler *et al.*[6] that the JA precursors OPDA and OPC-8:0 would induce tendril coiling at considerably lower levels

than JA, and that the response to OPDA or OPC-8:0 occured much faster. This kinetic effect was difficult to understand on the basis of their action as JA precursors.

2. It was found that the *Pseudomonas syringae* phytotoxin, coronatine (**6**) (Fig. 1) mimicked the dose-response curve and also the kinetics of OPDA or OPC-8:0 action on tendrils.[67] Coronatine did not induce JA accumulation in the tissue and is active *per se*. The toxin proved to be a close structural analog of OPDA and OPC-8:0.[67] Additionally, amidohydrolysis of coronatine yields coronafacic acid, which can be regarded as an analog of JA.[68] However, coronafacic acid is inactive in inducing tendril coiling.[67]

3. Recent work on signaling during mechanotransduction provided evidence for increases in OPDA-, but not (or else very delayed) of JA-levels in *Bryonia dioica* as well as *Phaseolus vulgaris*. OPDA levels, but not JA levels, correlated with the induction of the mechanoresponse.[26]

4. In parallel studies, Blechert et al.[64] showed that methyltrihomojasmonate (**10**) (Fig. 1) would induce the accumulation of defense metabolites in plant tissue cultures of *Eschscholtzia californica*. Since this compound cannot be converted to JA by β-oxidation, it is clear that β-oxidation is not necessary for an octadecanoid to be an active inducer of natural product accumulation.[64]

5. In a detailed study, a wide range of structural analogs of JA were examined for their activity in several different bioassays including *B. dioica* tendril coiling, elicitation of *E. californica* cell cultures, transpiration of tomato and barley, and tomato senescence. The message from these data is clear: structure-activity profiles were different for each of the assays under study.[69]

6. Our initial work on coronatine was taken further by Boland's group,[20,21] who showed that compounds, like JA, would induce volatile emission from leaves of a wide range of species. In this assay (as in many other JA bioassays), some JA-amino acid conjugates proved active, notably the L-isoleucine derivative.[21] The authors concluded from the observations that compounds such as 1-hydroxyindanoyl-L-isoleucine (**7**) (Fig. 1) were also active inducers of volatile emission, and that compounds like **7** and also coronatine (**6**) were to be regarded as mimics of JA-amino acid conjugates in this system. Compound **7**, interestingly, was inactive in the tendril coiling assay (Weiler, unpublished), proving that in *B. dioica*, coronatine does not act as a mimic of JA-amino acid conjugates.

The conclusions from these recent findings are: (a) OPDA, JA, and JA-amino acid conjugates are *per se* active octadecanoids. (b) It is probable that, while their biological profiles may overlap, the different active octadecanoids serve different biological functions. This is quite clear already in tendril mechanotransduction, where exogenously applied OPDA is, by at least a factor of 50, more active than JA, endogenous OPDA accumulates to higher levels than JA during the response, and, consequently, OPDA has a several-hundred-fold biological potency relative to the endogenous JA. The late accumulation of JA in tendrils correlates with the decline in OPDA levels when the organ has almost fully reacted. JA, thus, is to be regarded as an inactivation product of the endogenous signal transducer OPDA in this tissue.

These findings open the possibility of better understanding seemingly variant results in the literature. For example, it has been found that JA-inducible genes in rice are activated by pathogen attack without a concomitant increase in JA levels.[70] In the barley-*Erysiphe graminis* f.sp. *hordei* interaction, JA levels likewise did not rise and exogenous JA did not induce resistance to the pathogen.[71] OPDA should be analyzed in such cases as it is more than likely that in certain host-pathogen interactions, OPDA will be the signal and not JA. This may be a property of the host plant. We have shown, for example, that endogenous levels of OPDA may vary widely from species to species.[26] Barley belongs to the "high OPDA" class of species. Clearly, results based solely on analysis of JA can no longer be regarded as complete.

The picture that is now emerging is that of a structurally as well as functionally highly diverse system of lipid-derived signaling molecules. The individual roles of different octadecanoids can only in some cases be outlined today. Precise information about the octadecanoid receptors and transduction pathways will be required in order to better understand these roles. Unfortunately, this is an area of little knowledge, but the use of mutants and genetics[72] will undoubtedly shed more light on these processes in the future. The considerable overlap in the physiological responses to different octadecanoids is exemplified by the fact that coronatine-insensitive mutants of *A. thaliana* are also JA-insensitive,[72] and by the fact that JA, at sufficiently high levels, will give activity also in OPDA-responsive systems such as the *B. dioica* tendril.[67] This is most easily understood if one views the different physiological responses as governed by octadecanoid profiles. These may reflect both species specificity and process specificity in their composition and dynamics rather than any individual signal transducer. This would not only allow fine tuning of responses much more easily, but also improve networking with parallel signal processing pathways.

MOLECULAR PHYSIOLOGY OF OCTADECANOID ACTION

Plant–Herbivore Interactions

The pioneering observation of Farmer and Ryan[73] that airborne JAMe would induce, like wounding or herbivore attack, the accumulation of proteinase inhibitors in leaves of tomato plants has boosted work on the role of octadecanoids in herbivore defense. Much of this has been presented in several reviews.[1,74,75] Only recent findings will be discussed here. Definite proof for a role of octadecanoids in herbivore defense has been obtained for tomato and *Arabidopsis thaliana*. The tomato mutant, *defenseless1* (*def1*), accumulates reduced levels of proteinase inhibitors as well as of JA and is much more susceptible to herbivore foraging than the wild type.[65] The mutant can be normalized, i.e., its JA level, proteinase inhibitor accumulation, and herbivore defense restored by the addition of OPDA. Thus, the central role of octadecanoids in tomato herbivore defense is obvious. Likewise, an *A. thaliana* triple mutant (*fad3-2 fad7-2 fad8*) incapable of synthesizing α-linolenic acid showed extremely high mortality against the dipteran saprophagous fungal gnat, *Bradysia impatiens*, and negligible levels of JA.[76] Application of exogenous JAMe to this mutant resulted in substantial protection and reduced mortality from ca. 80% to ca. 12%. The mutant showed lack of inducibility by wounding of several, but not all, wound-inducible genes, suggesting that JA-independent wound responses occur besides JA-dependent ones. The latter appear to be the critical reactions for herbivore defense in *A. thaliana*.

Our current view of the role of octadecanoids in herbivore defense is summarized in Figure 5. Local wounding by the herbivore results in some initial production of octadecanoids due to the fact that in the wounded tissue, substrates and enzymes likely are mixed upon breakdown of the cellular architecture. The latter process results in the release of oligogalacturonides which exert local effects, among them the induction of octadecanoid accumulation.[77] In tomato, the octadecapeptide, systemin, is released proteolytically from its precursor, prosystemin.[78] Systemin is an inducer of octadecanoid biosynthesis[79] and may act both locally as well as systemically because the peptide can translocate via vascular bundles.[80] Octadecanoids themselves may be part of the systemic signal complex, and it has been shown that JA is translocated via the vascular system in tobacco.[63] The octadecanoid signal is propagated through the plant by systemin induction along its transport path (shown for tomato) and further by autocatalytic induction of allene oxide synthase, the rate-limiting enzyme of octadecanoid biosynthesis, by octadecanoids themselves (Laudert and Weiler, unpublished). This efficiently alerts the plant and initiates both local and systemic defenses. A role of hydraulic and/or electric signals in systemic defense, as proposed by several groups,[81,82] seems questionable.[83]

Chemical signaling in tritrophic interactions involves the production and emission of alarm volatiles from leaves of plants attacked by certain herbivores which attracts natural enemies of the herbivores.[84] Alarm volatiles are specific mixtures of compounds that differ from those emitted upon mechanical wounding, suggesting an elicitation process triggered by the herbivore itself. Recently, the first such elicitor was identified and shown to be N-(17-hydroxylinoleyl)-L-glutamine (**8**), named volicitin. Volicitin is found in the oral secretion of beet army

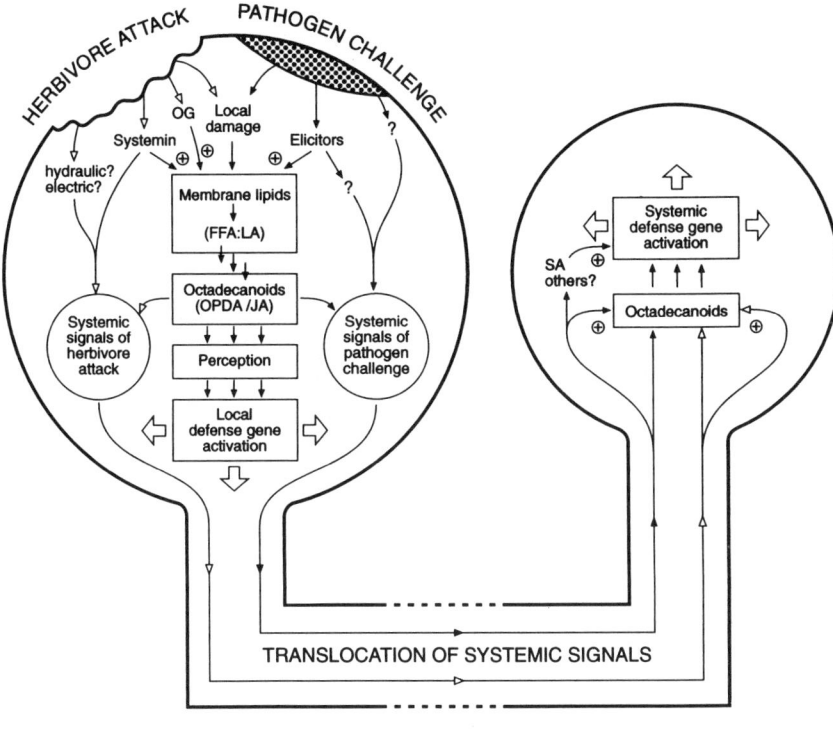

Figure 5. A synopsis of local and systemic reactions involved in herbivore and pathogen defenses. OG, oligogalacturonides; SA, salicylic acid, FFA, free fatty acids. All other abbreviations as in the text.

worms (*Spodoptera exigua* Hübner), and both natural as well as synthetic volicitin trigger the same blend of volatiles emitted from corn seedlings, as if they were attacked by the beet army worm larvae themselves.[12] It had earlier been shown that JA, JA-amino acid conjugates, coronatine, and synthetic analoges of coronatine such as compound **7** (1-hydroxyindanoyl-L-isoleucine) would induce volatile emissions from a number of diverse species.[21] Based on structure-activity data, it had been proposed earlier that the volatile emission was triggered by certain JA-amino acid conjugates rather than by JA itself.[21] The discovery of volicitin lends further support to the notion that octadecanoid-conjugates with amino acids may be active in triggering certain and presumably specific defense reactions.

Pathogen Defense

The situation is somewhat less clear with respect to the role of octadecanoids in pathogen defense. Cell cultures of different species accumulate toxic, low molecular mass metabolites upon challenge with fungi or fungal elicitors such as a yeast cell wall preparation.[52,53,64] Under these conditions, JA levels rise (sometimes drastically) in the cultures, and exogenous application of JAMe, its biosynthetic precursors, as well as structurally related compounds[64] or coronatine[67] induce the same sets of metabolites. As shown (Fig. 5), pathogen challenge is expected also in the intact plant to involve the octadecanoid signaling system to mount defense reactions. This may be a sub-set of the total repertoire of defense initiated by pathogen challenge,[70,71,85] but an important one that will result in the local accumulation of phytoalexins around infection sites and in induction of other local processes (*e.g.*, accumulation of phenolics leading to cell-wall reinforcements due to cross-linking processes and/or lignification, accelerated cell death characteristic for hypersensitive responses). In accord with this is the finding that elicitors act locally, but not systemically (for review, see ref. 86), and that ω-3 desaturases, supposedly involved in the biosynthesis of linolenic acid, are rapidly induced around infection sites in the *Phytophtora megasperma–Petroselinum crispum* interaction.[57]

Plants frequently mount a systemic defense response as a result of primary pathogen challenge, and develop systemic acquired resistance (SAR). There is strong evidence for an involvement of salicylic acid in SAR,[87–89] but the compound may act locally and is not the systemic signal of pathogen challenge.[90] Salicylic acid is known to be microbicidal and may act as a phytoalexin itself.[91] On the other hand, salicylic acid levels in plants under pathogen challenge are often highly transient, and the transient increase precedes the SAR response. Systemic signals of pathogen challenge must consequently occur, but their nature awaits elucidation. Locally released octadecanoids may be part of the systemic signal. A further uncertainty rests with the lack of data on octadecanoid levels in intact plants undergoing pathogen challenge. On the other hand, it has been shown several times that treatment of intact plants with JAMe will induce natural product accumulation[92–96] and may lead to pathogen resistance. This was demonstrated for the resistance of cotton plants to *Verticillium* wilt,[97] of tomato plants to *Fusarium*,[97] and of tomato and potato plants to *Phytophthora infestans*.[98]

Mechanotransduction

Work in our laboratory has shown, in parallel to the demonstration of the role of octadecanoids in plant defense, that mechanotransduction also involves octadecanoid signal transducers.[6,26,67,99] Tendril coiling in *Bryonia dioica* can be elicited by airborne JAMe without any mechanical contact,[99] and the chemically induced reaction is morphologically and biochemically indistinguishable from the mechanically elicited process.[100] Evidence has been obtained that the endogenous mechanotransducer is not JA, but rather its biosynthetic precursor, OPDA.[6,67] It has recently become possible to analyze the kinetics of OPDA levels in tendrils during the reaction[26] (Fig. 6). OPDA levels rise transiently, a process highly

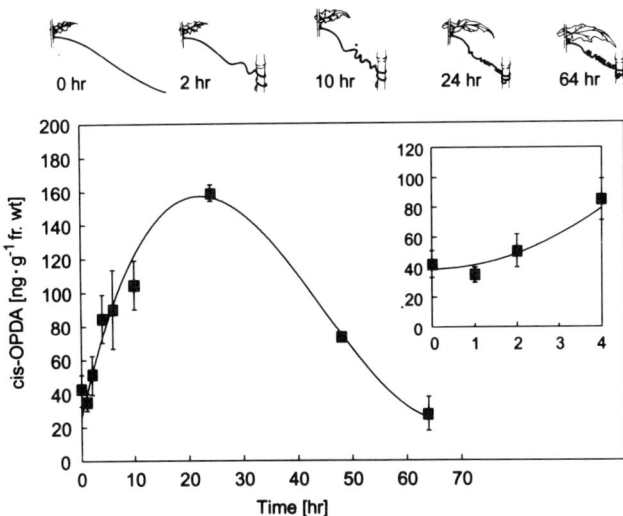

Figure 6. Levels of *cis*-OPDA in tendrils of *Bryonia dioica* during natural mechanical stimulation. The sketches show the progress of the response. Reprinted, with permission, from ref. 26.

correlated with the nastic growth response. JA levels remain low and constant during the phase of increasing levels of OPDA and start to rise late, during the phase of declining levels of OPDA.[26,67] Thus, there exists a situation when OPDA accumulates without giving rise to elevated levels of JA, which means uncoupling of the early and late reactions of JA biosynthesis. The thigmomorphogenetic response of bean internodes—a model system for a plant's reaction to mechanical stress—also proceeds with increases in the level of OPDA, but not JA,[26] well before growth inhibition takes place. Both in the tendril and in the bean internode system, coronatine is highly active,[26,67] and from the inactivity of 1-hydroxyindanoyl-L-isoleucine (7) on tendrils, the response is clearly not that of a JA-amino acid conjugate. We have proposed that coronatine acts like OPDA, and, in fact, coronatine and OPDA are very close structural analogs.[67] Growth inhibition was one of the earliest physiological properties attributed to JA.[101] It is likely that the development of stature in response to mechanical challenge (e.g., wind-forces, soil resistance, strain etc.) involves an interplay of auxins, ethylene, and octadecanoids. This will become better understood in the future, now that enzymes of auxin-, ethylene-, and octadecanoid biosynthesis of plants have been cloned.

Senescence

Higher levels of octadecanoids, predominantly of JA/JAMe, and also application of coronatine induce leaf senescence.[23,102,103] This process has been reviewed.[104] On a molecular basis, JA/JAMe induce loss of chlorophyll,[105] declining levels of ribulose-1,5-bisphosphate carboxylase,[106] inhibition of translation,[107] and many other processes.[108] It is not clear if endogenous levels of JA reach high enough to contribute significantly to senescence *in vivo*, but an interplay of declining levels of auxin and cytokinin and increasing levels of ethylene and jasmonates may well be decisive for the onset and progression of senescence in the intact plant. An interesting question in this context relates to a potential antagonism between cytokinins and octadecanoids in defining a plant's sink tissues, particularly during pathogen/herbivore challenge. Can it be that increases of octadecanoid levels around sites of pathogen challenge help to divert nutrient flow away from these sites thus decreasing a plant's nutritive value for a pathogen? This implication of the jasmonate-induced process of senescence may merit more attention.

Evolutionary Aspects

We have come to appreciate octadecanoids as highly potent, multi-faceted, signal transducers of higher plants. Their structures, biosynthesis, and even biological activities resemble somewhat the eicosanoids, arachidonic acid derived signaling compounds of animals. Whether these are analogous or homologous signaling systems cannot be answered. However, the physiology of octadecanoids in plants suggests them as representing evolutionarily quite an ancient system of control, perhaps even going back to the advent of oxygenic life. With the invention of photosynthesis, oxygen damage to biological membranes became a problem, and all aerobic organism evolved, and now dispose of, reactions to prevent the uncontrolled increase in reactive oxygen species and to repair oxidative damage. For membrane lipids, this repair function is exerted by A-type lipases which remove oxidized fatty acids from membrane lipids (see ref. 109 and refs. cited therein). Those are then subjected to various further metabolic conversions to yield less aggressive reaction products. It is conceivable that eventually products derived from oxidized fatty acids were used by cells as signaling compounds to detect oxidative stress and react appropriately to delimit damage. This could have been the origin of the octadecanoid pathway.

Many known reactions of octadecanoids indeed fit this hypothesis.[66] JA levels might increase as a response to photo-oxidative stress (this still needs to be determined) and under such circumstances would then down-regulate the amount of light absorbed by the photosynthetic apparatus by inducing loss of chlorophyll[104] and accumulation of light-absorbing, nonphotosynthetic pigments, notably anthocyanins.[110] Flavonoids are known potent antioxidants whose accumulation would protect a cell from oxidative damage.[111] At the same time, JA would also inhibit *rbcL* expression,[104,106] and thereby further reduce the capacity of the photosynthetic apparatus to fix CO_2. A regulatory link between photosynthesis and octadecanoid production is further suggested by co-localization of both pathways in chloroplasts, and also by the fact that many JA

FATTY ACID-DERIVED SIGNALING MOLECULES

responsive genes are not only regulated by light, but also by high levels of sugars.[112–114] A decreasing capacity of a chloroplast to remove reactive oxygen species may lead finally to increased accumulation of JA, resulting in accelerated senescence. A synopsis of this model is shown (Fig. 7).

The initial function acquired by oxylipins—originally representing simply oxidatively damaged membrane lipids—and a role still visible for octadecanoids today, may have been signaling oxidative damage to the cell to raise protective measures against such damage. There are a host of relations between these processes and pathogen and herbivore defense as well as mechanotransduction, all of which lead to plant protection (Fig. 8). They share common biochemical reactions, *e.g.*, an oxidative burst frequently occurs during hypersensitive pathogen defenses,[115] and also upon mechanostimulation,[116] and oxidative processes increase during wounding of a plant. The stimulation of phenolic metabolism leads to anti-oxidative flavonoids, to phytoalexins as microbial defense compounds, and to lignin as a structural reinforcement of cell walls in mechanically stressed plants and also in infected plants. By the erection of chemical and structural barriers, the plant acquires resistance to herbivores, pathogens, and mechanical forces. As in the case of oxidative stress, the result is protection, in one way or the other. This is the common physiological and, as we see it, evolutionary basis for the octadecanoids, a versatile group of plant defense regulators derived from fatty acids of structural lipids in membranes.

CONCLUSION

From a plant senescence factor and growth inhibitor, jasmonic acid research over only a few years has led to the appreciation of the octadecanoids

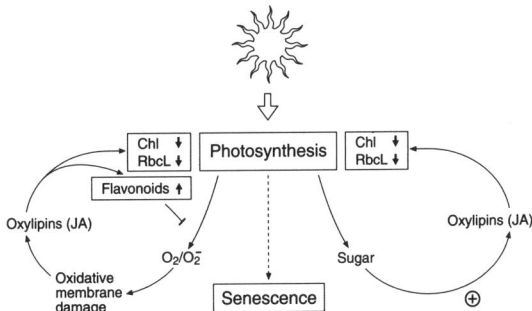

Figure 7. Known reactions of jasmonates (boxed or circled in) arranged in primordial regulatory circuits for coarse regulation of photosynthesis that might shed light on the evolution of the octadecanoid system (for details, see text). Based on ideas expressed in ref. 66. Oxylipins, oxidized metabolites of fatty acids; up arrows, increase in amounts; down arrows, decrease in amounts, ⊕, induction; ⊣, inhibition.

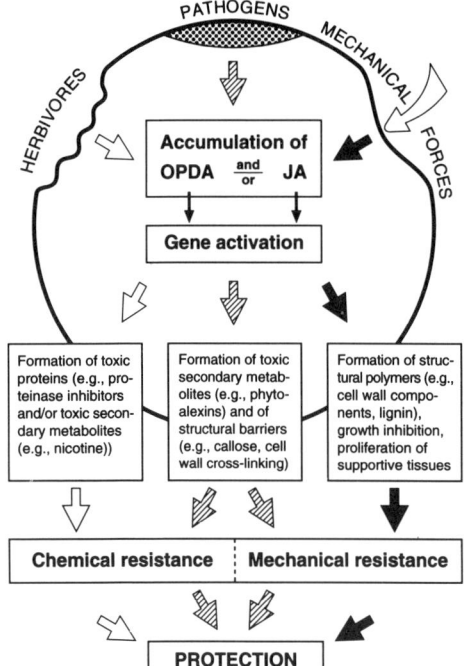

Figure 8. Integrated view on octadecanoid dependent defense reactions leading to natural plant protection.

as a novel and biologically highly diversified group of signal transducers regulating plant defenses against biotic and abiotic stressors. Yet, almost nothing is known about the molecular basis of octadecanoid action. The next task is to unveil the sites of octadecanoid perception and to answer the question of how the signal is relayed to the octadecanoid responsive genes. It will furthermore be necessary to better understand how the signals are integrated into the hormonal network of the plant and how their biosynthesis is regulated. Exciting discoveries are just ahead of us. One of these was recently reported by Farmer and colleagues,[117] who have presented conclusive evidence for the occurrence of dinor-oxo-phytodienoic acid (dnOPDA) in plants. This C16-analog of OPDA does not arise from β-oxidation of OPDA, but directly from the C16:3 fatty acid 7Z,10Z,13Z-hexadecatrienoic acid, a plastidic fatty acid occuring almost exclusively in the monogalactosyldiglyceride membrane lipid fraction of the so-called '16:3' plants such as potato or *A. thaliana*. Thus, there may exist parallel pathways from this fatty acid to dn-OPDA and from LA to OPDA in a cell. The biological activity of dnOPDA and its abundance in the plant[117] suggest it may be of biological relevance. This further stresses the importance of looking at the complexity of this signaling system rather than at any single compound alone.

Finally, in order to incorporate this new metabolite into the terminology, the term octadecanoids seems no longer valid as a general group term but rather only as one to denote the group of C18-derived metabolites. The term hexadecanoids may be appropriate for compounds such as dnOPDA (and possibly others still to be detected), if derived from hexadecatrienoic acid. Both groups would be cyclic oxylipins, and the term jasmonates could be used for this group of compounds irrespective of their biosynthetic origin whenever dealing with the more physiological aspects of the cyclic oxylipins.

ACKNOWLEDGMENTS

We thank all colleagues who have provided unpublished information or manuscripts in press in order to help keep this review timely. We apologize to all those whose work could not be included due to space constraints. The author's work was funded by the Deutsche Forschungsgemeinschaft, Bonn (Germany), Bundesminister für Bildung und Wissenschaft (BMBW), Bonn (Germany) and Fonds der Chemischen Industrie, Frankfurt (Germany). EWW is particularly indebted to Prof. Dr. M. H. Zenk, Munich, Germany, who persuaded him to start elucidating the processes of mechanotransduction in tendrils, work that led into the field of octadecanoid biology.

REFERENCES

1. SEMBDNER, G., PARTHIER, B. 1993. The biochemistry and the physiological and molecular actions of jasmonates. Annu. Rev. Plant Physiol. Plant Mol. Biol. 44:569–589.
2. MEYER, A., MIERSCH, O., BÜTTNER, C., DATHE, W., SEMBDNER, G. 1984. Occurrence of the plant growth regulator jasmonic acid in plants. J. Plant Growth Reg. 3:1–8.
3. ALDRIDGE, D.C., GALT, S., GILES, D., TURNER, W.B. 1971. Metabolites of *Lasiodiplodia theobromae*. J. Chem. Soc. 1623–1627.
4. MIERSCH, O., PREISS, A., SEMBDNER, G., SCHREIBER, K. 1987. (+)-7-Iso-jasmonic acid and related compounds from *Botryodiplodia theobromae*. Phytochemistry 26:1037–1039.
5. DEMOLE, E., LEDERER, E., MERCIER, D. 1962. Isolement et determination de la structure du jasmonate de methyle, constituant odorant caracteristique de l'essence de jasmin. Helvetica Chimica Acta XLV:675–685.
6. WEILER, E.W., ALBRECHT, T., GROTH, B., XIA, Z.-Q., LUXEM, M., LISS, H., ANDERT, L., SPENGLER, P. 1993. Evidence for the involvement of jasmonates and their octadecanoid precursors in the tendril coiling response of *Bryonia dioica*. Phytochemistry 32:591–600.
7. YOSHIHARA, T., OMER, E.-S.A., KOSHINO, H., SAKAMURA, S., KIKUTA, Y., KODA, Y. 1989. Structure of a tuber-inducing stimulus from potato leaves (*Solanum tuberosum* L.). Agric. Biol. Chem. 53:2835–2837.
8. XIA, Z.-Q., ZENK, M.H. 1993. Metabolism of jasmonic acid in suspension cultures of *Eschscholtzia*. Planta Medica 59: 575–576.
9. YOSHIHARA, T., AMANUMA, M., TSUTSUMI, T., OKUMURA, Y., MATSUURA, H., ICHIHARA, A. 1996. Metabolism and transport of [2-^{14}C](+/−) jasmonic acid in the potato plant. Plant Cell Physiol. 37:586–590.

10. BRÜCKNER, C., KRAMELL, R., SCHNEIDER, G., SCHMIDT, J., PREISS, A., SEMBDNER, G., SCHREIBER, K. 1988. N-[(−)-jasmonoyl]-S-tryptophan and a related tryptophan conjugate from *Vicia faba*. Phytochemistry 27:275–276.
11. SCHMIDT, J., KRAMELL, R., BRÜCKNER, C., SCHNEIDER, G., SEMBDNER, G., SCHREIBER, K., STACH, J., JENSEN, E. 1990. Gas chromatographic/mass spectrometric and tandem mass spectrometric investigations of synthetic amino acid conjugates of jasmonic acid and endogenously occuring related compounds from *Vicia faba* L. Biomed. Environ. Mass Spectrom. 19:327–338.
12. ALBORN, H.T., TURLINGS, T.C.J., JONES, T.H., STENHAGEN, G., LOUGHRIN, J.H., TUMLINSON, J.H. 1997. An elicitor of plant volatiles from beet armyworm oral secretion. Science 276:945–949.
13. Mason, H.S., DeWald, D.B., Mullet, J.E. 1993. Identification of a methyl jasmonate-responsive domain in the soybean *vspB* promoter. Plant Cell 5:241–251.
14. KIM, S.-R., KIM, Y., AN, G. 1993. Identification of methyl jasmonate and salicylic acid response elements from the nopaline synthase (nos) promotor. Plant Physiol. 103:97–103.
15. VICK, B.A., ZIMMERMAN, D.C. 1984. Biosynthesis of jasmonic acid by several plant species. Plant Physiol. 75:458–461.
16. BLEE, E., JOYARD, J. 1996. Envelope membranes from spinach chloroplasts are a site of metabolism of fatty acid hydroperoxides. Plant Physiol. 110:445–454.
17. MARÉCHAL, E., BLOCK, M.A., DORNE, A.J., DOUCE, R., JOYARD, J. 1997. Lipid synthesis and metabolism in the plastid envelope. Physiol. Plant. 100:65–77.
18. VICK, B.A., ZIMMERMAN, D.C. 1983. The biosynthesis of jasmonic acid: A physiological role for plant lipoxygenase. Biochem. Biophys. Res. Commun. 111:470–477.
19. QUINKERT, G., ADAM, F., DÜRNER, G. 1982. Asymmetrische Synthese von Methyljasmonat. Angew. Chem. 94:866–867.
20. BOLAND, W., HOPKE, J., DONATH, J., NÜSKE, J., BUBLITZ, F. 1995. Jasmonsäure- und Coronatin-induzierte Duftproduktion in Pflanzen. Angew. Chem. 107:1715–1717.
21. KRUMM, T., BANDEMER, K., BOLAND, W. 1995. Induction of volatile biosynthesis in the lima bean (*Phaseolus lunatus*) by leucine- and isoleucine conjugates of 1-oxo- and 1-hydroxyindan-4-carboxylic acid: Evidence for amino acid conjugates of jasmonic acid as intermediates in the octadecanoid signalling pathway. FEBS Lett. 377:523–529.
22. TAMOGAMI, S., RAKWAL, R., KODAMA, O. 1997. Phytoalexin production by amino acid conjugates of jasmonic acid through induction of naringenin-7-*O*-methyltransferase, a key enzyme of phytoalexin biosynthesis in rice (*Oryza sativa* L.). FEBS Lett. 401:239–242.
23. UEDA, J., KATO, J. 1980. Isolation and identification of a senescence-promoting substance from wormwood (*Artemisia absinthium* l.). Plant Physiol. 66:246–249.
24. HOPKE, J., DONATH, J., BLECHERT, S., BOLAND, W. 1994. Herbivore-induced volatiles: The emission of acyclic homoterpenes from leaves of *Phaseolus lunatus* and *Zea mays* can be triggered by a β-glucosidase and jasmonic acid. FEBS Lett. 352:146–150.
25. KOCH, T., BANDEMER, K., BOLAND, W. 1997. Biosynthesis of *cis*-jasmone: A pathway for the inactivation and the disposal of the plant stress hormone jasmonic acid to the gas phase? Helvetica Chimica Acta 80:838–850.
26. STELMACH, B.A., MÜLLER, A., HENNIG, P., LAUDERT, D., ANDERT, L., WEILER, E.W. 1997. Quantitation of the octadecanoid 12-oxo-phytodienoic acid, a signalling compound in mechanotransduction. Phytochemistry (in press).
27. BOHLMANN, F., JAKUPOVIC, J., AHMED, M., SCHUSTER, A. 1983. Sesquiterpene lactones and other constituents from *Schistostephium* species. Phytochemistry 22:1623–1636.
28. ZDERO, C., LEHMANN, L., BOHLMANN, F. 1991. Chemotaxonomy of *Athanasia* and related genera. Phytochemistry 30:1161–1163.
29. SIEDOW, J.N. 1991. Plant lipoxygenase: Structure and function. Annu. Rev. Plant. Physiol. 42:145–188.

30. BELL, E., MULLET, J.E. 1993. Characterization of an *Arabidopsis* lipoxygenase gene responsive to methyl jasmonate and wounding. Plant Physiol. 103:1133–1137.
31. BELL, E., CREELMAN, R.A., MULLET, J.E. 1995. A chloroplast lipoxygenase is required for wound-induced jasmonic acid accumulation in *Arabidopsis*. Proc. Natl. Acad. Sci. USA 92:8675–8679.
32. VICK, B.A., ZIMMERMANN, D.C. 1981. Lipoxygenase, hydroperoxide isomerase, and hydroperoxide cyclase in young cotton seedlings. Plant Physiol. 67:92–97.
33. VICK, B.A., ZIMMERMANN, D.C. 1987. Pathway of fatty acid hydroperoxide metabolism in spinach leaf chloroplasts. Plant Physiol. 85:1073–1078.
34. HAMBERG, M. 1987. Mechanism of corn hydroperoxide isomerase: Detection of 12,13(S)-oxido-9(Z),11-octadecadienoic acid. Biochim. Biophys. Acta 920:76–84.
35. HAMBERG, M., HUGHES, M.A. 1988. Fatty acid allene oxides. III. Albumin-induced cyclization of 12,13(S)-epoxy-9(Z),11-octadecadienoic acid. Lipids 23:469–475.
36. LAUDERT, D., PFANNSCHMIDT, U., LOTTSPEICH, F., HOLLÄNDER-CZYTKO, H., WEILER, E.W. 1996. Cloning, molecular and functional characterization of *Arabidopsis thaliana* allene oxide synthase (CYP 74), the first enzyme of the octadecanoid pathway to jasmonates. Plant Mol. Biol. 31:323–335.
37. SONG, W.-C., FUNK, C.D., BRASH, A.R. 1993. Molecular cloning of an allene oxide synthase: A cytochrome P450 specialized for the metabolism of fatty acid hydroperoxides. Proc. Natl. Acad. Sci. USA 90:8519–8523.
38. LAUDERT, D., HENNIG, P., STELMACH, B.A., MÜLLER, A., ANDERT, L., WEILER, E.W. 1977. Analysis of 12-oxo-phytodienoic acid enantiomers in biological samples by capillary gas chromatography-mass spectrometry using cyclodextrin stationary phases. Anal. Biochem. 246:211–217.
39. PAN, Z., DURST, F., WERCK-REICHHART, D., GARDNER, H.W., CAMARA, B., CORNISH, K., BACKHAUS, R.A. 1995. The major protein of guayule rubber particles is a cytochrome P450. J. Biol. Chem. 270:8487–8494.
40. SONG, W.-C., BRASH, A.R. 1991. Purification of an allene oxide synthase and identification of the enzyme as a cytochrome P-450. Science 253:781–784.
41. LAU, S.-M.C., HARDER, P.A., O'KEEFE, D.P. 1993. Low carbon monoxide affinity allene oxide synthase is the predominant cytochrome P450 in many plant tissues. Biochemistry 32:1945–1950.
42. BHAGWAT, S.S., HAMANN, P.R., STILL, W.C., BUNTING, S., FITZPATRICK, F.A. 1985. Synthesis and structure of the platelet aggregation factor thromboxane A2. Nature 315:511–513.
43. NELSON, D.R., KOYMANS, L., KAMATAKI, T., STEGEMAN, J.J., FEYEREISEN, R., WAXMAN, D.J., WATERMAN, M.R., GOTOH, O., COON, M.J., ESTABROOK, R.W. 1997. P450 superfamily: Update on new sequences, gene mapping, accession numbers and nomenclature. Pharmacogenetics 6:1–42.
44. HAMBERG, M. 1988. Biosynthesis of 12-oxo-10,15(Z)-phytodienoic acid: Identification of an allene oxide cyclase. Biochem. Biophys. Res. Commun. 156:543–550.
45. HAMBERG, M., FAHLSTADIUS, P. 1990. Allene oxide cyclase: A new enzyme in plant lipid metabolism. Arch. Biochem. Biophys. 276:518–526.
46. ZIEGLER, J., HAMBERG, M., MIERSCH, O., PARTHIER, B. 1997. Purification and characterization of allene oxide cyclase from dry corn seeds. Plant Physiol. 114:565–573.
47. SCHALLER, F., WEILER, E.W. 1997. Enzymes of octadecanoid biosynthesis in plants: 12-Oxo-phytodienoate-10,11-reductase. Eur. J. Biochem. 245:294–299.
47a. SCHALLER, F., WEILER, E.W. 1997. Molecular cloning and characterization of 12-Oxophytodienoate reductase, an enzyme of the octadecanoid signaling pathway from *Arabidopsis thaliana*. Structural and fuctional relationship to yeast old yellow enzyme. J. Biol. Chem. (in press).
48. FRENCH, C.E., BRUCE, N.C. 1995. Bacterial morphinone reductase is related to old yellow enzyme. Biochem. J. 312:671–678.

49. TSERNG, K.Y., JIN, S.J. 1990. NADPH-dependent reductive metabolism of cis-5 unsaturated fatty acids. J. Biol. Chem. 266:11614–11620.
50. VICK, B.A., ZIMMERMAN, D.C. 1995. Thermal alteration of a cyclic fatty acid produced by a flaxseed extract. Lipids 14:734–740.
51. ALBRECHT, T., KEHLEN, A., STAHL, K., KNÖFEL, H.-D., SEMBDNER, G., WEILER, E.W. 1993. Quantification of rapid, transient increases in jasmonic acid in wounded plants using a monoclonal antibody. Planta 191:86–94.
52. GUNDLACH, H., MÜLLER, M.J., KUTCHAN, T.M., ZENK, M.H. 1992. Jasmonic acid is a signal transducer in elicitor-induced plant cell cultures. Proc. Natl. Acad. Sci. USA 89:2389–2393.
53. MUELLER, M.J., BRODSCHELM, W., SPANNAGL, E., ZENK, M.H. 1993. Signaling in the elicitation process is mediated through the octadecanoid pathway leading to jasmonic acid. Proc. Natl. Acad. Sci. USA 90:7490–7494.
54. HERTEL, S.C., KNÖFEL, H.D., KRAMELL, R., MIERSCH, O. 1997. Partial purification and characterization of a jasmonic acid conjugate cleaving amidohydrolase from the fungus *Botryodiplodia theobromae*. FEBS Lett. 407:105–110.
55. MUELLER, M.J., BRODSCHELM, W. 1994. Quantification of jasmonic acid by capillary gas chromatography-negative chemical ionization-mass spectrometry. Anal. Biochem. 218:425–435.
56. FARMER, E.E., RYAN, C.A. 1992. Octadecanoid precursors of jasmonic acid activate the synthesis of wound-inducible proteinase inhibitors. Plant Cell 4:129–134.
57. KIRSCH, C., TAKAMIYA-WIK, M., REINOLD, S., HAHLBROCK, K., SOMSSICH, I.E. 1997. Rapid, transient, and highly localized induction of plastidial ω-3 fatty acid desaturase mRNA at fungal infection sites in *Petroselinum crispum*. Proc. Natl. Acad. Sci. USA 94:2079–2084.
58. CONCONI, A., MIQUEL, M., BROWSE, J.A., RYAN, C.A. 1996. Intracellular levels of free linolenic and linoleic acids increase in tomato leaves in response to wounding. Plant Physiol. 111:797–803.
59. CHANDRA, S., HEINSTEIN, P.F., LOW, P.S. 1996. Activation of phospholipase A by plant defense elicitors. Plant Physiol. 110:979–986.
60. ROY, S., POUENAT, M.-L., CAUMONT, C., CARIVEN, C., PREVOST, M.-C., ESQUERRE-TUGAYE, M.-T. 1995. Phospholipase activity and phospholipid patterns in tobacco cells treated with fungal elicitor. Plant Sci. 107:17–25.
61. LEE, S.-S., KAWAKITA, K., TSUGE, T., DOKE, N. 1992. Stimulation of phospholipase A2 in strawberry cells treated with AF-toxin 1 produced by *Alternaria alternata* strawberry phenotype. Physiol. Molec. Plant Pathol. 41:283–294.
62. HARMS, K., ATZORN, R., BRASH, A., KÜHN, H., WASTERNACK, C., WILLMITZER, L., PENA-CORTÉS, H. 1995. Expression of a flax allene oxide synthase cDNA leads to increased endogenous jasmonic acid (JA) levels in transgenic potato plants but not to a corresponding activation of JA-responding genes. Plant Cell 7:1645–1654.
63. BALDWIN, I.T., ZHANG, Z.P., DIAB, N., OHNMEISS, T.E., McCLOUD, E.S., LYNDS, G.Y., SCHMELZ, E.A. 1997. Quantification, correlations and manipulations of wound-induced changes in jasmonic acid and nicotine in *Nicotiana sylvestris*. Planta 201:397–404.
64. BLECHERT, S., BRODSCHELM, W., HÖLDER, S., KAMMERER, L., KUTCHAN, T.M., MUELLER, M.J., XIA, Z.-Q., ZENK, M.H. 1995. The octadecanoic pathway: Signal molecules for the regulation of secondary pathways. Proc. Natl. Acad. Sci. USA 92:4099–4105.
65. HOWE, G.A., LIGHTNER, J., BROWSE, J., RYAN, C.A. 1996. An octadecanoid pathway mutant (JL5) of tomato is compromised in signaling for defense against insect attack. Plant Cell 8:2067–2077.
66. CREELMAN, R.A., MULLET, J.E. 1995. Jasmonic acid distribution and action in plants: Regulation during development and response to biotic and abiotic stress. Proc. Natl. Acad. Sci. USA 92:4114–4119.

67. WEILER, E.W., KUTCHAN, T.M., GORBA, T., BRODSCHELM, W., NIESEL, U., BUBLITZ, F. 1994. The *Pseudomonas* phytotoxin coronatine mimics octadecanoid signalling molecules of higher plants. FEBS Lett. 345:9–13.
68. KODA, Y., TAKAHASHI, K., KIKUTA, Y., GREULICH, F., TOSHIMA, H., ICHIHARA, A. 1996. Similarities of the biological activities of coronatine and coronafacic acid to those of jasmonic acid. Phytochemistry 41:93–96.
69. BLECHERT, S., BRÜMMER, O., FÜSSLEIN, M., HÖLDER, S., BOCKELMANN, C., KUTCHAN, T., WEILER, E.W., ZENK, M.H. 1997. Structural separation of biological activities of jasmonates and related compounds. J. Chem. Soc. Perkin I (in press).
70. SCHWEIZER, P., BUCHALA, A., SILVERMAN, P., SESKAR, M., RASKIN, I., MÉTRAUX, J.P. 1997. Jasmonate-inducible genes are activated in rice by pathogen attack without a concomitant increase in endogenous jasmonic acid levels. Plant Physiol. 114:79–88.
71. HAUSE, B., KOGEL, K.H., PARTHIER, B., WASTERNACK, C. 1997. In barley leaf cells, jasmonates do not act as a signal during compatible or incompatible interactions with the powdery mildew fungus (*Erysiphe graminis* f sp *hordei*). J. Plant Physiol. 150:127–132.
72. FEYS, B.J.F., BENEDETTI, C.E., PENFOLD, C.N., TURNER, J.G. 1994. *Arabidopsis* mutants selected for resistance to the phytotoxin coronatine are male sterile, insensitive to methyl jasmonate, and resistant to a bacterial pathogen. Plant Cell 6:751–759.
73. FARMER, E.E., RYAN, C.A. 1990. Interplant communication: Airborne methyl jasmonate induces synthesis of proteinase inhibitors in plant leaves. Proc. Natl. Acad. Sci. USA 87:7713–7716.
74. FARMER, E.E., RYAN, C.A. 1992. Octadecanoid-derived signals in plants. Trends Cell Biol. 2:236–241.
75. CREELMAN, R.A., MULLET, J.E. 1997. Biosynthesis and action of jasmonates in plants. Annu. Rev. Plant. Physiol. Plant Mol. Biol. 48:355–381.
76. McCONN, M., CREELMAN, R.A., BELL, E., MULLET, J.E., BROWSE, J. 1997. Jasmonate is essential for insect defense in *Arabidopsis*. Proc. Natl. Acad. Sci. USA 94:5473–5477.
77. DOARES, S.H., SYROVETS, T., WEILER, E.W., RYAN, C.A. 1995. Oligogalacturonides and chitosan activate plant defensive genes through the octadecanoid pathway. Proc. Natl. Acad. Sci. USA 92:4095–4098.
78. SCHALLER, A., RYAN, C.A. 1995. Systemin—a polypeptide defense signal in plants. BioEssays 18:27–33.
79. CONSTABEL, C.P., BERGEY, D.R., RYAN, C.A. 1995. Systemin activates synthesis of wound-inducible tomato leaf polyphenol oxidase via the octadecanoid defense signaling pathway. Proc. Natl. Acad. Sci. USA 92:407–411.
80. NARVAEZ-VASQUEZ, J., PEARCE, G., OROZCO-CARDENAS, M.L., FRANCESCHI, V.R., RYAN, C.A. 1995. Autoradiographic and biochemical evidence for the systemic translocation of systemin in tomato plants. Planta 195:593–600.
81. MALONE, M., ALARCON, J.-J. 1995. Only xylem-borne factors can account for systemic wound signalling in the tomato plant. Planta 196:740–746.
82. WILDON, D.C., THAIN, J.F., MINCHIN, P.E.H., GUBB, I.R., REILLY, A.J., SKIPPER, Y.D., DOHERTY, H.M., O'DONNELL, P.J., BOWLES, D.J. 1992. Electrical signalling and systemic proteinase inhibitor induction in the wounded plant. Nature 360:62–64.
83. BERGEY, D.R., HOWE, G.A., RYAN, C.A. 1996. Polypeptide signaling for plant defensive genes exhibits analogies to defense signaling in animals. Proc. Natl. Acad. Sci. USA 93:12053–12058.
84. TAKABAYASHI, J., DICKE, M. 1996. Plant-carnivore mutualism through herbivore-induced carnivore attractants. Trends Plant Sci. 1:109–113.
85. RICKAUER, M., BRODSCHELM, W., BOTTIN, A., VERONESI, C., GRIMAL, H., ESQUERRÉ-TUGAYÉ, M.T. 1997. The jasmonate pathway is involved differentially in the regulation of different defence responses in tobacco cells. Planta 202:155–162.

86. YOSHIKAWA, M., YAMAOKA, N., TAKEUCHI, Y. 1993. Elicitors: Their significance and primary modes of action in the induction of plant defense reactions. Plant Cell Physiol. 34:1163–1173.
87. MALAMY, J., KLESSIG, D.F. 1992. Salicylic acid and plant disease resistance. Plant J. 2:643–654.
88. DURNER, J., SHAH, J., KLESSIG, D.F. 1997. Salicylic acid and disease resistance in plants. Trends Plant Sci. 2:266–274.
89. HUNT, M.D., NEUENSCHWANDER, U.H., DELANEY, T.P., WEYMANN, K.B., FRIEDRICH, L.B., LAWTON, K.A., STEINER, H., RYALS, J.A. 1996. Recent advances in systemic acquired resistance research - a review. Gene 179:89–95.
90. VERNOOIJ, B., FRIEDRICH, L., MORSE, A., REIST, R., KOLDITZ-JAWHAR, R., WARD, E., UKNES, S., KESSMANN, H., RYALS, J. 1994. Salicylic acid is not the translocated signal responsible for inducing systemic acquired resistance but is required in signal transduction. Plant Cell 6:959–965.
91. RÜFFER, M., STEIPE, B., ZENK, M.H. 1995. Evidence against specific binding of salicylic acid to plant catalase. FEBS Lett. 377:175–180.
92. BODNARYK, R.P. 1994. Potent effect of jasmonates on indole glucosinolates in oilseed rape and mustard. Phytochemistry 35:301–305.
93. DOUGHTY, K.J., KIDDLE, G.A., PYE, B.J., WALLSGROVE, R.M., PICKETT, J.A. 1995. Selective induction of glucosinolates in oilseed rape leaves by methyl jasmonate. Phytochemistry 38:347–350.
94. BALDWIN, I.T. 1996. Methyl jasmonate-induced nicotine production in *Nicotiana attenuata*: Inducing defenses in the field without wounding. Entomol. Exp. Appl. 80:213–220.
95. AERTS, R.J., GISI, D., DE CAROLIS, E., DE LUCA, V., BAUMANN, T.W. 1994. Methyljasmonate vapor increases the developmentally controlled synthesis of alkaloids in *Catharanthus* and *Cinchona* seedlings. Plant J. 5:635–643.
96. MIKSCH, M., BOLAND, W. 1996. Airborne methyl jasmonate stimulates the biosynthesis of furanocoumarins in the leaves of celery plants (*Apium graveolens*). Experientia 52:739–743.
97. LI, J., ZINGEN-SELL, I., BUCHENAUER, H. 1996. Induction of resistance of cotton plants to *Verticillium* wilt and of tomato plants to *Fusarium* wilt by 3-aminobutyric acid and methyl jasmonate. J. Plant Dis. Prot. 103:288–299.
98. COHEN, Y., GISI, U., NIDERMAN, T. 1993. Local and systemic protection against *Phytophthora infestans* induced in potato and tomato plants by jasmonic acid and jasmonic acid methyl ester. Phytopathology 83:1054–1062.
99. FALKENSTEIN, E., GROTH, B., MITHÖFER, A., WEILER, E.W. 1991. Methyljasmonate and α-linolenic acid are potent inducers of tendril coiling. Planta 185:316–322.
100. KAISER, I., ENGELBERTH, J., GROTH, B., WEILER, E.W. 1994. Touch-and methyl jasmonate-induced lignification in tendrils of *Bryonia dioica* Jacq. Bot. Acta 107:24–29.
101. DATHE, W., RÖNSCH, H., PREISS, A., SCHADE, W., SEMBDNER, G., SCHREIBER, K. 1981. Endogenous plant hormones of the broad bean, *Vicia faba* L. (−)-Jasmonic acid, a plant growth inhibitor in pericarp. Planta 153:530–535.
102. UEDA, J., KATO, J., YAMANE, H., TAKAHASHI, N. 1981. Inhibitory effect of methyl jasmonate and its related compounds on kinetin-induced retardation of oat leaf senescence. Physiol. Plant. 52:305–309.
103. MITCHELL, R.E., YOUNG, H. 1978. Identification of a chlorosis-inducing toxin of *Pseudomonas glycinea* as coronatine. Phytochemistry 17:2028–2029.
104. PARTHIER, B. 1990. Jasmonates: Hormonal regulators or stress factors in leaf senescence? J. Plant Growth Reg. 9:57–63.
105. KOVAC, M., RAVNIKAR, M. 1994. The effect of jasmonic acid on the photosynthetic pigments of potato plants grown in vitro. Plant Sci. 103:11–17.
106. POPOVA, L.P., VAKLINOVA, S.G. 1988. Effect of jasmonic acid on the synthesis of ribulose-1,5-bisphosphate carboxylase-oxygenase in barley leaves. J. Plant Physiol. 133:210–215.

107. REINBOTHE, S., REINBOTHE, C., PARTHIER, B. 1993. Methyl jasmonate represses translation initiation of a specific set of mRNAs in barley. Plant J. 4:459–467.
108. REINBOTHE, C., PARTHIER, B., REINBOTHE, S. 1997. Temporal pattern of jasmonate-induced alterations in gene expression of barley leaves. Planta 201:281–287.
109. McLEAN, L.R., HAGAMAN, K.A., DAVIDSON, W.S. 1993. Role of lipid structure in the activation of phospholipase A_2 by peroxidized phospholipids. Lipids 28:505–509.
110. FRANCESCHI, V.R., GRIMES, H.D. 1991. Induction of soybean vegetative storage proteins and anthocyanins by low-level atmospheric methyl jasmonate. Proc. Natl. Acad. Sci. USA 88:6745–6749.
111. RICE-EVANS, C.A., MILLER, N.J., PAGANGA, G. 1997. Antioxidant properties of phenolic compounds. Trends Plant Sci. 2:152–159.
112. BERGER, S., BELL, E., SADKA, A., MULLET, J.E. 1995. *Arabidopsis thaliana Atvsp* is homologous to soybean *VspA* and *VspB*, genes encoding vegetative storage protein acid phosphatases, and is regulated similarly by methyl jasmonate, wounding, sugars, light and phosphate. Plant Mol. Biol. 27:933–942.
113. SADKA, A., DEWALD, D.B., MAY, G.D., PARK, W.D., MULLET, J.E. 1994. Phosphate modulates transcription of soybean *VspB* and other sugar-inducible genes. Plant Cell 6:737–749.
114. JOHNSON, R., RYAN, C.A. 1990. Wound-inducible potato inhibitor II genes: Enhancement of expression by sucrose. Plant Mol. Biol. 14:527–536.
115. DOKE, N., MIURA, Y., SANCHEZ, L.M., PARK, H., NORITAKE, T., YOSHIOKA, H., KAWAKITA, K. 1996. The oxidative burst protects plants against pathogen attack: Mechanism and role as an emergency signal for plant bio-defence—a review. Gene 179:45–51.
116. YAHRAUS, T., CHANDRA, S., LEGENDRE, L., LOW, P.S. 1997. Evidence for a mechanically induced oxidative burst. Plant Physiol. 109:1259–1266.
117. WEBER, A., VICK, B.A., FARMER, E.E. 1997. Dinor-oxo-phytodienoic acid: A new hexadecanoid signal in the jasmonate family. Proc. Natl. Acad. Sci. USA 94:10473–10478.

In nature, *A.tumefaciens* induces crown galls (plant tumors) on dicotyledonous plants (Fig. 1). The molecular mechanisms underlying this transformation process are now known in some detail, and will be discussed in this review.

An overview of the transformation process is shown in Figure 2. Virulent *Agrobacterium* strains harbor a large plasmid, the tumor inducing (Ti) plasmid, which carries most of the genes involved in the virulence of the bacterium (see 1 for a recent review). Part of the Ti plasmid, the T-DNA, is transferred from the bacterium to plant cells during tumor formation. The T-DNA is targetted to the plant cell nucleus where it integrates randomly into the plant genome. Genes present on the T-DNA encode proteins involved in production of plant hormones (auxins and cytokinins). Expression of these genes results in uncontrolled plant cell proliferation and formation of a tumor. Similarly, *A. rhizogenes* causes hairy root disease on the aerial parts of plants. It transfers the T-DNA of its Ri (root inducing) plasmid to plant cells at an infection site. The Ri T-DNA induces the transformed cells to divide and differentiate into roots, leading to the production of hairy roots on plant stems.

Figure 1. Crown gall tumours induced by *A. tumefaciens*.

Chapter Eleven

INTERACTIONS BETWEEN *AGROBA TUMEFACIENS* AND PLANT CELLS

Paul Bundock and Paul Hooykaas

Institute for Molecular Plant Sciences
Leiden University
Wassenaarseweg 64
2333 AL, Leiden, The Netherlands

Introduction . . . .
The Ti Plasmid . . . .
The Virulence Region . . . .
    VirA/VirG: A Two Component Regulatory System . . . .
    Border Processing: The VirD1/VirD2 Endonuclease Action
        T-DNA Borders . . . .
    The T-Complex . . . .
    Transport of T-DNA to the Plant Cell: Role of VirB and VirD
        Proteins . . . .
    Protein Transport . . . .
    T-DNA Integration . . . .
Applications of the *Agrobacterium*/Plant Interaction . . . .

INTRODUCTION

In the last 20 years we have seen tremendous progress in th molecular biology. This has been due largely to our ability to tr plant cells, which can then be regenerated into complete plants. I possible to genetically engineer plants with specific desired tr transformation can be achieved by chemical or physical means widely used transformation protocols utilize the elegant transforr of the gram negative soil bacterium, *Agrobacterium tumefaciens*.

Phytochemical Signals and Plant–Microbe Interactions, edited by Romeo *et al.*
Plenum Press, New York, 1998.

INTERACTIONS BETWEEN *AGROBACTERIUM TUMEFACIENS* AND PLANT CELLS   209

Tumor formation is advantageous for *Agrobacterium* because tumor cells produce a class of compounds called opines which the bacterium can catabolize. These compounds are normally not found in plant cells. Their production is catalyzed by enzymes called opine synthases, which are encoded by the T-DNA. Most opines are formed by condensation of an amino acid and a keto acid, while others, such as the agrocinopines, are phosphorylated saccharides. The structures of some of these opines are shown in Figure 3. Each opine catabolase gene can

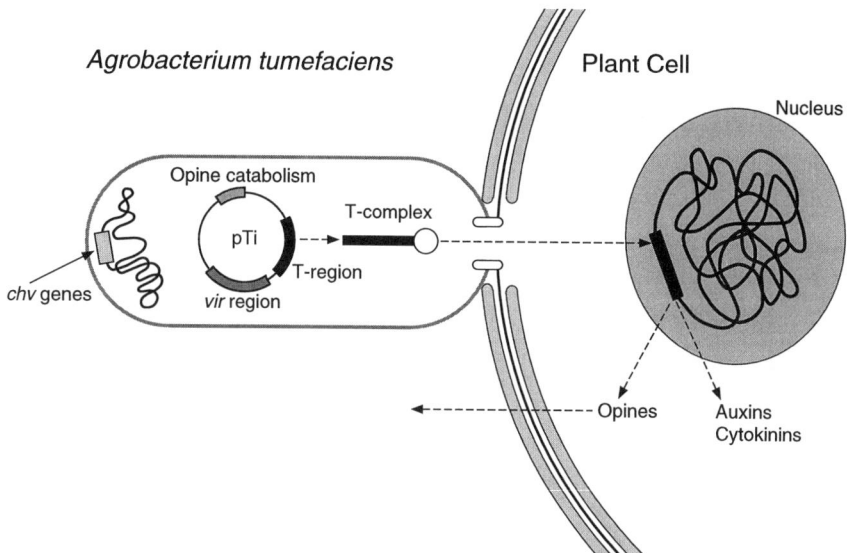

Figure 2. A model summarizing the interaction of *Agrobacterium* with plant cells.

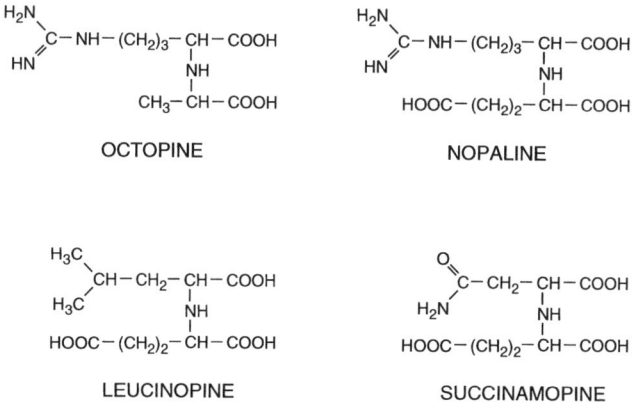

Figure 3. Four common opines and their structural formulae.

only degrade a specific opine. For instance, octopine catabolase is able to utilize octopine, but not nopaline. Ti plasmids are classified as either octopine, nopaline, succinamopine, or leucinopine types, depending upon the opine synthase and catabolase genes they carry. The opines secreted from the tumor can be catabolized by *Agrobacterium*, but not by most other soil organisms. By inducing these tumors, *Agrobacterium* is thought to gain a competitive advantage over other soil organisms in the rhizosphere. This process has been termed 'genetic colonization'.

THE Ti PLASMID

A. tumefaciens possesses a complex genome made up from a circular chromosome (3 Mbps), a linear chromosome (2.1 Mbps), a cryptic plasmid (450 kbps), and the Ti plasmid (200 kbps).[2] The Ti plasmid carries most of the genes important in tumorigenesis.

Ti plasmids are extrachromosomal elements of approximately 200 kb with a host range restricted to the Rhizobiaceae. The structure of a typical octopine type Ti plasmid is shown in Figure 4. Transfer of Ti to other bacteria, such as *Rhizobium leguminosarum* b.v. *trifolii*, renders these bacteria tumorigenic.[3] However, the related bacterium, *R. meliloti*, remains avirulent when it contains the Ti plasmid.[4] This can be explained by the presence of chromosomal virulence genes (*chv*) in *Agrobacterium*, some of which mediate attachment of the bacteria to their plant hosts. Mutations at these loci lead to avirulence on most plant species.[5–7] Homologs of these genes may be absent in other bacteria such as *R.meliloti*.

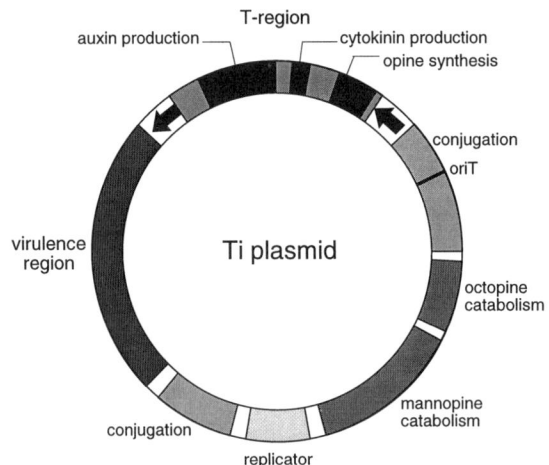

Figure 4. Genetic map of a Ti plasmid.

Ti plasmids carry transfer (*tra*) genes involved in the conjugal transfer of Ti plasmids between different *Agrobacterium* cells.[8] The structural genes of the Ti *tra* region are related to those of the wide host range *incP* plasmids.[9] Opines released from tumors not only induce the opine catabolism genes but also the *tra* genes of the Ti plasmids.[10–12] Therefore, in nature, conjugative transfer of Ti plasmids probably occurs only in the proximity of crown galls. Indeed, transfer of virulence properties from a virulent to an avirulent *Agrobacterium* strain both present in a crown gall tumor was the first indication of a possible extrachromsomal location for the virulence genes.[13]

Induction of the *tra* genes requires two different signals. First, a specific opine stimulates the transcription of a *tra* gene regulator, *traR*.[14] The TraR protein in turn activates the *tra* gene promoters, provided that a 3-oxo-octanoyl homoserine lactone, known as Agrobacterium Auto Inducer (AAI), is present.[15] AAI is synthesized by the TraI protein encoded by the Ti plasmid[16–17] and is able to diffuse across the *Agrobacterium* membrane. A high intracellular concentration of AAI is required for *tra* gene induction. In this way, AAI allows *A. tumefaciens* to regulate expression of the *tra* genes in a cell density dependent manner. Cell density dependent gene regulation (quorum sensing) is used by many other bacteria, *e.g. Vibrio fisheri* uses a similar system for regulation of its light producing *lux* genes.[18]

Despite occupying half of the Ti plasmid, the Ti plasmid *tra* genes and opine catabolism genes are not involved in tumorigenesis. This is determined by (1) the *onc* genes on the T-DNA, which confer the tumorous phenotype on plant cells, and (2) the products of the virulence (*vir*) genes present on the Ti plasmid, which process and transport the T-DNA to the plant cell.

During tumor induction, T-DNA copies are formed from the T-region of the Ti plasmid. Sequencing of the T-regions of several Ti plasmids revealed that these regions are surrounded by conserved 24 bp direct repeats known as the border sequences.[19] As the borders are located at either end of the T-region, they are referred to as the left or right border (LB and RB), respectively. Nopaline Ti plasmids only contain one T-region, while octopine Ti plasmids contain two T-regions (T_L-DNA and T_R-DNA), which can be independantly transferred to the plant cell.

The T-region of the Ti plasmid carries genes that confer the tumorous phenotype on the plant cells. Mutations in three different genes on the octopine T_L-DNA were identified which lead to non-oncogenicity on certain plant species (*e.g.* tomato) and aberrant tumor formation on others, such as tobacco and kalanchoe.[20–21] Two of these genes (*iaaM* and *iaaH*) mediate an auxin effect in the transformed plants, while the remaining gene is involved in a cytokinin effect and termed the *cyt* or *ipt* gene. Expression of these genes in *E. coli* showed the *iaaM* gene to be involved in the conversion of tryptophan to indole acetamide (IAM), which can then be converted into the active auxin indole acetic acid via the action of *iaaH*.[22–23] Interestingly, this pathway of auxin synthesis, which is also used by other gall inducing pathogenic bacteria, such as *Pseudomonas syringae* p.v. *savastanoi*,[24] is not normally used in plant cells. Synthesis of indole acetic acid in plant cells usually proceeds via the production of indole pyruvic acid as an intermediate. The T-DNA genes have well known 5' and 3' eukaryotic expression signals, such as the TATA box, for transcription initiation and the AATAAA box involved in transcription termination and poly-adenylation.[25] The T-DNA genes are also surrounded by various plant regulatory sequences, leading to a different expression of each gene depending on the transformed tissue type and/or the local concentrations of phytohormones. The *A. rhizogenes* T-DNA gene *rolA* has a unique mechanism of posttranscriptional control. *RolA* contains an intron which is removed in the plant cell via pre-mRNA splicing.[26] As this form of intron has not been previously discovered in prokaryotes, it raises the intriguing possibility that at least some of the genes present on the T-DNA may be eukaryotic in origin.

The deletion of the *onc* genes from the T-region leads to the loss of oncogenicity. Nevertheless, such non-oncogenic T-DNAs are still transferred, as shown by the local production of opines by plant cells at infection sites.[27-28] Further analysis has shown that none of the T-region genes are involved in transfer, but that the border repeats surrounding the T-region are *cis* acting signals for the transfer apparatus, which is determined by the virulence genes located adjacent to the T-DNA on the Ti plasmid. Below, we discuss in detail which chemical signals derived from plants induce *Agrobacterium* T-DNA transfer and how the transfer is accomplished.

THE VIRULENCE REGION

Mutagenesis of an area of the Ti plasmid adjacent to the T-region showed that it contained functions essential for *in vivo* tumor formation, and it was therefore named the virulence (*vir*) region. This *vir* region is not transferred to plant cells by *Agrobacterium*, but is involved in processing the T-DNA and transferring it to the plant cell. The *vir* region of the octopine type Ti plasmid is approximately 40 kb in size and consists of nine operons *virA* to *virJ* (Fig. 5).

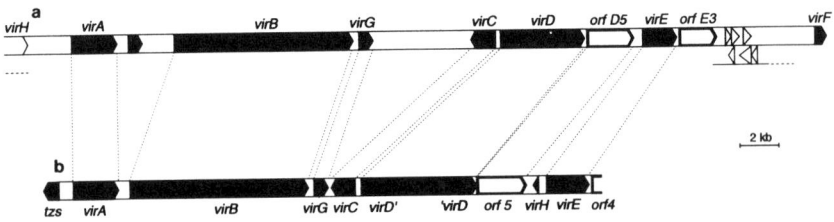

Figure 5. The virulence regions of (a) octopine and (b) nopaline type Ti plasmids.

INTERACTIONS BETWEEN *AGROBACTERIUM TUMEFACIENS* AND PLANT CELLS

Together the *vir* operons form a regulon, *i.e.* a set of operons that are coregulated by the same regulatory proteins.[29]

VirA/VirG: A Two Component Regulatory System

Wounding of plants facilitates plant tumor induction by *Agrobacterium*. Plant wound sites not only allow penetration of the bacterium, but also secrete a wide range of compounds which trigger the bacterium to prepare for T-DNA transfer. These signal compounds are detected by the VirA receptor protein, which together with the VirG protein forms a two component regulatory system which controls the *vir* regulon.[30] By fusing a reporter gene to the promoters of *vir* genes, the compounds which mediate induction of the *vir* system were identified.[31]

The first compounds purified from wounded tobacco tissue that led to induction of the *vir* genes were the phenolic compounds, acetosyringone (AS) and hydroxy-acetosyringone (Fig.6).[32] Since then, several other inducing compounds have also been identified, including other plant cell wall precursors and degradation products, such as coniferyl alcohol and sinapic acid.[33-35]

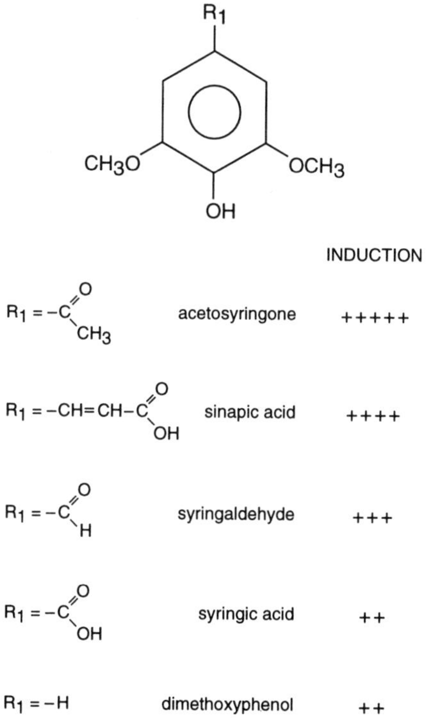

Figure 6. Acetosyringone and related phenolic compounds known to induce *vir* genes.

Some *Agrobacterium* isolates from Korea do not respond to the chemical inducer acetosyringone. Rather they are induced by related compounds, such as 4-hydroxyacetophone, which are only weak inducers of the laboratory strains normally used.[36] Genetic evidence suggests that this requirement for different inducers is dependent upon the *virA* gene. For infection of Douglas fir, specific *Agrobacterium* strains are needed. These strains produce high levels of a β-glucosidase, which converts the non-inducing compound coniferin into coniferyl alcohol.[37] Plants may also excrete compounds that are inhibitory to *vir* gene induction. For instance, certain maize varieties have an inhibitor in their root exudate, 2,4-dihydroxy-7-methoxy-1,4-benzoxazin-3-one (DIMBOA).[38]

A pH of 5.0–5.8 is also critical for efficient *vir* gene induction. This correlates well with the pH of plant sap, which presumably acidifies the soil after plant cell wounding.[30–31] Also, *vir* gene induction only occurs at a temperature below 30°C.

The VirA protein detects the low pH of the medium and acts as a receptor for the plant phenolic compounds. It is localized in the inner membrane of *Agrobacterium* and consists of two transmembrane domains with the C terminus protruding into the cytoplasm and the N terminus into the periplasm.[39] The topology of VirA is similar to many other sensor proteins from *E. coli*, including EnvZ, Tar, and Tsr. Surprisingly, the periplasmic domain is not the sensor domain for phenolic compounds but it does have a role in sensing external temperature and pH conditions. Upon sensing an inducing compound such as AS, VirA acts as an autokinase, phosphorylating itself on a conserved histidine residue in the C-terminal cytoplasmic part.[40–41] The phosphorylated form of the VirA protein can then transfer its phosphate moiety directly to a conserved aspartate residue in the response regulator, the VirG protein.[42] Phosphorylation of VirG is thought to activate its DNA binding activity, perhaps by inducing a conformational change in the protein. VirG then activates transcription of the *vir* genes by binding as a dimer or multimer to a specific 12bp sequence in the 5' non coding regions of the *vir* genes known as the 'vir box'.[43–44]

Monomers of plant cell wall polysaccharides, such as D-galacturonic acid, D-glucuronic acid, or the sugars glucose, arabinose, and galactose, act synergistically with AS in *vir* gene induction.[45] This effect is mediated by the chromosomally located *chvE* gene,[46] which shares homology with genes encoding periplasmically located sugar binding proteins in *E. coli*. ChvE binds monosaccharides and then interacts with the periplasmic domain of VirA, thus making VirA supersensitive to induction by phenolic compounds.[47]

Border Processing: The VirD1/VirD2 Endonuclease Action on T-DNA Borders

After *vir* gene induction, single stranded (ss) copies of the T-region (T-strands) can be detected in *Agrobacterium*.[48] Figure 7 summarizes the predicted molecular mechanism of T-strand production. The formation of T-strands

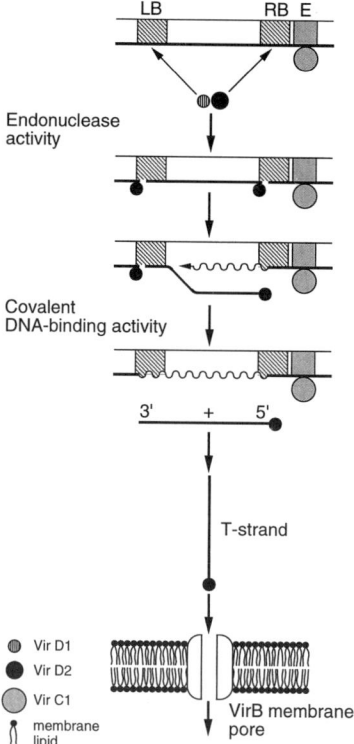

Figure 7. A model for T-strand synthesis in *Agrobacterium*.

is dependent on the activity of two proteins, VirD1 and VirD2.[49] Together, VirD1 and VirD2 form an endonuclease that introduces nicks (ss breaks) at a precise site in each border sequence. Purified VirD2 protein is able to specifically nick and ligate ss border sequences *in vitro*,[50] but requires VirD1 for nicking similar borders in a double stranded form.[51] The 3' OH ends of border nicks may serve as starting points for DNA synthesis, thereby releasing the T-strand from the T-region by a strand displacement mechanism.

Processing of the T-region by VirD1 and VirD2 to form T-strands is comparable to the processing of the origin of transfer (*oriT*) by mobilization proteins (Mob) during the transfer of plasmids between bacteria. Components involved in the two systems are also similar at the DNA sequence level. The LB and RB and the *oriT* sequence from certain wide host range plasmids show significant sequence homology,[52] as do VirD2 and the TraI protein of *incP* type plasmids.[53] These similarities suggest that the *Agrobacterium* T-DNA transfer system and the conjugation system for transferring plasmids between bacteria have common evolutionary origins. Indeed, the mobilizable *incQ* plasmid, RSF1010, can be transferred by the *Agrobacterium vir* system to plant cells.[54] This transfer is VirD2 independent, showing that the *incQ* Mob proteins are able to replace, at least partially, the function of VirD2 in transfer.

The VirC1 protein binds to the overdrive sequence located next to the RB.[55] This sequence strongly enhances T-strand formation in *Agrobacterium*,[56-57] perhaps by targetting the VirD1/VirD2 endonuclease to the RB, or by enhancing DNA synthesis from the RB nick site.

The T-Complex

The VirD2 and VirE2 proteins are bound to the T-strand and facilitate its transport to the plant cell nucleus. This complex of T-strand and Vir proteins is called the T-complex. During nicking of the RB, VirD2 becomes covalently attached to the T-DNA 5' end *in vivo*[58] and *in vitro*[50] at amino acid Tyr29. *Agrobacterium* strains producing only the N terminal half of VirD2 are unable to elicit tumors[59] but do generate T-strands.[60] Therefore, the N terminal part of VirD2 is involved in nicking the borders, and the C terminal part of VirD2 must play another role in tumorigenesis. Sequencing of VirD2 has revealed that the C terminus contains a bipartite nuclear localization signal (NLS) as found in many nuclear proteins.[61-62] Such NLSs consist of two basic regions with a KR/KXR/K motif separated by at least four amino acids. Fusion of this C terminal region of VirD2 to a β-glucuronidase (GUS) reporter protein can mediate its transport to the plant cell nucleus.[59-61] Therefore, the C terminal part of VirD2 has a role in targetting the T-strand to the plant cell nucleus. The VirD2 protein also protects the T-DNA from 5'-3' exonucleases.[63]

VirE2 is a 60kD protein which binds to ssDNA in a non sequence specific manner. VirE2 protein can convert T-strands from a compact folded structure into an extended linear form *in vitro*.[64] These extended structures may be more efficiently transferred into the plant cell nucleus.[65-67] *Agrobacterium* strains lacking *virE2* show reduced virulence on plants, and fewer T-strands accumulate in the plant cells infected by *virE2* mutants.[68] Recently, it was found that transgenic plants expressing the VirE2 protein can complement for the absence of VirE2 in the infecting agrobacteria, as shown by wild type tumor formation by *virE2* mutants on such plants.[69] Apparently, binding of VirE2 to the T-strand is not essential for transfer of the T-strand from the bacterium to plant cells. It may be that binding of VirE2 to the T-strand occurs within the plant cell and is important for transfer to the nucleus. Indeed, the C terminus of VirE2 contains two bipartite NLSs which are able to target β-glucuronidase into tobacco nuclei.[69] Microinjection studies have shown that VirE2 can mediate the import of ssDNA into the nucleus; fluorescently labelled ssDNA is unable to enter the plant cell nucleus unless complexed with VirE2.[70] Alternatively, VirE2 may have another function in the transfer process, *e.g.* in conversion of the T-strand into a double stranded form or in integration. T-DNA copies delivered from a *virE2* mutant strain which had integrated into the plant genome had large deletions at the LB end.[71] Therefore, VirE2 protects the T-DNA 3' end from degradation by plant exonucleases.

Transport of T-DNA to the Plant Cell: Role of VirB and VirD4 Proteins

Using a sensitive PCR based method, T-strands can be observed disappearing from AS induced *A. tumefaciens* cells within thirty minutes of bacterial co-cultivation with tobacco protoplasts.[68] At the same time point, the first T-strands can be detected in the tobacco protoplasts. This does not occur when the *Agrobacterium* strain has a mutation in the *virB* or *virD4* genes.[68]

The large *virB* operon (9600 bp) of several Ti plasmids has been sequenced and contains eleven genes.[72–75] All the *virB* genes, except *virB1*, are essential for T-DNA transfer to plant cells.[76] Nearly all the predicted proteins contain hydrophobic stretches and many have signal sequences. Thus, all of these VirB proteins are probably associated with the membrane and have periplasmic and/or extracellular domains, or are exported to the periplasm. Biochemical evidence for their presence as a complex associated with the membrane has been obtained. The VirB1 protein shares significant homology with a family of hydrolases which cleave β-1,4-glycosidic bonds found in the bacterial murein layer.[77–78] Therefore, this protein may assist in the formation of the transport apparatus by weakening the peptidoglycan layer. The VirB4 and VirB11 proteins are ATPases,[79–80] which reside at the cytoplasmic face of the cytoplasmic membrane. VirB6 is thought to be a cytoplasmic membrane protein with multiple membrane spanning domains, while VirB7 and VirB8 to VirB10 seem to be membrane associated proteins with large periplasmic domains. Studies on the assembly of the VirB transport apparatus have led to a model in which VirB7, an outer membrane-associated lipoprotein, acts with VirB9 to promote accumulation of the remaining VirB proteins to wild type levels.[81] The VirB7/VirB9 heterodimer may form the core around which the remaining VirB proteins assemble.[82]

The *virB* encoded transport apparatus of *Agrobacterium* can promote the transfer of a mobilizable RSF1010 plasmid not only to plant cells, but also to recipient bacterial cells.[83] This indicates a relatedness between the T-DNA transport system and the pilus/pore system determined by conjugative plasmids. Indeed, it has been found that the VirB proteins share homology with Tra proteins, especially those encoded by broad host range plasmids of the IncN, IncP, and IncW incompatibilty groups.[84–86]

Under the electron microscope, pili can be seen accumulating on the surface of induced *Agrobacterium* cells.[87] This suggests that pili may play an important role in T-DNA transfer, perhaps for attachment to plant cells, or by serving as a 'tunnel' through which T-DNA can travel into the plant cell. The

VirB2 protein shares some similarity with the TraA pilin protein of the *E. coli* F plasmid and the putative pilin subunit encoded by the IncN plasmid pKM101 and may, therefore, form the building block of the pilus structure. Processing of VirB2 seems to occur in a similar fashion to that of the TraA protein.[88]

The VirD4 protein is essential for T-DNA transfer and tumorigenisis.[89] The N-terminal region anchors VirD4 in the bacterial inner membrane, and the C terminus probably protrudes into the periplasm.[90] VirD4 has homology with a number of bacterial conjugation proteins, including TrwB of plasmid R388, TraD of F, and the TraG protein of RP4 and R751.[91] These proteins form part of the pilus/pore structure and facilitate transport of the nucleoprotein complex produced during conjugation/T-strand formation via the transmembrane pore to the recipient.

Protein Transport

When *virE* or *virF* mutants infect plants, only small tumors are formed. When such mutants are co-inoculated with a helper strain containing a complete Vir region, but lacking a T-region, normal tumors are formed by extra-cellular complementation.[92–93] Initially, it was thought that the *vir* loci mediated production of compounds which were secreted into the medium and which were necessary for tumor induction. However, further work found no evidence for this. For extra-cellular complementation, the helper strain needs to contain not only the *vir* regulatory genes, *virA* and *virG*, and the *virE* and *virF* genes, respectively, but also the *virB* and *virD4* genes determining the T-DNA transport apparatus.[94] Regensburg-Tuink, and Hooykaas[95] reasoned that the T-DNA transport apparatus might mediate transport of the VirE2 and VirF proteins directly into the plant cells of the host. To obtain evidence for this, the *virF* gene was cloned between plant regulatory sequences and expressed in transgenic plants. Such plants turned out to be infected effectively by *virF* mutants, showing that the VirF protein was active within the plant cell. Recently, it was found that the genes of the *virB* operon share sequence similarities with components of the *Bordetella pertussis* Ptl transporter, which directs the export of the six-subunit pertussis toxin to mammalian cells.[96] This corroborates the idea that the T-DNA transport system may be a specific type of protein transport system capable of transporting not only proteins, but also protein-DNA complexes.

T-DNA Integration

T-DNA integration sites seem random at the genome level. Analysis of T-DNA integration sites in *Arabidopsis thaliana* revealed an even distribution.[97] No T-DNA preferential integration sites could be found for *Crepis capilaris*,[98] *Lycopersicum esculentum*,[99] or *Petunia hybrida*.[100] T-DNA integrates preferentially into transcriptionally active areas.[101] The chromatin structure must be relaxed before transcription can take place, likely making the area more accessable for the T-DNA. Also, many of the enzymes perhaps necessary for T-DNA integration, such as DNA helicases, ligases, and polymerases, are abundant around transcription bubbles.

T-DNA integration has been studied also at the DNA level.[102–104] This has revealed that T-DNA integrates via an illegitimate recombination (IR) mechanism (random integration) at the sequence level. By comparing T-DNA/plant DNA junctions with the pre-integration genomic sites, certain features of T-DNA integration can be identified. T-DNA integration results in small deletions of the plant DNA at the site of insertion (13–73 bp).[104] The LB end of the T-DNA is less well conserved than the RB end, and 'filler' DNA of unknown origin is often found linking T-DNA and plant sequences.[104] Some integration events show short regions of homology (microhomology) between the T-DNA LB and the plant DNA.[103] Microhomology may be responsible for stabilizing the interaction between T-DNA and plant DNA before integration can occur.

Unlike other mobile DNA elements such as transposons, T-DNA does not encode any gene important for the integration process. Therefore, in order to integrate into the plant genome, the T-DNA must use existing host mechanisms. *A. tumefaciens* is able to transfer T-DNA to the yeast, *Saccharomyces cerevisiae*.[105] In *S. cerevisiae*, T-DNA sharing homology with the yeast genome integrates via homologous recombination. Even when T-DNA carries large homologous sequences with the plant genome, it still preferentially integrates into the plant genome via IR. Therefore, the dominant recombination mechanism of the host determines T-DNA integration. We have developed *S. cerevisiae* as a model organism for the study of host factors involved in the integration of T-DNA by IR.[106] Illegitimate recombination events in *S. cerevisiae* can be detected when the transforming DNA lacks homology with the yeast genome. By exploiting this, we have demonstrated that T-DNA can integrate via IR in *S. cerevisiae*. Furthermore, sequencing of the T-DNA/yeast DNA junctions has revealed that IR of T-DNA in yeast shows the same deletions of target DNA, loss of nucleotides from the LB, microhomology, and filler DNA found in the plant studies. Therefore, the IR mechanisms of higher and lower eukaryotes seem conserved. Several *S. cerevisiae* genes involved in IR have already been identified. These include the product of the *rad50* gene,[107] topoisomerase I,[108] and the Ku70/80 proteins.[109–111] A study of T-DNA integration in yeast may, therefore, lead to the identification of plant proteins involved.

APPLICATIONS OF THE *AGROBACTERIUM*/PLANT INTERACTION

The use of *Agrobacterium* in an efficient transformation system depends upon the introduction of new genes into the T-region. Cloning genes into the

T-region is made complicated by the large size of the Ti plasmid. An important discovery was made when it was found that no physical linkage was necessary between the T-region and the rest of the Ti plasmid for T-DNA transfer to occur.[112–113] From this, binary systems were developed in which an artificial T-DNA is present on a WHR plasmid, *i.e.* a minimal *incP* type plasmid with the RK2 origin of replication.[114] Binary plasmids carry a bacterial selection marker and can be maintained in both *E. coli* and *Agrobacterium*. Cloning of new genes into the T-DNA is done in *E. coli*, and the binary plasmid is subsequently transformed to *Agrobacterium* using either a tri-parental mating or electroporation protocol.[115] The Ti plasmids in such binary systems are called 'helper' Ti plasmids. They retain the *vir* region but lack the T-region so that only the T-DNA present on the binary plasmid is transferred. A wide range of binary vectors is now available carrying different combinations of bacterial replicons and selection markers.

There is an ever increasing list of plant species which can be transformed by *Agrobacterium*. This includes such plants as tobacco,[116] rapeseed,[117] apple,[118] and elm.[119] Since tumor formation is not observed on monocotyledonous plant species, these have not been considered as hosts for *Agrobacterium*. However, tumor formation is the end result of a complex process which depends upon many discrete steps, and, therefore, T-DNA transfer may sometimes not be accompanied by a phenotype. To obtain evidence that *Agrobacterium* was indeed able to transfer T-DNA to monocots, a biochemical assay for opine synthases was used.[120] Crude extracts from infection sites of *Agrobacterium* on the monocots *Chlorophytum capense* and *Narcissus* cv. Paperwhite were assayed for opine synthase activity. In this way, for the first time, evidence was obtained for T-DNA transfer to monocots. Another even more sensitive reporter system, termed 'agroinfection', was later developed.[121] In this assay, a virus is introduced into plants by the T-DNA, and successful T-DNA transfer is then visualized by the occurrence of symptoms of viral infection. Using this system, it was possible to demonstrate T-DNA transfer to maize,[122] and wheat and barley.[123] Although T-DNA transfer could thus be demonstrated, no stable transformants containing integrated T-DNA copies were obtained in these experiments. Recently, however, efficient T-DNA transfer and integration was demonstrated for rice by using scutellum tissue as the target for transformation with an *Agrobacterium* strain containing a so-called 'super binary' vector.[124] Similar approaches were used to obtain stable transformants of maize[125] and cassava.[126]

The *Agrobacterium* vector system also is used extensively for the study of gene function in plants. Due to the random nature of T-DNA integration, it has been used for insertional inactivation screens for many traits in the model plant *Arabidopsis thaliana*.[127] Such T-DNA tagged loci can be subsequently cloned and characterized. To date, at least 40 novel genes have been characterized in this way.[128] By placing a strong promoter near the RB, T-DNA can be used to activate the expression of genes near the T-DNA integration site. Such 'gain of function' mutants are easier to select for than 'loss of function' mutants, and often are of greater importance in understanding the physiology of plants.[127] Finally, T-DNA constructs containing a plant reporter gene lacking a promoter have been used to tag endogenous plant promoters. Expression of the reporter gene can be detected if the T-DNA integrates near a plant promoter.[127] Such an approach has been used successfully to identify plant genes upregulated by plant hormones (R. Offringa, personal communication).

Agrobacterium also has been used extensively for the transfer of traits to crop plants, confering resistance to many chemicals and pathogens. The bacterium, *Bacillus thuringiensis*, produces proteins which are toxic to a variety of insect species. Transgenic tobacco plants expressing the toxin gene *bt2* are protected from feeding damage by larvae of the tobacco hookworm.[129] Similarly, tobacco plants expressing the tobacco mosaic virus coat protein show resistance to this virus.[130]

Tomatoes with an extended shelf-life have been produced by specifically inhibiting a gene involved in ripening by using an 'antisense' construct.[131] Essentially, the gene of interest is isolated from the plant and then cloned behind a strong plant promoter, but in an inverted orientation. *Agrobacterium* is then used to integrate this antisense construct into the plant genome. Expression of the antisense construct has a negative effect on the endogenous gene and leads to decreased production of the protein.

Secondary metabolism also may be modified by the introduction of transgenes or antisense constructs by *Agrobacterium*. By introduction of a construct leading to overexpression of the enzyme, tryptophan decarboxylase (TDC), an enhanced level of tryptamine was found in *Brassica napus*, with a corresponding decrease in levels of the unpalatable indole glucosinolates derived from tryptophan.[132] Introduction of the *Erwinia carotovora* gene *crtI* for phytoene desaturase into tobacco led to increased production of β-carotenoids and resistance to the herbicide norflurazon but at the expense of α-carotenoid levels.[133] By introduction of an antisense construct for chalcone synthase (CHS), it was possible to modify flower color in petunia and chrysanthemum. Plants expressing the antisense CHS construct produced white flowers due to the inhibition of anthocyanin production.[134]

Components of the *Agrobacterium*/plant interaction can be tailored for specific tasks. For instance, plants containing opine synthase genes but not the *onc* genes have been constructed.[135-136] Bacteria in the rhizosphere containing opine catabolase genes are able to utilize the opines secreted by plant roots as a carbon and nitrogen source. It was shown that production of opines led to the selection of bacteria in the rhizosphere which are capable of degrading the substrate. This system may be useful in the future for tailoring microbes to plant hosts and helping genetically manipulated bacteria, such as nitrogen fixing *Rhizobium* sp., to compete with resident microflora.

A good understanding of the processes underlying the *Agrobacterium*-plant interaction has provided insight into a vast range of important bacterial and plant processes. Further study of this unique interaction will undoubtedly lead to the development of even more powerful and sophisticated tools for the genetic modification of plants.

REFERENCES

1. HOOYKAAS, P.J.J., BEIJERSBERGEN, A.G.M. 1994. The virulence system of *Agrobacterium tumefaciens*. Annu. Rev. Phytopathol. 32: 157–179.
2. ALLARDET-SERVENT, A., MICHAUX-CHARACHON, S., JUMAS-BILAK, E., KARAYAN, L., RAMUZ, M. 1993. Presence of one linear and one circular chromosome in the *Agrobacterium tumefaciens* C58 genome. J. Bacteriol. 175: 7869–7874.
3. HOOYKAAS, P.J.J., KLAPWIJK, P.M., NUTI, M.P., SCHILPEROORT, R.A., RORSCH, A. 1977. Transfer of the *Agrobacterium tumefaciens* Ti plasmid to avirulent agrobacteria and to *Rhizobium ex planta*. J. Gen. Microbiol. 98: 477–484.
4. VAN VEEN, R.J.M., DEN DULK-RAS, H., SCHILPEROORT, R.A., HOOYKAAS, P.J.J. 1989. Ti plasmid containing *Rhizobium meliloti* are non-tumorigenic on plants, despite proper virulence gene induction and T-strand formation. Arch. Microbiol. 153: 85–89.
5. DOUGLAS, C.J., STANELONI, R.J., RUBIN, R.A., NESTER, E.W. 1985. Identification and genetic analysis of an *Agrobacterium tumefaciens* chromosomal virulence region. J. Bacteriol. 161: 850–860.
6. MATTHYSSE, A.G. 1987. Characterization of non-attaching mutants of *Agrobacterium tumefaciens*. J. Bacteriol. 169: 313–323.
7. THOMASHOW, M.F., KARLINSEY, J.E., MARKS, J.R., HURLBERT, R.E. 1987. Identification of a new virulence locus in *Agrobacterium tumefaciens* that affects polysaccharide composition and plant cell attachment. J. Bacteriol. 169: 3209–3216.
8. FARRAND, S.K. 1993. Conjugal transfer of Agrobacterium plasmids. In: Bacterial Conjugation. (D.B. Clewell, ed.) Plenum Press, New York. pp. 255–291.
9. ALT-MORBE, J., STRYKER, J.L., FUQUA, C., LI, P.L., FARRAND, S.K., WINANS, S.C. 1996. The conjugal transfer system of *Agrobacterium tumefaciens* octopine type Ti plamsids is closely related to the transfer system of an IncP plasmid and distantly related to Ti plasmid vir genes. J. Bacteriol. 14: 4233–4247.
10. KLAPWIJK, P.M., SCHEULDERMAN, T., SCHILPEROORT, R.A. 1978. Coordinated regulation of octopine degradation and conjugative transfer of Ti plasmids in *Agrobacterium tumefaciens*: Evidence for a common regulatory gene and separate operons. J. Bacteriol. 136: 775–785.
11. PETIT, A., TEMPE, J., KERR, A., HOLSTERS, M., VAN MONTAGU, M., SCHELL, J. 1978. Substrate induction of conjugative activity of *Agrobacterium tumefaciens* Ti plasmids. Nature 271: 570–572.
12. ELLIS, J.G., KERR, A., PETIT, A., TEMPE J. 1982. Conjugal transfer of nopaline and agropine Ti plasmids. The role of agrocinopines. Mol. Gen. Genet. 186: 269–274.
13. KERR, A. 1969. Transfer of virulence between isolates of *Agrobacterium*. Nature 223: 1175–1176.
14. FUQUA, W.C., WINANS, S.C. 1994. A LuxR-LuxI type regulatory system activates Agrobacterium Ti plamsid conjugal transfer in the presence of a plant tumor metabolite. J. Bacteriol. 176: 2796–2806.
15. FUQUA, W.C., WINANS, S.C., GREENBERG, E.P. 1994. Quorum sensing in bacteria: The LuxR-LuxI family of cell-density responsive transcriptional regulators. J. Bacteriol. 176: 269–275.

16. HWANG, I., COOK, D.M., FARRAND, S.K. 1995. A new regulatory element modulates homoserine lactone mediated autoinduction of Ti plasmid conjugal transfer. J. Bacteriol. 177: 449–458.
17. MORÉ, M.I., FINGER, D.L., STRYKER, J.L., FUQUA, C., EBERHARD, A., WINANS, S.C. 1996. Enzymatic synthesis of a quorum sensing autoinducer through use of defined substrates. Science 272: 1655–1658.
18. MEIGHEN, E.A. 1991. Molecular biology of bacterial luminescence. Microbiol. Rev. 55: 123–142.
19. YADAV, N.S., VANDERLEYDEN, J., BENNETT, D.R., BARNES, W.M., CHILTON, M-D. 1982. Short direct repeats flank the T-DNA on a nopaline Ti plasmid. Proc. Natl. Acad. Sci. USA 79: 6322–6326.
20. GARFINKEL, D.J., SIMPSON, R.B., REAM, L.W., WHITE, F.F., GORDON, M.P., NESTER, E.W. 1981. Genetic analysis of crown gall: Fine structure map of the T-DNA by site directed mutagenesis. Cell 27: 143–153.
21. OOMS, G., HOOYKAAS, P.J.J., MOOLENAAR, G., SCHILPEROORT, R.A. 1981. Crown gall plant tumors of abnormal morphology, induced by *Agrobacterium tumefaciens* carrying mutated octopine Ti plasmids: Analysis of T-DNA functions. Gene 14: 33–50.
22. SCHRÖDER, G., WAFFENSCHMIDT, S., WEILER, E.W., SCHRÖDER, J. 1984. The T region of Ti plasmids codes for an enzyme synthesizing indole-3-acetic acid. Eur. J. Biochem. 138: 387–391.
23. THOMASHOW, M.F., HUGLY, S., BUCHHOLZ, W.G., THOMASHOW, L.S. 1986. Molecular basis for the auxin independent phenotype of crown gall tumor tissues. Science 231: 616–618.
24. YAMADA, T., PALM, C.J., BROOKS, B., KOSUGE, T. 1985. Nucleotide sequences of the *Pseudomonas savastanoi* indole acetic acid gene show homology with *Agrobacterium tumefaciens* T-DNA. Proc. Natl. Acad. Sci. USA 82: 6522–6526.
25. BARKER, R.F., IDLER, K.B., THOMPSON, D.V., KEMP, J.D. 1983. Nucleotide sequence of the T-DNA region from *Agrobacterium tumefaciens* octopine Ti plasmid pTi15955. Plant Mol. Biol. 2: 335–350.
26. MAGRELLI, A., LANGENKEMPER, K., DEHIO, C., SCHELL, J., SPENA, A. 1994. Splicing of the rolA transcript of *Agrobacterium rhizogenes* in *Arabidopsis*. Science 266: 1986–1988.
27. LEEMANS, J., DEBLAERE, R., WILLMITZER, L., DE GREVE, H., HERNALSTEENS, J.P., VAN MONTAGU, M., SCHELL, J. 1982. Genetic identification of functions of TL-DNA transcripts in octopine crown galls. EMBO J. 1: 147–152.
28. HILLE, J., WULLEMS, G., SCHILPEROORT, R.A. 1983. Non-oncogenic T-region mutants of *Agrobacterium tumefaciens* do transfer T-DNA into plant cells. Plant Mol. Biol. 2: 155–163.
29. WINANS, S.C. 1992. Two way chemical signalling in *Agrobacterium*-plant interactions. Microbiol. Rev. 56: 12–31.
30. STACHEL, S.E., ZAMBRYSKI, P.C. 1986. *VirA* and *VirG* control the plant induced activation of the T-DNA transfer process of *Agrobacterium tumefaciens*. Cell 46: 325–333
31. TURK, S.C.H.J., MELCHERS, L.S., DEN DULK-RAS, H., REGENSBURG-TUINK, A.J.G., HOOYKAAS, P.J.J. 1991. Environmental conditions differentially affect *vir* gene induction in different *Agrobacterium* strains. Role of the *VirA* sensor protein. Plant Mol. Biol. 16: 1051–1059.
32. STACHEL, S.E., MESSENS, E., VAN MONTAGU, M., ZAMBRYSKI, P. 1985. Identification of signal molecules produced by wounded plant cells that activate T-DNA transfer in *Agrobacterium tumefaciens*. Nature 318: 624–629.
33. SPENCER, P.A., TOWERS, G.H.N. 1988. Specificity of signal compounds detected by *Agrobacterium tumefaciens*. Phytochemistry 27: 2781–2785.
34. MELCHERS, L.S., REGENSBURG-TUINK, A.J.G., SCHILPEROORT, R.A., HOOYKAAS, P.J.J. 1989. Specificity of signal molecules in the activation of *Agrobacterium* virulence gene expression. Mol. Microbiol. 3: 969–977.

35. SONG, Y-N., SHIBUYA, M., EBRIZUKA, Y., SANKAWA, U. 1991. Identification of plant factors inducing virulence gene expression in *Agrobacterium tumefaciens*. Chem. Pharm. Bull. 39: 2347–2350.
36. LEE, Y-W., SHOUGUANG, J., SIM, W-S., NESTER, E.W. 1995. Genetic evidence for direct sensing of phenolic compounds by the VirA protein of *Agrobacterium tumefaciens*. Proc. Natl. Acad. Sci. USA 92: 12245–12249.
37. MORRIS, J.W., MORRIS, R.O. 1990. Identification of an *Agrobacterium tumefaciens* virulence gene inducer from the pinaceous gymnosperm *Pseudotsuga menziesii*. Proc. Natl. Acad. Sci. USA 87: 3614–3618.
38. SAHI, S.V., CHILTON, M-D., CHILTON, W.S. 1990. Corn metabolites affect growth and virulence of *Agrobacterium tumefaciens*. Proc. Natl. Acad. Sci. USA 87: 3879–3883.
39. MELCHERS, L.S., REGENSBURG-TUINK, A.J.G., BOURRET, R.B., SEDEE, N.J.A., SCHILPEROORT, R.A., HOOYKAAS, P.J.J. 1989. Membrane topology and functional analysis of the sensory protein VirA of *Agrobacterium tumefaciens*. EMBO J. 8: 1919–1925.
40. HUANG, Y., MOREL, P., POWELL, B., KADO, C.I. 1990. VirA, a coregulator of Ti-specified virulence genes, is phosphorylated in vitro. J. Bacteriol. 172: 1142–1144.
41. JIN, S., ROITSCH, T., ANKENBAUER, R.G., GORDON, M.P., NESTER, E.W. 1990. The VirA protein of *Agrobacterium tumefaciens* is autophosphorylated and is essential for *vir* gene regulation. J. Bacteriol. 172: 525–530.
42. JIN, S., PRUSTI, R.K., ROITSCH, T., ANKENBAUER, R.G., NESTER, E.W. 1990. The VirG protein of *Agrobacterium tumefaciens* is phosphorylated by the autophosphorylated VirA protein and this is essential for its biological activity. J. Bacteriol. 172: 4945–4950.
43. PAZOUR, G.J., DAS, A. 1990. VirG, an *Agrobacterium tumefaciens* transcriptional activator, initiates translation at a UUG codon and is a sequence specific DNA-binding protein. J. Bacteriol. 172: 1241–1249.
44. TAMAMOTO, S., AOYAMA, T., TAKANAMI, M., OKA, A. 1990. Binding of the regulatory protein VirG to the phased signal sequences upstream from virulence genes on the hairy root inducing plasmid. J. Mol. Biol. 215: 537–547.
45. SHIMODA, N., TOYODA-YAMAMOTO, A., NAGAMINE, J., USAMI, S., KATAYAMA, M. 1990. Control of expression of *Agrobacterium vir* genes by synergistic actions of phenolic signal molecules and monosaccharides. Proc. Natl. Acad. Sci. USA 87: 6684–6688.
46. CANGELOSI, G.A., ANKENBAUER, R.G., NESTER, E.W. 1990. Sugars induce the *Agrobacterium* virulence genes through a periplasmic binding protein and a transmembrane signal protein. Proc. Natl. Acad. Sci. USA 87: 6708–6712.
47. SHIMODA, N., TOYODA-YAMAMOTO, A., AOKI, S., MACHIDA, Y. 1993. Genetic evidence for an interaction between the VirA sensor protein and the ChvE sugar binding protein of *Agrobacterium*. Proc. Natl. Acad. Sci. USA 268: 26552–26558.
48. STACHEL, S.E., TIMMERMAN, B., ZAMBRYSKI, P. 1986. Generation of single-stranded T-DNA molecules during the initial stages of T-DNA transfer from *Agrobacterium tumefaciens* to plant cells. Nature 322: 706–712.
49. YANOFSKY, M.F., PROTER, S.G., YOUNG, C., ALBRIGHT, L.M., GORDON, M.P., NESTER, N.W. 1986. The *virD* operon of *Agrobacterium tumefaciens* encodes a site-specific endonuclease. Cell 47: 471–477.
50. PANSEGRAU, W., SCHOUMACHER, F., HOHN, B., LANKA, E. 1993. Site-specific cleavage and joining of single-stranded DNA by VirD2 protein of *Agrobacterium tumefaciens* Ti plamsids: Analogy to bacterial conjugation. Proc. Natl. Acad. Sci. USA 90: 11538–11542.
51. SCHEIFFELE, P., PANSEGRAU, W., LANKA, E. 1995. Initiation of *Agrobacterium tumefaciens* T-DNA processing. J. Bio. Chem. 270: 1269–1276.
52. PANSEGRAU, W., LANKA, E. 1991. Common sequence motifs in DNA relaxases and nick regions from a variety of DNA transfer systems. Nuc. Acids Res. 19: 3455.

53. ILYINA, T.V., KOONIN, E.V. 1992. Conserved sequence motifs in the initiator proteins for rolling circle DNA replication encoded by diverse replicons from eubacteria, eukaryotes and archaebacteria. Nuc. Acids Res. 20: 3279–3285.
54. BUCHANON-WOLLASTON, V., PASSIATORE, J.E., CANNON, F. 1987. The mob and oriT mobilization functions of a bacterial plasmid promote its transfer to plants. Nature 328: 172–175.
55. TORO, N., DATTA, A., CARMI, O.A., YOUNG, C., PRUSTI, R.K., NESTER, E.W. 1989 The *Agrobacterium tumefaciens virC1* gene product binds to overdrive, a T-DNA transfer enhancer. J. Bacteriol. 171: 6845–6849.
56. PERALTA, E.G., HELLMISS, R., REAM, W. 1986. Overdrive, a T-DNA transmission enhancer on the *A.tumefaciens* tumour inducing plasmid. EMBO J. 5: 1137–1142.
57. VAN HAAREN, M.J.J., SEDEE, N.J.A., SCHILPEROORT, R.A., HOOYKAAS, P.J.J. 1987. Overdrive is a T-region transfer enhancer which stimulates T-strand production in *Agrobacterium tumefaciens*. Nuc. Acids. Res. 15: 8983–8997.
58. VOGEL, A.M., DAS, A. 1992. Mutational analysis of *Agrobacterium tumefaciens* VirD2: tyrosine 29 is essential for endonuclease activity. J. Bacteriol. 174: 303–308.
59. KOUKOLIKOVA-NICOLA, Z., RAINERI, D., STEPHENS, K., RAMOS, C., TINLAND, B., HOHN, B. 1993. Genetic analysis of the *virD* operon of *Agrobacterium tumefaciens*: A search for functions involved in transport of T-DNA into the plant cell nucleus and in T-DNA integration. J. Bacteriol. 175: 723–731.
60. STACHEL, S.E., TIMMERMAN, B., ZAMBRYSKI, P. 1987. Activation of *Agrobacterium tumefaciens vir* gene expression generates multiple single stranded T-strand molecules from the pTiA6 T-region: Requirement for 5' *virD* gene products. EMBO J. 6: 857–863.
61. HOWARD, E.A., ZUPAN, J.R., CITOVSKY, V., ZAMBRYSKI, P. 1992. The VirD2 protein of *Agrobacterium tumefaciens* contains a C terminal bipartite nuclear localization signal: Implications for nuclear uptake of DNA in plant cells. Cell 68: 109–118.
62. ROSSI, L., HOHN, B., TINLAND, B. 1993. The VirD2 protein of *Agrobacterium tumefaciens* carries nuclear localization signals important for transfer of T-DNA to plants. Mol. Gen. Genet. 239: 345–353.
63. DÜRRENBERGER, F., CRAMERI, A., HOHN, B., KOUKOLIKOVA-NICOLA, Z. 1989. Covalently bound VirD2 protein of *Agrobacterium tumefaciens* protects the T-DNA from exonucleolytic degradation. Proc. Natl. Acad. Sci. USA 86: 9154–9158.
64. CITOVSKY, V., WONG, M.L., ZAMBRYSKI, P. 1989. Cooperative interaction of *Agrobacterium* VirE2 protein with single stranded DNA: Implications for the T-DNA transfer process. Proc. Natl. Acad. Sci. USA 86: 1193–1197.
65. CHRISTIE, P.J., WARD, J.E., WINANS, S.C., NESTER, E.W. 1988. The *Agrobacterium tumefaciens virE2* gene product is a single stranded DNA binding protein that associates with T-DNA. J. Bacteriol. 170: 2584–2591.
66. SEN, P., PAZOUR, G.J., ANDERSON, D., DAS, A. 1989. Cooperative binding of *Agrobacterium tumefaciens* VirE2 protein to single-stranded DNA. J. Bacteriol. 171: 2573–2580.
67. HOWARD, E., CITOVSKY, V. 1990. The emerging structure of the *Agrobacterium* T-DNA transfer complex. BioEssays 12: 103–108.
68. YUSIBOV, V.M., STECK, T.R., GUPTA, V., GELVIN, S.B. 1994. Association of single-stranded transferred DNA from *Agrobacterium tumefaciens* with tobacco cells. Proc. Natl. Acad. Sci. USA 91: 2994–2998.
69. CITOVSKY, V., ZUPAN, J., WARNICK, D., ZAMBRYSKI, P. 1992. Nuclear localization of *Agrobacterium* VirE2 protein in plant cells. Science 256: 1802–1805.
70. ZUPAN, J.R., CITOVSKY, V., ZAMBRYSKI, P. 1996. *Agrobacterium* VirE2 protein mediates nuclear uptake of single-stranded DNA in plant cells. Proc. Natl. Acad. Sci. USA 93: 2392–2397.
71. ROSSI, L., HOHN, B., TINLAND, B. 1996. Integration of complete transferred DNA units is dependant on the activity of virulence E2 protein of *Agrobacterium tumefaciens*. Proc. Natl. Acad. Sci. USA 93: 126–130.

72. THOMPSON, D.V., MELCHERS, L.S., IDLER, K.B., SCHILPEROORT, R.A., HOOYKAAS, P.J.J. 1988. Analysis of the complete nucleotide sequence of the *Agrobacterium tumefaciens virB* operon. Nuc. Acids Res. 16: 4621–4636.
73. WARD, J.E., AKIYOSHI, D.E., REGIER, D., DATTA, A., GORDON, M.P., NESTER, E.W. 1988. Characterization of the *virB* operon from an *Agrobacterium tumefaciens* Ti plasmid. J. Biol. Chem. 263: 5804–5814.
74. KULDAU, G.A., DE VOS, G., OWEN, J., McCAFFERY, G., ZAMBRYSKI, P. 1990. The *virB* operon of *Agrobacterium tumefaciens* pTiC58 encodes 11 open reading frames. Mol. Gen. Genet. 221: 256–266.
75. SHIRASU, K., MOREL, P., KADO, C.I. 1990. Characterization of the virB operon of an *Agrobacterium tumefaciens* Ti plasmid: nucleotide sequence and protein analysis. Mol. Microbiol. 4: 1153–1163.
76. BERGER, B.R., CHRISTIE, P.J. 1994. Genetic complementation analysis of the *Agrobacterium tumefaciens virB* operon: *virB2* through *virB11* are essential virulence genes. J. Bacteriol. 176: 3646–3659.
77. BAYER, M., EFERL, R., ZELLNIG, G., TEFERLE, K., DIJKSTRA, A., KORIAMANN, G. 1995. Gene 19 of plasmid R1 is required for both efficient conjugative DNA transfer and bacteriophage R17 infection. J. Bacteriol. 177: 4279–4288.
78. MUSHEGIAN, A.R., FULLNER, K.J., KOONIN, E.V., NESTER, E.W. 1996. A family of lysozyme like virulence factros in bacterial pathogens of plants and animals. Proc. Natl. Acad. Sci. USA 93: 7321–7326.
79. CHRISTIE, P.J., WARD, J.E., GORDON, M.P., NESTER, E.W. 1989. A gene required for transfer of T-DNA to plants encodes a ATPase with autophosphorylating activity. Proc. Natl. Acad. Sci. USA 86: 9677–9681.
80. JONES, A.L., SHIRASU, D., KADO, C.I. 1994. The product of the virB4 gene of *Agrobacterium tumefaciens* promotes accumulation of the VirB3 protein. J. Bacteriol. 176: 5255–5261.
81. FERNANDEZ, D., SPUDICH, G.M., ZHOU, X,-R., CHRISTIE, P.J. 1996. The *Agrobacterium tumefaciens* VirB7 lipoprotein is required for stabilization of VirB proteins during assembly of the T-complex transport apparatus. J. Bacteriol. 178: 3168–3176.
82. SPUDICH, G.M., FERNANDEZ, D., ZHOU, X,-R., CHRISTIE, P.J. 1996. Intermolecular disulphide bonds stabilize VirB7 homodimers and VirB7/VirB9 heterodimers during biogenesis of the *Agrobacterium tumefaciens* T-complex transport apparatus. Proc. Natl. Acad. Sci. USA 93: 7512–7517.
83. BEIJERSBERGEN, A., DULK-RAS, A.D., SCHILPEROORT, R.A., HOOYKAAS, P.J.J. (1992). Conjugative transfer by the virulence system of *Agrobacterium tumefaciens*. Science 256: 1324–1327.
84. LESSL, M., BALZER, D., PANSEGRAU, W., LANKA, E. 1992. Sequence similarities between the RP4 Tra2 and the virB regions strongly support the conjugation model for T-DNA transfer. J. Biol. Chem. 267: 20471–20480.
85. KADO, C.I. 1993. *Agrobacterium* mediated transfer and stable incorporation of foreign genes in plants. In: Bacterial Conjugation. (D.B. Clewell, ed.) Plenum, New York. pp. 243–254.
86. POHLMAN, R.F., GENETTI, H.D., WINANS, S.C. 1994. Common ancestry between IncN conjugal transfer genes and macromolecular export systems of plant and animal pathogens. Mol. Microbiol. 14: 655–668.
87. FULLNER, K.J., LANA, J.C., NESTER, E.W. 1996. Pilus assembly by *Agrobacterium* T-DNA transfer genes. Science 273: 1107–1109.
88. JONES, A.L., ERH-MIN, L., SHIRASU, K., KADO, C.I. 1996. VirB2 is a processed pilin-like protein encoded by the *Agrobacterium tumefaciens* Ti plasmid. J. Bacteriol. 178: 5706–5711.
89. LIN, T.-S., KADO, C.I. 1993. The *virD4* gene is required for virulence while the *virD3* and *orf5* are not required for virulence of *Agrobacterium tumefaciens*. Mol. Microbiol. 9: 803–812.

90. OKAMOTO, S., TOYODA-YAMAMOTO, A., ITO, K., TAKEBE, I., MACHIDA, Y. 1991. Localization and orientation of the VirD4 protein of *Agrobacterium tumefaciens* in the cell membrane. Mol. Gen. Genet. 288: 24–32.
91. LESSL, M., PANSEGRAU, W., LANKA, E. 1992. Relationship of DNA transfer systems: Essential transfer functions of plasmids RP4, Ti and F share common sequences. Nucleic Acid Res. 20: 6099–6100.
92. OTTEN, L.A.B.M., DEGREVE, H., LEEMANS, J., HAIN, R., HOOYKAAS, P.J.J., SCHELL, J. (1984). Restoration of virulence of vir region mutants of *Agrobacterium tumefaciens* strain B6S3 by coinfection with normal and mutant *Agrobacterium* strains. Mol. Gen. Genet. 195: 159–163.
93. OTTEN, L.A.B.M., PIOTROWIAK, G., HOOYKAAS, P.J.J., DUBOIS, M., SZEGEDI, E., SCHELL, J. 1985. Identification of an *Agrobacterium tumefaciens* pTiB6S3 vir region fragment that enhances the virulence of pTiC58. Mol. Gen. Genet. 199: 189–193.
94. MELCHERS, L.S., MARONEY, M.J., DEN DULK-RAS, A., THOMPSON, D.V., VAN VUUREN, H.A.J., SCHILPEROORT, R.A. HOOYKAAS, P.J.J. 1990. Octopine and nopaline strains of *Agrobacterium tumefaciens* differ in virulence: Molecular characterization of the *virF* locus. Plant Mol. Biol. 14: 249–259.
95. REGENSBERG-TUINK, A.J.G., HOOYKAAS, P.J.J. 1993. Transgenic *N. glauca* plants expressing bacterial virulence gene *virF* are converted into hosts for nopaline strains of *A. tumefaciens*. Nature 363: 69–70.
96. WEISS, A.A., JOHNSON, F.D., BURNS, D.L. 1993. Molecular characterization of an operon required for pertussis toxin secretion. Proc. Natl. Acad. Sci. USA 90: 2970–2974.
97. TINLAND, B. 1996. The integration of T-DNA into plant genomes. Trends In Plant Science 1: 178–184.
98. AMBROS, P.F., MATZKE, A.J.M., MATZKE, M.A. 1986. Localization of *Agrobacterium rhizogenes* T-DNA in plant chromosomes by in situ hybridization. EMBO J. 5: 2073–2077.
99. THOMAS, C.M., JONES, D.A., ENGLISH, J.J., CARROLL, B.J., BENNETZEN, J.L., HARRISON, K., BURBIDGE, A., BISHOP, G.J., JONES, J.D.G. 1994. Analysis of the chromsomal distribultion of transposon carrying T-DNAs in tomato using the inverse polymerase chain reaction. Mol. Gen. Genet. 242: 573–585.
100. WALLROTH M, GERATS, A.G.M., ROGERS, S.G., FRALEY, R.T., HORSCH, R.B. 1986. Chromosomal localization of foreign genes in *Petunia hybrida*. Mol. Gen. Genet. 202: 6–15.
101. HERMAN, L., JACOBS, A., VAN MONTAGU, M., DEPICKER, A. 1990. Plant chromosome/marker gene fusion assay for study of normal and truncated T-DNA integration events. Mol. Gen. Genet. 224: 248–256.
102. GHEYSEN, G., VILLAROEL, R., VAN MONTAGU, M. 1991. Illegitimate recombination in plants: A model for T-DNA integration. Genes Dev. 5: 287–297.
103. MATSUMOTO, S., ITO, Y., HOSOI, T., TAKAHASHI, Y., MACHIDA, Y. 1990. Integration of *Agrobacterium* T-DNA into a tobacco chromosome: Possible involvement of DNA homology between T-DNA and plant DNA. Mol. Gen. Genet. 224: 309–316.
104. MAYERHOFER, R., KONCZ-KALMAN, Z., NAWRATH, C., BAKKEREN, G., CRAMERI, A., ANGELIS, K., REDEI, G.P., SCHELL, J., HOHN, B., KONCZ, C. 1991. T-DNA integration: A mode of illegitimate recombination in plants. EMBO J. 10: 697–704.
105. BUNDOCK, P., DEN DULK-RAS, A., BEIJERSBERGEN, A.G.M., HOOYKAAS, P.J.J. 1995. Trans-kingdom T-DNA transfer from *Agrobacterium tumefaciens* to *Saccharomyces cerevisiae*. EMBO J. 14: 3206–3214.
106. BUNDOCK, P., HOOYKAAS, P.J.J. 1996. Integration of *Agrobacterium tumefaciens* T-DNA in the *Saccharomyces cerevisiae* genome by illegitimate recombination. Proc. Natl. Acad. Sci. USA 93: 15272–15275.
107. SCHIESTL, R.H., ZHU, J., PETES, T.D. 1994. Effect of mutations in genes affecting homologous recombination on restriction enzyme mediated and illegitimate recombination in *Saccharomyces cerevisiae*. Mol. Cell Biol. 14: 4493–4500.

108. ZHU, J., SCHIESTL, R.H. 1996. Topoisomerase I involvement in illegitimate recombination in *Saccharomyces cerevisiae*. Mol. Cell Biol. 16: 1805–1812.
109. BOLTON, S.J., JACKSON, S.P. 1996. *Saccharomyces cerevisiae* Ku70 potentiates illegitimate double strand break repair and serves as a barrier to error prone DNA repair pathways. EMBO J. 15: 5093–5103.
110. BOLTON, S.J., JACKSON, S.P. 1996. Identification of a *Saccharomyces cerevisiae* Ku80 homologue: Roles in DNA double strand break rejoining and in telomeric maintainance. Nuc. Acids Res. 24: 4639–4648.
111. FELDMAN, H., DRILLER, L., MEIER, B., MAGES, G., KELLERMAN, J., WINNACKER, E.L. 1996. HDF2, the second subunit of the Ku homologue from *Saccharomyces cerevisiae*. J. Biol. Chem. 271: 27765–27769.
112. DE FRAMOND, A.J., BARTON, K.A., CHILTON, M-D. 1983. Mini Ti: A new vector strategy for plant genetic engineering. Biotechnology 1: 262–269.
113. HOEKEMA, A., HIRSCH, P.R., HOOYKAAS, P.J.J., SCHILPEROORT, R.A. 1983. A binary plant vector strategy based on separation of *vir* and T-region of the *Agrobacterium tumefaciens* Ti plasmid. Nature 303: 179–180.
114. BEVAN, M. 1984. Binary *Agrobacterium* vectors for plant transformation. Nuc. Acids Res. 12: 8711–8721.
115. DEN DULK-RAS, A., HOOYKAAS, P.J.J. 1995. Electroporation of *Agrobacterium tumefaciens*. In: Methods in Molecular Biology Vol 55: Plant Cell Electroporation and Electrofusion Protocols. (J.A.Nickoloff, ed.) Humana Press Inc, Totowa, NJ. pp. 63–72.
116. HORSCH, R.B., FRY, J.E., HOFFMANN, N.L., EICHHOLTZ, D., ROGERS, S.G., FRALEY, R.T. 1985. A simple and general method for transferring genes into plants. Science 227: 1229–1231.
117. CHAREST, P.J., HOLBROOK, L.A., GABARD, J., IYER, V.N., MIKI, B.L. 1988. *Agrobacterium* mediated transformation of thin cell layer explants from *Brassica napus* L. Theor. Appl. Genet. 75: 438–445.
118. PUITE, K.J., SCHAART, J.G. 1996. Genetic modification of the commercial apple cultivars gala, golden delicious and elstar via an *Agrobacterium tumefaciens* mediated transformation method. Plant Science 119: 125–133.
119. FENNING, T.M., TYMENS, S.S., GARTLAND, J.S., BRASIER, C.M., GARTLAND, K.M.A. 1996. Transformation and regeneration of english elm using wild type *Agrobacterium tumefaciens*. Plant Science 116: 37–46.
120. HOOYKAAS-VAN SLOGTEREN, G.M.S., HOOYKAAS, P.J.J., SCHILPEROORT, R.A. 1984. Expression of Ti plasmid genes in monocotyledonous plants infected with *Agrobacterium tumefaciens*. Nature 311: 763–764.
121. GRIMSLEY, N., HOHN, B., HOHN, T., WALDEN, R. 1986. Agroinfection' an alternative route for viral infection of plants by using the Ti plasmid. Proc. Natl. Acad. Sci. USA 83: 3282–3286.
122. GRIMSLEY, N., HOHN, T., DAVIES, J.W., HOHN, B. 1987. *Agrobacterium* mediated delivery of infectious maize streak virus into maize plants. Nature 325: 177–179.
123. BOULTON, M.I., BUCHHOLZ, W.G., MARKS, M.S., PARKHAM, P.G., DAVIES, J.W. 1989. Specificity of *Agrobacterium* mediated delivery of maize streak virus DNA to members of the Gramineae. Plant Mol. Biol. 12: 31–40.
124. HIEI, Y., OHTA, S., KOMARI, T., KUMASHIRO, T. 1994. Efficient transformation of rice (*Oryza sativa* L.) mediated by *Agrobacterium* and sequence analysis of the boundaries of the T-DNA. Plant Journal 6: 271–282.
125. ISHIDA, Y., SAITO, H., OHTA, S., HIEI, Y., KOMARI, T., KUMASHIRO, T. 1996. High efficiency transformation of maize (*Zea mays* L.) mediated by *Agrobacterium tumefaciens*. Nature Biotechnology 14: 745–750.

126. LI, H-Q., SAUTTER, C., POTRYKUS, I., PUONTI-KAERLAS, J. 1996. Genetic transformation of cassava (*Manihot esculenta* Crantz). Nature Biotechnology 14: 736–739.
127. KONCZ, C., NEMETH, K., REDEI, G.P., SCHELL, J. 1992. T-DNA insertional mutagenesis in *Arabidopsis*. Plant. Mol. Biol. 20: 963–976.
128. AZIPIROZ-LEEHAN, R., FELDMANN, K.A. 1997. T-DNA insertion mutagenesis in *Arabidopsis*: Going back and forth. Trends in Gen. 13: 152–156.
129. VAECK, M., REYNAERTS, A., HÖFTE, H., JANSENS, S., DE BEUCKELEER, M., DEAN, C., ZABEAU, M., VAN MONTAGU, M., LEEMANS. 1987. Transgenic plants protected from insect attack. Nature 328: 33–37.
130. ABEL, P.P., NELSON, R.S., HOFFMAN, N., ROGERS, S.G., FRALEY, R.T., BEACHY, R.N. 1986. Delay of disease development in transgenic plants that express the tobacco mosaic virus coat protein gene. Science 232: 738–743.
131. SMITH, C.J.S., WATSON, C.F., RAY, J., BIRD, C.R., MORRIS, P.C., SCHUCH, W., GRIERSON, D. 1988. Antisense RNA inhibition of polygalacturonase gene expression in transgenic tomatoes. Nature 334: 724–726.
132. CHAVADEJ, S., BRISSON, N., McNEIL, J.N., DE LUCA, V. 1994. Redirection of tryptophan leads to production of low indole glucosinolate canola. Proc. Natl. Acad. Sci. USA 91: 2166–2170.
133. MISAWA, N., MASAMOTO, K., HORI, T., OHTANI, T., BÖGER, P., SANDMANN, G. 1994. Expression of an *Erwinia* phytoene desaturase gene not only confers multiple resistance to herbicides interfering with carotenoid biosynthesis but also alters xanthophyll metabolism in transgenic plants. Plant J. 6: 481–489.
134. COURTNEY-GUTTERSON, N., NAPOLI, C., LEMIEUX, C., MORGAN, A., FIROOZABADY, E., ROBINSON, E.P. 1994. Modification of flower color in florists Chrysanthemum: Production of a white flowering variety through molecular genetics. Biotechnology 12: 268–271.
135. OGER, P., PETIT, A., DESSAUX, Y. 1997. Genetically engineered plants producing opines alter their biological environment. Nature Biotechnology 15: 369–372.
136. SAVKA, M.A., FARRAND, S.K. 1997. Modification of rhizobacterial populations by engineering bacterium utilization of a novel plant-produced resource. Nature Biotechnology 15: 363–368.

Chapter Twelve

WOUND AND DEFENSE RESPONSES IN CASSAVA AS RELATED TO POST-HARVEST PHYSIOLOGICAL DETERIORATION

John R. Beeching, Yuanhuai Han, Rocío Gómez-Vásquez,
Robert C. Day, and Richard M. Cooper

Department of Biology and Biochemistry
University of Bath
Bath BA2 7AY, United Kingdom

Introduction 231
Post-Harvest Physiological Deterioration 233
   Inhibition of PPD 238
   Potential for Genetic Improvement 240
Experimental Strategies 241
   Isolation and Characterization of PPD-Related Clones 241
   Use of Cassava Cell Suspension Cultures 242
Conclusions 243

INTRODUCTION

Cassava (*Manihot esculenta* Crantz), a member of the Euphorbiaceae, is a perennial shrub of one to three meters high, which originates in Latin America where there are two centers of diversity of the genus *Manihot*, a major one in Brazil and a minor one in Mexico and Guatemala. After the colonization of America by Spain and Portugal, the cultivation of cassava spread during the sixteenth and subsequent centuries from Latin America to Africa, the Philippines and beyond.[1] Today cassava is grown throughout the humid tropics principally as a root crop; world production in 1991 was 161.5 million metric tons of which 75 million was in Africa.[2] In these regions, it is the staple food of over 500 million people; in some African countries, such as Zaire, cassava can provide over 50% of the dietary carbohydrate.[3] However, in other countries such as Thailand, cassava is increasingly being grown as animal feed for export or as input into the starch industry.

Cassava is cultivated for its large starchy roots. It is one of the most efficient producers of starch and is the fourth most important producer of carbohydrate after rice, sugar cane, and maize. However, while 85% of the storage root dry weight is starch, only 0.7–2.6% is protein, in which cysteine is in limiting quantities.[1] The storage roots develop by a process of secondary thickening to form a structure consisting of an outer periderm and a thin cortex surrounding a core of mainly starch parenchyma which forms the bulk of the root. It is important to realize that these swollen roots do not possess bud primordia and are incapable of acting as organs of propagation. They simply serve as stores of carbohydrate which may be used in times of climatic adversity, such as drought-induced defoliation, to enable the plant as a whole to survive. Detached from the plant, the roots serve no natural function, and, as a result, unlike other crops, such as potatoes or yams, cassava storage roots do not enter a period of dormancy and cannot serve as propagules.

As a crop, cassava is propagated vegetatively by placing mature stem cuttings of about 30 cm in the soil either vertically or horizontally. While moisture is required for the establishment of the plant, cassava will withstand periods of drought. It will grow on impoverished soil and on soils which contain high levels of toxic compounds, such as aluminium and magnesium, which would inhibit the growth of other crops such as maize or beans.[1] The roots can be harvested within nine to 12 months after planting. The minimal technology required for its cultivation, coupled to its hardiness and robustness, makes cassava a vitally important crop to the poorest sections of society, especially in Africa. It is a critical crop for survival in the event of famine, war, or other social disruptions. With increasing population pressures and resulting demands for food and resources, it is anticipated that cassava cultivation and production will increase world-wide.

Though of vital importance to millions, as a crop of the poor in developing countries cassava has not been subject to the intense research afforded temperate species such as potato or model systems such as *Arabidopsis*. Until recently, research has been intermittent and to a certain extent haphazard. However, during the last few years, the importance of cassava has been recognized, and research in national and international centers both in developing and developed countries is now linked via the Cassava Biotechnology Network (CBN). This network sets priorities for and brings together those working in research and development. As a result, progress has been made towards the genetic improvement of cassava, for example, by establishing a molecular map of the cassava genome[4] and developing transformation systems.[5,6]

Cassava suffers from a variety of pests, arthropod, viral, and microbial; the most important of which are cassava mealy bug (*Phanacoccus manihoti*), African cassava mosaic virus (ACMV), and bacterial blight (*Xanthomonas campestris* pv *manihotis*). These pests cause considerable losses and they, their control, and the response which they induce in the plant are the subject of current research. This paper, however, will concentrate largely on an abiotic stress response of cassava known as post-harvest physiological deterioration (PPD). It will draw analogies with defense responses which we are also studying here. Within 24 to 72 hours after harvest, cassava roots start to deteriorate, which renders them unpalatable and unmarketable. This phenomenon severely limits the versatility of the crop. The farmer responds by keeping the crop in the field until required for consumption, which reduces the quality of the root with time and ties up otherwise usable land,[7,8] or by processing the roots into a storable dry product. Because of distance and transport difficulties, good quality fresh cassava is scarce and commands a high mark-up in the expanding towns, thereby encouraging imports of carbohydrate alternatives such as wheat. As a result, cash income to national small farmers is decreased.[9] In addition, unreliable and variable quality input poses problems to small and larger scale industrial processors of cassava for starch and other purposes. Thus, the control of PPD is a priority of the CBN. Although the subject of research over a number of years, the problem is far from understood, and, while post-harvest treatments have been developed to control PPD, these are often impracticable for economic and other reasons.

A comparative evaluation of cassava PPD with defense and stress responses in other better studied plant systems provides the context for potentially creative insights into this problem of an under-studied tropical root crop. These insights stimulate new biochemical and molecular experimentation on PPD which will assist in the illumination of key aspects of the stress response, and may lead to the identification of mechanisms for its modulation or even control. These themes are explored in this paper.

POST-HARVEST PHYSIOLOGICAL DETERIORATION

Post-harvest deterioraton of cassava occurs in two stages: primary or post-harvest physiological deterioration (PPD), which is abiotic and in many ways resembles aspects of wound responses observed in other better studied plant systems; and secondary, or microbial, deterioration, which involves the invasion and proliferation of fungi and other micro-organisms within the root. It is the vascular discoloration and associated physiological and biochemical changes of PPD which cause the changes in consumer acceptability of cassava roots. While microbial deterioration, rotting, occurs later, it has been shown that the changes observed during PPD are not the result of microbial attack.[10-12] Secondary or microbial deterioration, which occurs four to five days after harvesting, commonly involves the pathogens and saprotrophs, *Botryodiplodia theobromae*, *Pythium butleri*, and *Rhizopus* sp..[13]

During harvesting, cassava roots are broken off from the parental plant, the tip is often damaged, and the root as a whole suffers abrasions. Handling and transport can further damage the roots. It is from these wound sites that post-harvest physiological deterioration (PPD) is initiated and spreads through the root.[10] PPD is initially observed as a blue-black discoloration of vascular tissue within the root, known as vascular streaking. There is a strong fluorescence under ultra violet light, prior to the appearance of a more general tissue discoloration. For this reason, cassava on sale in markets is often broken open in order to reveal the internal quality of the roots. Microscopic observations show, associated with this discoloration, the formation of colored occlusions and tyloses from the xylem parenchyma which block adjacent xylem vessels.[12,14]

Harvesting and injury of cassava roots induce an increase in respiration which accompanies the development of PPD.[15-17] During PPD, acid invertase is active, and there is some mobilization of starch to free sugars.[18] The phytohormone ethylene is produced in cassava roots during PPD. Increased levels of ethylene are detected in root slices prior to the appearance of tissue discoloration and continue during the development of PPD.[15,19,20] These data show that the deteriorating root is metabolically active and suggest that ethylene may be playing a co-ordinating role as has been observed in wound responses, senescence phenomena, and fruit ripening in other plant species.[21-23] However, preharvest pruning, which inhibits the development of PPD in roots post-harvest, does not prevent the synthesis of ethylene in the roots after harvesting.[24]

Cyclohexamide, an inhibitor of protein synthesis, prevents PPD and the appearance of the blue fluorescence, suggesting that PPD is an active process requiring the synthesis of novel or increased amounts of enzymes.[18,25] Experiments involving *in vivo* labeling of proteins in cassava root disks indicate that there is a massive synthesis of proteins, including novel ones, as a response to post-harvest wounding.[26] These data are strong evidence that PPD is an active process involving changes in gene expression and protein synthesis. Certainly, there is an increase in activity of a range of enzymes during PPD, including acid invertase, catalase, dehydrogenases, peroxidases, phenylalanine ammonia lyase (PAL), and polyphenol oxidase.[14,18,19,27-29]

Plant defense against wounding or pathogen attack involves several different layers of activity or responses. There are physical barriers which are either pre-existing or are produced in response to the damage. These include cell walls and the associated polymers, lignin, suberin, and callose, and occlusion of xylem vessels by tyloses and polysaccharide gels.[30] There is also the synthesis of proteins involved in defense, such as proteinase inhibitors, chitinases, and extracellular glucanases.[31] In addition, there are low molecular weight defense molecules. These natural products can be pre-existing constitutive components, such as simple phenols and tannins, which tend to have a diffuse or peripheral location in the plant. They also include compounds, such as antimicrobial phytoalexins, produced locally in response to the stress with the end of detaining the spread of and preventing further damage from the wound or infection.[32,33]

Playing a central role in many of these responses is general phenylpropanoid metabolism,[34,35] the key entry enzyme of which is phenylalanine ammonia lyase (PAL),[36] which increases in activity during PPD. General phenylpropanoid metabolism, though constitutive at low levels in most plant tissues, is induced by damage or attack.[35] Branch pathways from the core reactions of phenylpropanoid metabolism provide components of many defense responses. These include barriers (suberin and lignin), low molecular weight anti-microbial molecules (phytoalexins), and signalling molecules (salicylic acid). The increase in phenolic and polyphenolic compounds in the cassava root during PPD are products of general phenylpropanoid metabolism. The main components identified are scopoletin, scopolin, esulentin, esculin, (+)-catechin and gallocatechin, flavanols, leucoanthocyanidins and condensed tannins (Fig. 1).[27-29,37,38] Some of these compounds are pigmented or fluoresce under ultra-violet light, and they probably are constituents of the observed vascular streaking of the root. Polyphenol oxidase and peroxidase activities increase during PPD, though only the former has been found to be directly associated with the colored deposits found in the deteriorating roots.[14] These two enzymes are associated with browning of damaged plant tissues due to the oxidation of released phenols. This oxidation can lead to the production of quinones, which can polymerize to produce melanin, or phenols, which may combine with amino groups in protein to produce brown precipitates. Polyphenol oxidases, involved in browning in plants, form a small multi-gene family which in potato are spatially and temporally differentially expressed, in response to wounding.[39,40] Peroxidases play multiple roles in plant defense responses and can function in the removal of hydrogen peroxide and in the formation of protective barriers via the biosynthesis of suberin and lignin and the cross-linking of hydroxyproline-rich glycoproteins (HRGPs).[31] Catalases, like peroxidases, are involved in the removal of toxic hydrogen peroxide from plants.[41]

A diffusible substance produced in deteriorated cassava roots has been shown to be closely associated with the deterioration process in fresh root tissues.[42] The coumarin, scopoletin (Fig. 1), which fluoresces bright blue under ultra violet light, has been isolated from deteriorating cassava root tissue, where it peaks 24 hours after harvesting before the appearance of any visible signs of PPD. The application of exogenous scopoletin to fresh cassava root tissue produces a rapid and intense discoloration of the vascular and surrounding parenchymal tissues within 12 hours. In other words, scopoletin appears to accelerate the process of PPD. While the mechanism by which it may act as a trigger of PPD is unknown, scopoletin has been shown to have anti-fungal properties and to be induced in *Ulmus* species cell suspension cultures in response to *Ophiostoma ulmi* spores, and it may function as a phytoalexin in the related euphorb, *Hevea brasilensis*.[43,44] Several unidentified compounds, separable by thin layer chromatography from cassava roots undergoing PPD, which probably included scopoletin, also showed anti-fungal activity towards *Alternaria alternata*.[45] Therefore, the implication is that scopoletin plays a role as a phytoalexin in cassava roots, though it may also have signaling properties.

Other classes of stress metabolites detected in the deteriorating root include steroids and diterpenoids.[46] The former are probably the product of the

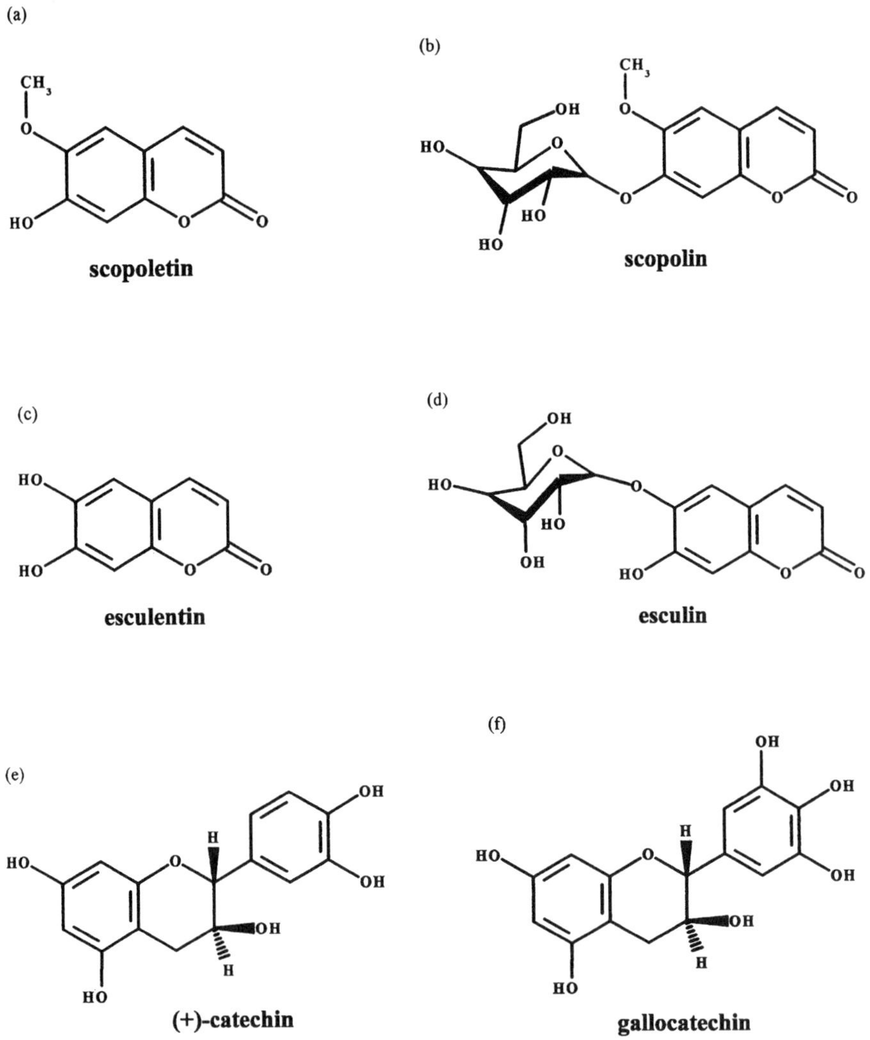

Figure 1. Phenolic and polyphenolic compounds identified as increasing in concentration in the cassava root during post-harvest physiological deterioration. (a) scopoletin, (b) scopolin, (c) esculentin, (d) esculin, (e) (+)-catechin, (f) gallocatechin.

enzymatic oxidation of original membrane sterols. Diterpenes are not common as stress metabolites in plants, and most of the 22 found in cassava are novel. Some may play a role as phytoalexins, such as casbene (Fig. 2) in the related euphorb, *Ricinus communis*,[47] and may also contribute to the unpalatability of cassava undergoing PPD.[48]

Total phospholipids and glyceroglycolipids decrease progressively during PPD, while there is an increase in sterol containing lipids.[49] These changes could result from the breakdown of membrane lipids in the damaged tissue. In other plants, it has been shown that membrane damage results in the production of long chain fatty acids, including linolenic acid, by the action of fatty acyl hydrolase on membrane lipids.[50] Jasmonic acid, synthesized from linolenic acid, plays a key role in the signal transduction pathway from wounding or infection to the biosynthesis of proteinase inhibitors and other defense-related molecules including phytoalexins.[51,52] The membrane degradation observed in cassava may similarly provide the molecules for the initiation of signal transduction cascades leading to the phenomenon of PPD.

Cassava contains cyanogenic glucosides,[53,54] and the roots have been shown to exhibit increased cyanide content during PPD. Although linamarase, which releases hydrogen cyanide from cyanogenic glucosides, is activated by wounding, there is no evidence of its *de novo* synthesis during PPD, as its activity decreased over a three day period post-harvest.[55,56] The hydrogen cyanide released by mechanical damage is believed to act as a deterrent to herbivores; in many parts of Africa, high cyanide varieties are grown around plots of low cyanide cassava.[57]

The available information on PPD suggests that there are strong parallels between this phenomenon in cassava roots and wound or pathogen induced responses in other crops and plant model systems. Certainly, similar enzymes, metabolic pathways, and secondary metabolites are induced or synthesized. However, in most plants, the co-ordinated cascade of defense responses triggered

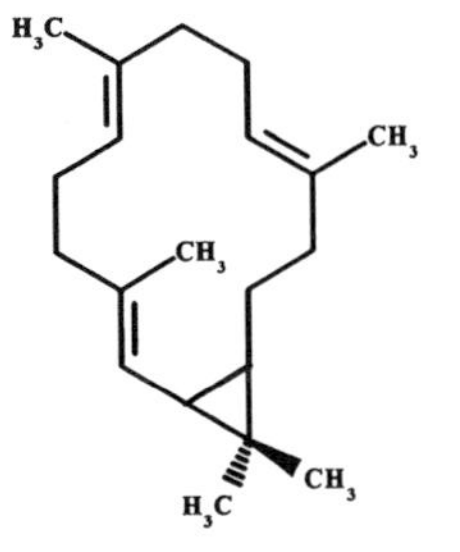

Casbene

Figure 2. Casbene, an anti-fungal from the castor-bean plant, *Ricinus communis* (Euphorbiaceae).

by wounding or pathogen attack lead up to and include the sealing and repair of the exposed tissue by barrier materials such as suberin, callose, lignin, and periderm formation. Such wound healing, or boundary sealing, prevents the continual production of signal molecule precursors, such as the products of membrane damage, which trigger the wound response cascade in the first instance, leading to a down modulation of defensive responses and a return to normal plant growth. During PPD in cassava roots, we appear to be observing a continuous production of these defensive cascades without the negative feed back loop of wound healing which normally turns off the triggering signals. While such wound repair occurs in other harvested root crops such as potato, sweet potato, and yam, it does not normally occur in the harvested cassava root. This difference in response could indicate that there is something unique about cassava. The fact that the detached storage root cannot serve as a propagule suggests that wound repair is a superfluous aspect of its defensive response. Alternatively, because traditionally cassava roots are harvested as required to be eaten or processed, and because the roots are not used for propagation, there has been no selection for post-harvest conservation of the root during the 7,000 and more years since cassava was first domesticated.[58] Therefore, wound repair, after detachment from the plant, could have been lost from the cassava root's defensive repertoire. In an attempt to shed light on this question, it is worthwhile to consider those circumstances in which wound repair can be induced or PPD inhibited in the cassava storage root.

Inhibition of PPD

The PPD response in cassava can be inhibited or delayed by "pruning", "curing", and by various storage methods. Pruning, cutting off the plant 20-30 cm from the surface of the soil 2-3 weeks prior to harvesting, reduces susceptibility to PPD. However, it has been observed that re-growth of new shoots breaks this inhibition. Foliage re-growth probably involves the mobilization of starch from the storage root until such time as the foliage reaches a size to become self sufficient via photosynthesis in carbohydrate production. Certainly, storage root starch decreases during re-growth.[25] "Natural" defoliation due to the biotic or abiotic stresses of pests or drought also leads to reduced susceptibility to PPD.[25] "Curing", treating cassava at high temperatures (35°C) and humidity (80-85% relative humidity), for a period after harvesting inhibits PPD by promoting wound healing via suberization and meristem formation in the cortex. These "cured" roots have a longer shelf-life, less weight loss, and a lower deterioration index than control roots. However, their re-damage promotes PPD in a similar manner to freshly-harvested roots.[10] Storage of cassava roots in polythene bags, especially if pre-treated with a fungicide (thiabendazole-based) sufficient to control microbial deterioration which can develop in the humid atmosphere of the bag, can effectively prevent PPD and enable the storage of the roots for two or more weeks. Under these conditions, the wounds seal and are repaired.[8,9] Dipping roots in wax, preferably containing a fungicide, similarly inhibits PPD. Re-wounding of roots stored in polyethylene bags will re-initiate PPD; this was the basis of many experiments on cassava PPD with air-freighted material.[59]

Pruning affects PPD in a manner different from "curing" or polyethylene bag storage. While the last two permit wound healing but retain sensitivity to subsequent damage, no wound healing has been described in the roots from pruned plants. However, the roots exhibit reduced sensitivity to PPD despite the damage normally associated with harvesting. These results with pruned cassava plants suggest that in some way "pre-stressing" the plant before harvest reduces susceptibility of the roots to PPD. Supportive evidence comes from the interesting observation that non-locally adapted varieties of cassava are less susceptible to PPD than local varieties. However, the former do not grow so well and are less productive under these abnormal agro-ecological conditions.[25] This could be conceived as an environmental "stress". It is tempting to speculate that "pre-stressing" might operate in a manner analogous to the effects on a plant of prior exposure to a pathogen as in systemic acquired resistance (SAR).[60]

Under the "curing" and polythene bag storage conditions, several environmental parameters may be involved in the observed effects of wound sealing and lack of PPD in the absence of further damage; these include temperature, humidity, and oxygen tension. Damage to the periderm during harvesting disrupts the main barrier to oxygen diffusion into cassava roots. It is possible that this increased access of oxygen promotes PPD. Indeed, storing cassava discs in atmospheres depleted of oxygen inhibits the development of vascular streaking compared to controls for up to 70 hours.[11,61] High humidity, however, suppresses the rate of vascular discoloration compared to low humidity with the same levels of aeration, suggesting that water stress is a pe-requisite for initation of vascular streaking subsequent to wounding.[16] High humidity will reduce water loss from the harvested root, thereby preventing the extensive pernetration of air, including oxygen, into the root. An interpretation of these data is that while exposure to the stress of atmospheric oxygen may precipitate PPD, high humidity, by reducing oxygen penetration, can down-modulate PPD sufficiently to permit wound healing and thereby pervent the extensive wound response cascades discussed above.

Oxygen and reactive oxygen species (ROS) derived from it are potentially damaging to all life forms including plants. Organisms counter such oxidative stress by restricting their boundaries in order to control access of oxygen to their tissues and by synthesizing enzymes and other molecules which can detoxify ROS. Such detoxifying mechanisms include enzymes such as superoxide dismutases (SOD), catalases and peroxidases, and antioxidant compounds such as glutathione, phenolic compounds, ascorbate, β-carotene, and citric acid.[62,63] Oxidative stress is a double-edged sword in that ROS, as well as damaging the plant and requiring detoxification,[64] play a defensive role. For example, ROS are involved in the cross-linking of cell wall proteins, lignin biosynthesis, the induction of defense-related genes, the stimulation of phytoalexin biosynthesis, and the promotion of the hypersensitive response.[65] The exposure of wounded cassava tissues to atmospheric oxygen and its further penetration into the root due to water loss would expose these tissues to the stress of higher than normal oxygen tensions and the potentially extremely damaging effects of reactive oxygen species. While there is not conclusive evidence that such oxidative stress plays a significant role in PPD, or in precipitating the condition, strong circumstantial evidence suggests that it does. Peroxidases, and especially catalases, are involved in the detoxification of the reactive oxygen species, hydrogen peroxide. The activity of both enzymes increase during PPD.[14,18,19] Membrane lipid degradation, probably including lipid peroxidation, occurs during PPD.[46,49] The synthesis of phenolic compounds increases during PPD. Several of these are capable of reacting with hydrogen peroxide to produce pigmented compounds which could account for the vascular streaking observed.[28,29] β-carotene (Fig. 3), which has anti-oxidant properties, is found in cassava and has been shown to decrease in concentration during PPD.[66] Yellow cassava roots generally contain much more β-carotene than white, and there are data indicating a correlation between β-carotene content and reduced PPD response.[67,68] While the cassava plant, including the roots, contain relatively high amounts of ascorbate (35 mg per 100 g),[69] there is no information as to any correlation between concentrations of this antioxidant and PPD sensitivity. The potential ability of cassava suspension cells to generate a rapid, intensive oxidative burst is described later.

Potential for Genetic Improvement

Although some cassava varieties have been identified which show ranges of sensitivity to PPD, the phenomenon is far from understood from a genetic perspective. The problem is compounded by the large influence of environmental growth conditions and pre-harvest stress on the PPD response.[70] The availability of biochemical and molecular markers for components of the PPD response may assist with the genetic definition of the problem and with breeding programs designed to improve cassava with respect to this trait. However, manipulation of a key aspect of PPD via genetic modification has an elegant simplicity which

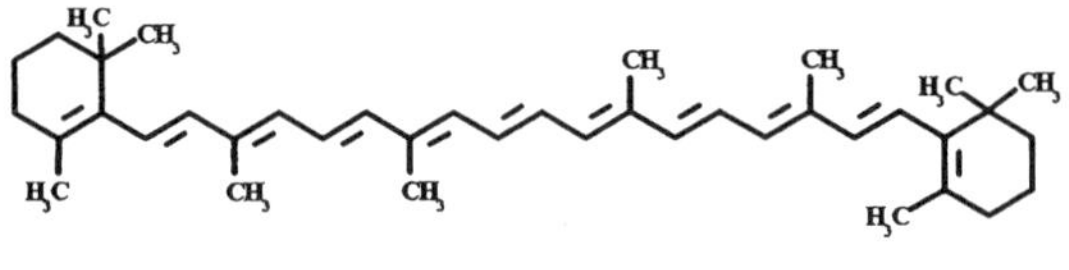

β-carotene

Figure 3. The antioxidant β-carotene is correlated with reduced PPD-response.

makes this approach attractive. Certainly, the use of anti-sense RNA constructs or developmentally specific promoters driving genes whose products have regulatory effects have proven powerful tools in manipulating developmental pathways in other plants.[71,72] While detailed knowledge of the regulation of PPD is not yet sufficiently advanced to be able to identify key regulatory points, the development of the experimental work described here should enable such a stage to be approached.

EXPERIMENTAL STRATEGIES

Research on stress responses in other plant species provides the theoretical background and the practical tools with which to approach PPD from molecular and biochemical perspectives. The core reactions of general phenylpropanoid metabolism and its branch pathways which lead to the synthesis of defense-related molecules are at the heart of many stress responses. So too are the enzymes and other components of oxidative stress responses. Therefore, we are cloning genes for enzymes involved in controlling key steps in the above reactions. By analyzing the profiles of expression of these genes during PPD, we can gain insights into their relative importance and roles in this process. This approach is assisted by the availability of cassava varieties from the Centro Internacional de Agricultura Tropical (CIAT), Colombia exhibiting differences in susceptibility to PPD, and by the ability to control environmental parameters such as atmospheric gasses and humidity.

Cell suspension cultures have proved successful in dissecting aspects of defensive responses in several other plants.[35] Therefore, in addition to using material and tissues from whole plants, we are using cassava cell suspension cultures, which provide a simple model system in which many parameters can be controlled.

Isolation and Characterization of PPD-Related Clones

Using heterologous probes[73-75] at low stringency, we have isolated clones from a cDNA library constructed with poly(A)$^+$ RNA purified from cassava root discs 48 hours after harvest, a time point identified by *in vivo* labeling as showing a substantial increase in protein synthesis.[26] The identity of the isolated clones has been confirmed by DNA sequencing and alignment of the corresponding DNA and deduced amino acid sequences with the appropriate sequences obtained from the EMBL and GenBank databases. These clones include phenylalanine ammonia lyase (PAL), β-1,3-glucanase, and a hydroxyproline-rich glycoprotein (HRGP); indicating that these genes are active during PPD (Han, Cooper, and Beeching, unpublished). PAL is the key entry enzyme to the core reactions of general phenylpropanoid metabolism. Its activity controls the flux into the pathway and thereby influences outputs from the several branch pathways.[33,35] Enzyme activity profiles and the accumulation of phenolic compounds during PPD indicates that general phenylpropanoid metabolism is active during PPD. Results show that increased activity includes PAL gene expression. PAL is encoded by a small multi-gene family in most plants, usually of about four members, although potato PAL has been shown to have over 40 members[76] and loblolly pine appears to consist of only a single gene.[77] Southern blots of cassava genomic DNA and restriction mapping of cassava PAL genomic clones suggest that cassava PAL contains at least two family members (Li and Beeching, unpublished).

β-1,3-glucanases are hydrolases involved in the depolymerization of β-glucan, a significant component of fungal cell walls. As such they are believed to play a role in defense due to their anti-fungal activity. By their release of cell wall fragments they are capable of triggering more extensive defensive responses. The expression of β-1,3-glucanase genes has been shown to be induced by pathogens, elicitors, and phytohormones.[74,78,79] The expression of a β-1,3-glucanase gene during PPD is consistent with PPD being a normal wound response involving anti-microbial activity and the induction of defensive cascades.

Hydroxyproline-rich glycoproteins (HRGPs), or extensins, are highly basic structural components of the normal primary cell wall which are also expressed during defense responses, including those to wounding and pathogen attack.[75,80] The expression of a protein involved in cell wall strengthening and repair during PPD, and the increased activity of peroxidase[15,19,20] which is involved in forming cross-links between HRGPs using hydrogen peroxide, is significant in that it indicates that PPD is more than an unstoppable defensive response. It also includes components associated with boundary sealing and wound repair. HRGPs are generally believed to be encoded by a multi-gene family.[31] However, while the cassava HRGP cDNA clone has extensive sequence similarity to, and structural features common with, published HRGP sequences, it only hybridizes at high stringency to a single band on Southern blots of genomic DNA (Li and Beeching, unpublished). While this result could be interpreted as indicating only one family member, it is more likely to mean that there is sufficient sequence diversity in the cassava HRGP gene family to prevent cross hybridization at high stringency.

Use of Cassava Cell Suspension Cultures

Cell suspension cultures and elicitors have proved excellent systems for studying components of defense responses, including those of general phenylpropanoid metabolism in parsley, bean, and other plants.[35,81,82] While the exploitation of cassava cell suspension cultures is perceived as having an indirect bearing on the specific problem of PPD, it will illustrate the range of versatility of stress responses in this plant. The cell suspension system enables the biochemical and molecular responses of cassava to biotic and abiotic stresses, including microbial and chemical elicitors, to be firmly placed within the context of the extensive and detailed work using similar systems in other plant species. While responses to elicitors measured in cell suspension cultures are not proof of parallel responses occurring in the cassava root during PPD, they show what the plant is genetically and metabolically capable of and provide strong evidence that particular biochemical or molecular events are avenues worth exploring in the root system. Conversely, the insights gained and the molecular tools generated by the PPD work can benefit the cell suspension system. A creative synergistic interplay occurs between the two systems which is already proving fruitful.

In common with other plant cell suspension culture systems, cassava shows an increase in phenylalanine ammonia-lyase (PAL) activity and an increase in extracellular pH in response to elicitors such as *Colletotrichum lindemuthianum* glucan cell wall fraction (Gómez-Vásquez, Beeching, and Cooper, unpublished). This implies that the responses of the cassava system to elicitors is not unusual compared to those of cell suspension cultures from other plants.[83-85] Reactive oxygen species (ROS), while potentially damaging, can also be produced in high concentrations by the plant itself as an important part of its defense response. This oxidative burst can be triggered by elicitors or pathogens in cell suspension cultures and *in vivo*, and probably plays a key role in the hypersensitive response induced by an incompatible pathogen. In the case of incompatible pathogens, the oxidative burst is biphasic, with a short acute peak a few minutes after induction and a more prolonged peak several hours later.[65,86] Typical, very rapid, oxidative bursts have been observed in cassava cell suspension cultures in response to elicitors (*C. lindemuthianum*, oligogalacturonic acid, and yeast) and to the causative agent of cassava bacterial blight, *Xanthomonas campestris* pv. *manihotis*. A biphasic burst also occurs in response to incompatible bacterial pathogens such as *Erwinia amylovora*. The enzyme xanthine oxidoreductase has the capacity to produce ROS. While it has been well characterized in mammalian systems, its precise biological function is not completely understood.[87] The activity of xanthine oxidoreductase can increase in response to a pathogen in tobacco,[88] but it remains an understudied enzyme in plants. Using a sensitive fluorometric assay, we have shown xanthine oxidoreductase activity in non-elicited cassava cell suspension cultures, and preliminary results suggest a response to elicitation by *C. lindemuthianum*, yeast, and lipopolysaccharide from *Pseudomonas syringae* and *X. campestris* (Gómez-Vásquez, Beeching, and Cooper, unpublished).

CONCLUSIONS

Although not exhaustive, the literature on PPD in cassava and results from our laboratory summarized here imply that cassava's reactions to biotic and abiotic stresses encompass the normal repertoire of defense responses of other more extensively studied plant species. The implication for PPD is that cassava has a similar capacity to respond to wounding or pathogen damage, as do other root crops such as potato, yam, or sweet potato. Certainly, the responses of the cassava plant to wounds, other than those due to root harvesting, and to pathogens are comparable to those of other plants. However, except under extraordinary circumstances, such as low oxygen tensions and high humidity, the full range of responses required for both defense and repair are inadequately expressed in the detached cassava root. During PPD, there is an imbalance in the expression of these two components. For defense, they appear to be excessive, while for repair, they are insufficient or incomplete. Inadequate wound repair leads to the continual production of the signals, such as those from membrane damage. These trigger the activation of defensive cascades which are perceived as the symptoms of PPD. As wound healing can occur in detached cassava roots under certain conditions, such as polyethylene bag storage, and as at least some components of the wound healing response are synthesized during PPD, such as HRGPs, the conclusion is that the repair aspects of the wound response are insufficiently active to promote healing. This condition is probably exacerbated by the penetration of atmospheric oxygen into the root tissue due to water loss provoking oxidative stress responses.

AKNOWLEDGMENTS

We would like to thank Professor M. Hughes, University of Newcastle, for the generous gift of the cassava genomic library; Professors W. Schuch, Zeneca Plant Science, M. van Montagu, University of Gent, and K. Hahlbrock, Max-Planck Institute, for the kind gifts of their clones for PAL, glucanase and HRGP, respectively, and Hongying Li for isolating the genomic clones. Y.H. would like to thank the Department of Biology and Biochemistry, University of Bath, and R. G-V. COLCIENCIAS, Colombia, for studentships. The research described here was in part supported by ODA grants R6726(H) and R6105, to J.R.B. and R.M.C., respectively.

REFERENCES

1. COCK, J.H. 1985. Cassava: New Potential for a Neglected Crop, Westfield Press, Boulder.
2. WENHAM, J.E. 1995. Post-harvest Deterioration of Cassava. A Biotechnological Perspective, FAO, Rome.
3. CIAT 1992. Cassava Programme 1987–1991. Working Document 116, CIAT, Cali, Colombia.
4. GOMEZ, R., ANGEL, F., BONIERBALE, M.W., RODRIQUEZ, F., TOHME, J., ROCA, W.M. 1996. Inheritance of random amplified polymorphic DNA markers in cassava (*Manihot esculenta* Crantz). Genome 39: 1039–1043.

5. LI, H-Q., SAUTTER, C., POTRYKUS, I., PUONTI-KAERLAS, J. 1996. Genetic transformation of cassava (*Manihot esculenta* Crantz). Nature Biotechnology 14: 736–740.
6. SCHOPKE, C., TAYLOR, N., CARCAMO, R., KONAN, N.K., MARMEY, P., HENSHAW, G.G., BEACHY, R.N., FAUQUET, C. 1996. Regeneration of transgenic cassava plants (*Manihot-esculenta* Crantz) from microbombarded embryogenic suspension-cultures. Nature Biotechnology 14: 731–735.
7. WHEATLEY, C., GOMEZ, G. 1985. Evaluation of some quality characteristics in cassava storage roots. Qual. Plant Foods Hum. Nutr. 35: 121–129.
8. RICKARD, J.E., COURSEY, D.G. 1981. Cassava storage. Part 1: Storage of fresh cassava roots. Tropical Science 23: 1–32.
9. WHEATLEY, C.C., BEST, R. 1991. How can traditional forms of nutrition be maintained in urban centers: The case for cassava. Entwicklung Und Laendlicher Raum 91: 13–16.
10. BOOTH, R.H. 1976. Storage of fresh cassava (*Manihot esculenta*). I. Post-harvest deterioration and its control. Experimental Agriculture 12: 103–111.
11. NOON, R.A., BOOTH, R.H. 1977. Nature of post-harvest deterioration of cassava roots. Trans. British Myc. Soc. 69: 287–290.
12. RICKARD, J.E., MARRIOTT, J., GAHAN, P.B. 1979. Occlusions in cassava xylem vessels associated with vascular discoloration. Ann. Bot. 43: 523–526.
13. TANIGUCHI, T., DATA, E.S., BURDEN, O.J., URITANI, I., GORGONIO, M., UMERES, E. 1984. The appearance of antifungal activity in cassava root tissues in response to physiological and microbial deterioration. Ann. Phytopath. Soc. Japan 50: 286–288.
14. RICKARD, J.E., GAHAN, P.B. 1983. The development of occlusions in cassava (*Manihot esculenta* Crantz) root xylem vessels. Ann. Bot. 52: 811–821.
15. HIROSE, S. 1986. Physiological studies on postharvest deterioration of cassava plants. Jap. Agric. Res. Quart. 19: 241–252.
16. MARRIOTT, J., BEEN, B.O., PERKINS, C. 1978. The aetiology of vascular discoloration in cassava roots after harvesting: Association with water loss from wounds. Physiol. Plant. 44: 38–42.
17. MARRIOTT, J., BEEN, B.O., PERKINS, C. 1979. The aetiology of vascular discoloration in cassava roots after harvesting: Development of endogenous resistance in stored roots. Physiol. Plant. 45: 51–56.
18. URITANI, I., DATA, E.S., TANAKA, Y. 1984. Biochemistry of postharvest deterioration of cassava and sweet potato roots. In: Tropical Root Crops: Postharvest Physiology and Processing. (I. Uritani and E.D. Reyes, eds.) JSSP, Tokyo, pp. 61–75.
19. PLUMBLEY, R.A., HUGHES, P.A., MARRIOT, J. 1981. Studies on peroxidases and vascular discoloration in cassava root tissues. J. Sci. Food Agric. 32: 723–731.
20. HIROSE, S., DATA, E.S., QUEVEDO, M.A. 1984. Changes in respiration and ethylene production in cassava roots in relation to postharvest deterioration. In: Tropical Root Crops: Postharvest Physiology and Processing. (I. Uritani and E.D. Reyes, eds.) JSSP, Tokyo, pp. 83–98.
21. ECKER, J.R., DAVIS, R.W. 1987. Plant defense genes are regulated by ethylene. Proc. Nat. Acad. Sci. USA 84: 5202–5206.
22. FLUHR, R., MATTOO, A.K. 1996. Ethylene—biosynthesis and perception. Critical Reviews in Plant Sciences 15: 479–523.
23. PICTON, S., GRAY, J.E., GRIERSON, D. 1995. Ethylene genes and fruit ripening. In: Plant Hormones: Physiology, Biochemistry and Molecular Biology. (P.J. Davies, ed.) 2nd ed., Kluwer, Netherlands, pp. 327–394.
24. HIROSE, S., DATA, E.S., MATURAN, E. 1984. Relation of respiration and ethylene production to postharvest deterioration in cassava roots from pruned and unpruned plants. In: Tropical Root Crops: Postharvest Physiology and Processing. (I. Uritani and E.D. Reyes, eds.) JSSP, Tokyo, pp. 99–107.
25. WHEATLEY, C. 1980. Studies related with the nature of post-harvest physiological deterioration in cassava roots. CIAT Seminarios Internos SE-16–80.

26. BEECHING, J.R., DODGE, A.D., MOORE, K.G., WENHAM, J.E. 1995. Physiological deterioration in cassava: An incomplete wound response? The Cassava Biotechnology Network: Proceedings of the Second International Scientific Meeting, (A.M. Thro, W. Roca, eds.) Aug 22-26, 1994; Bogor, Indonesia. Cali, Colombia: CIAT, pp. 729–736.
27. RICKARD, J.E. 1981. Biochemical changes involved in the post-harvest deterioration of cassava roots. Tropical Science 23: 235–237.
28. RICKARD, J.E. 1985. Physiological deterioration of cassava roots. J. Sci. Food Agric. 36: 167–176.
29. TANAKA, Y., DATA, E.S., HIROSE, S., TANIGUCHI, T., URITANI, I. 1983. Biochemical changes in secondary metabolites in wounded and deteriorated cassava roots. Agric. Biol. Chem. 47: 693–700.
30. COOPER, R.M. 1981. Pathogen-induced changes in host ultrastructure. Plant Disease Control: Resistance and Susceptibility. (B. Solheim, J. Raa, eds.) Wiley, New York, pp. 105–142.
31. BOWLES, D.J. 1990. Defense-related proteins in higher plants. Annu. Rev. Biochem. 59: 873–907.
32. COOPER, R.M., RESENDE, M.L.V., FLOOD, J., ROWAN, M.G., BEALE, M.H., POTTER, U. 1996. Detection and cellular-localization of elemental sulfur in disease-resistant genotypes of *Theobroma cacao*. Nature 379: 159–162.
33. BENNETT, R.N., WALLSGROVE, R.M. 1994. Secondary metabolites in plant defense-mechanisms. New Phytologist 127: 617–633.
34. DIXON, R.A., PAIVA, N.L. 1995. Stress-induced phenylpropanoid metabolism. Plant Cell 7: 1085–1097.
35. HAHLBROCK, K., SCHEEL, D. 1989. Physiology and molecular biology of phenylpropanoid metabolism. Annu. Rev. Plant Physiol. Plant Mol. Biol. 40: 347-369.
36. JONES, D.H. 1984. Phenylalanine ammonia-lyase: Regulation of its induction and its role in plant development. Phytochemistry 23: 1349–1359.
37. LALAGUNA, F. 1993. Purification of fresh cassava root polyphenols by solid-phase extraction with amberlite XAD-8 resin. J. Chromatogr. A 657: 445–449.
38. URITANI, I., DATA, E.S., VILLEGAS, R.J., FLORES, P., HIROSE, S. 1983. Relationship between secondary metabolism changes in cassava root tissue and physiological deterioration. Agric. Biol. Chem. 47: 1591–1598.
39. THIPYAPONG, P., HUNT, M.D., STEFFENS, J.C. 1995. Systemic wound induction of potato (*Solanum tuberosum*) polyphenol oxidase. Phytochemistry 40: 673–676.
40. THYGESEN, P.W., DRY, I.B., ROBINSON, S.P. 1995. Polyphenol oxidase in potato—a multigene family that exhibits differential expression patterns. Plant Physiol. 109: 525–531.
41. WILLEKENS, H., INZE, D., VAN MONTAGU, M., VAN CAMP, W. 1995. Catalases in plants. Molecular Breeding 1: 207–228.
42. WHEATLEY, C.C., SCHWABE, W.W. 1985. Scopoletin involvement in post-harvest physiological deterioration of cassava root (*Manihot esculenta* Crantz). J. Exp. Bot. 36: 783–791.
43. GIESEMANN, A., BIEHL, B., LIEBEREI, R. 1986. Identification of scopoletin as a phytoalexin of the rubber tree *Hevea brasiliensis*. J. Phytopath. 117: 273–276.
44. VALLE, T., LOPEZ, J.L., HERNANDEZ, J.M., CORCHETE, P. 1997. Antifungal activity of scopoletin and its differential accumulation in *Ulmus pumila* and *Ulmus campestris* cell suspension cultures infected with *Ophiostoma ulmi* spores. Plant Sci. 125: 97–101.
45. TANIGUCHI, T., DATA, E.S., BURDEN, O.J., GORGONIO, M.A., UMERES, E. 1984. Production of antifungal substances in cassava roots in response to physiological and microbial deterioration. In: Tropical Root Crops: Postharvest Physiology and Processing. (I. Uritani and E.D. Reyes, eds.) JSSP, Tokyo, pp. 145–149.
46. SAKAI, T., NAKAGAWA-MURATA, Y., URITANI, I., DATA, E.S. 1994. Occurrence and characteristics of stress metabolites in cassava roots. In: Postharvest Biochemistry of Plant

Food-Materials in the Tropics. (I. Uritani, V.V. Garcia, E.M.T. Mendoza, eds.), Japan Sci. Soc. Press, Tokyo, pp. 95–110.
47. MAU, C.J.D., WEST, C.A. 1994. Cloning of casbene synthase cDNA: Evidence for conserved structural features among terpenoid cyclases in plants. Proc. Nat. Acad. Sci. USA 91: 8497–8501.
48. SAKAI, T., NAKAGAWA, Y. 1988. Diterpenic stress metabolites from cassava roots. Phytochemistry 27: 3769–3779.
49. LALAGUNA, F., AGUDO, M. 1989. Relationship between changes in lipid with ageing of cassava roots and senescence parameters. Phytochemistry 28: 2059–2062.
50. MARANGONI, A.G., PALMA, T., STANLEY, D.W. 1996. Membrane effects in postharvest physiology. Postharvest Biol. Technol. 7: 193–217.
51. FARMER, E.E., RYAN, C.A. 1992. Octadecanoid precursors of jasmonic acid activate the synthesis of wound-inducible proteinase inhibitors. Plant Cell 4: 129–134.
52. NOJIRI, H., SUGIMORI, M., YAMANE, H., NISHIMURA, Y., YAMADA, A., SHIBUYA, N., KODAMA, O., MUROFUSHI, N., OMORI, T. 1996. Involvement of jasmonic acid in elicitor-induced phytoalexin production in suspension-cultured rice cells. Plant Physiol. 110: 387–392.
53. LYKKESFELDT, J., MOLLER, B.L. 1994. Cyanogenic glycosides in cassava, *Manihot esculenta* Crantz. Acta Chemica Scandinavica 48: 178–80.
54. PANCORO, A., HUGHES, M.A. 1992. In-situ localization of cyanogenic β-glucosidase (linamerase) gene expression in leaves of cassava (*Manihot esculenta* Crantz) using non-isotopic riboprobes. Plant J. 2: 821–827.
55. IWATSUKI, N., KOJIMA, M., DATA, E.S., VILLEGAS-GODOY, C.D.V. 1984. Changes in cyanide content and linamarase activity in cassava roots after harvest. In: Tropical Root Crops: Postharvest Physiology and Processing. (I. Uritani, E.D. Reyes, eds.) JSSP, Tokyo, pp. 151–161.
56. KOJIMA, M., IWATSUKI, N., DATA, E.S., VILLEGAS, C.D.V., URITANI, I. 1983. Changes of cyanide content and linamarase activity in wounded cassava roots. Plant Physiol. 72: 186–189.
57. KAPINGA, R. 1996. Selection of cassava varieties by producers in the lake zone of Tanzania. The Cassava Biotechnology Network: Third International Scientific Meeting; Aug. 26–31, 1996, Kampala, Uganda.
58. LATHRAP, D.W. 1970. The Upper Amazon, Thames and Hudson, London.
59. RICKARD, J.E. 1982. Investigation into Post-Harvest Behaviour of Cassava Roots and Their Response to Wounding. Ph.D. thesis, University of London.
60. RYALS, J.A., NEUENSCHWANDER, U.H., WILLITS, M.G., MOLINA, A., STEINER, H.Y., HUNT, M.D. 1996. Systemic acquired resistance. Plant Cell 8: 1809–1819.
61. PASSAM, H.C., NOON, R.A. 1977. Deterioration of yams and cassava during storage. Ann. App. Biol. 85: 436–440.
62. LARSON, R.A. 1995. Plant defenses against oxidative stress. Arch. Insect Biochem. Physiol. 29: 175–186.
63. SCANDALIOS, J.G. 1990. Response of plant antioxidant defense genes to environmental stress. Adv. Genet. 28: 1–41.
64. BOWLER, C., VAN CAMP, W., VAN MONTAGU, M., INZE, D. 1994. Superoxide-dismutase in plants. Crit. Revi. Plant Sci. 13: 199–218.
65. LOW, P.S., MERIDA, J.R. 1996. The oxidative burst in plant defense - function and signal-transduction. Physiol. Plant. 96: 533–542.
66. GLORIA, L.A., URITANI, I. 1984. Changes in β-carotene content of golden yellow cassava in relation to physiological deterioration. In: Tropical Root Crops: Postharvest Physiology and Processing. (I. Uritani, E.D. Reyes, eds.) JSSP, Tokyo. pp. 163–168.
67. ADEWUSI, S.R.A., BRADBURY, J.H. 1993. Carotenoids in cassava—comparison of open-column and HPLC methods of analysis. J. Sci. Food Agric. 62: 375–383.
68. IGLESIAS, C., MAYER, J.E., CHAVES, A.L., CALLE, F. 1995. Exploring the genetic potential and stability of β-carotene content in cassava roots. In: Biotechnology Research Unit Annual Report 1995, CIAT, Cali, pp. 33–38.

69. ONWUEME, I. 1978. The Tropical Tuber Crops, John Wiley, Chichester, U.K.
70. WHEATLEY, C. 1982. Studies on Cassava (*Manihot esculenta* Crantz) Root Post-Harvest Deterioration. PhD thesis. University of London.
71. HAMILTON, A.J., LYCETT, G.W., GRIERSON, D. 1990. Antisense gene that inhibits synthesis of the hormone ethylene in transgenic plants. Nature 346: 284–286.
72. GAN, S., AMASINO, R.M. 1995. Inhibition of leaf senescence by autoregulated production of cytokinin. Science 270: 1986–1968.
73. CRAMER, C.L., EDWARDS, K., DRON, M., LIANG, X., DILDINE, S.L., BOLWELL, G.P., DIXON, R.A., LAMB, C.J., SCHUCH, W. 1989. Phenylalanine ammonia-lyase gene organization and structure. Plant Mol. Biol. 12: 367–383.
74. DE LOOSE, M., ALLIOTTE, T., GHEYSEN, G., GENETELLO, C., GIELEN, J., SOETAERT, P., VAN MONTAGU, M., INZE, D. 1988. Primary structure of a hormonally regulated β-glucanase of *Nicotiana plumbaginifolia*. Gene 70: 13–23.
75. KAWALLECK, P., SCHMELZER, E., HAHLBROCK, K., SOMSSICH, I.E. 1995. Two pathogen-responsive genes in parsley encode a tyrosine-rich hydroxyproline-rich glycoprotein (HRGP) and an anionic peroxidase. Mol. Gen. Genet. 247: 444–452.
76. JOOS, H.J., HAHLBROCK, K. 1992. Phenylalanine ammonia-lyase in potato (*Solanum-tuberosum* L)—genomic complexity, structural comparison of 2 selected genes and modes of expression. Eur. J. Biochem. 204: 621–629.
77. WHETTEN, R.W., SEDEROFF, R.R. 1992. Phenylalanine ammonia-lyase from loblolly-pine—purification of the enzyme and isolation of complementary-DNA clones. Plant Physiol. 98: 380–386.
78. CASTRESANA, C., DE CARVALHO, F., GHEYSEN, G., HABETS, M., INZE, D., VAN MONTAGU, M. 1990. Tissue-specific and pathogen-induced regulation of a *Nicotiana plumbaginifolia* β-1,3-glucanase gene. Plant Cell 2: 1131–1143.
79. STINTZI, A., HEITZ, T., PRASAD, V., WIEDEMANNMERDINOGLU, S., KAUFFMANN, S., GEOFFROY, P., LEGRAND, M., FRITIG, B. 1993. Plant pathogenesis-related proteins and their role in defense against pathogens. Biochimie 75: 687–706.
80. CHEN, J., VARNER, J.E. 1985. Isolation and characterization of cDNA clones for carrot extensin and a proline-rich 33-kDa protein. Proc. Nat. Acad. Sci. USA 82: 4399–4403.
81. BOLWELL, G.P., ROBBINS, M.P., DIXON, R.A. 1985. Metabolic changes in elicitor-treated bean cells: Enzymatic responses associated with rapid changes in cell wall components. Eur. J. Biochem. 148: 571–578.
82. DOUGLAS, C.J., DANGL, J.L., HOFFMANN, H., LIPPHARDT, S., HAHLBROCK, K. 1987. Analysis of fungal elicitor—a UV light-induced gene expression in parsley cells. In: Plant Genes Systems and Their Biology, (J.L. Key, L. McIntosh, eds.) Alan R. Liss Inc., New York, pp. 139–151.
83. DALE, K.L., NOVACKY, A. 1989. The initiation of membrane lipid peroxidation during bacteria-induced hypersensitive response. Physiol. Mol. Plant Path. 2: 233–245.
84. DIXON, R.A., DEY, M.P., MURPHY, D.L., WHITEHEAD, I.M. 1981. Dose responses for *Colletotrichum lindemuthianum* elicitor-mediated enzyme induction in French bean cell suspension cultures. Planta 151: 272–280.
85. HAHLBROCK, K., LAMB, C.J., PURVIN, C., EBEL, J., FAUTZ, E., SCHAFER, E. 1981. Rapid response of suspension-culture parsley cells to the elicitor from *Phytophthora megasperma* var. *sojae*. Plant Physiol. 67: 768–773.
86. DOKE, N., MIURA, Y., SANCHEZ, L.M., PARK, H.J., NORITAKE, T., YOSHIOKA, H., KAWAKITA, K. 1996. The oxidative burst protects plants against pathogen attack—mechanism and role as an emergency signal for plant bio-defense - a review. Gene 179: 45–51.
87. HARRISON, R. 1997. Human xanthine oxidoreductase: in search of a function. Biochem. Soc. Trans. 25: 786-790.
88. MONTALBINI, P. 1993. Xanthine-oxidase activity in the susceptible and hypersensitive responses of tobacco-leaves to tobacco mosaic-virus infection. J. Phytopath. 139: 177–186.

INDEX

Abscisic acid, 103, 185
Acanthaceae, 73
Acetosyringone, 59, 213–214
Acetyltransferase, 22, 25
Active oxygen, 31, 33–34, 37, 40–41, 51, 120, 196; see also Reactive oxygen species
Aflatoxins, 18, 22
Agrobacterium, 11, 59, 141, 207–214, 216–217, 219–222
 A. tumefaciens, 11, 59, 147, 207
Alfalfa, 101, 148, 152–156
Allelochemicals, 73, 75–76, 78–81, 88–89
Allene oxide cyclase, 182, 184, 187–189, 192
Allene oxide synthase, 182–184, 187–189, 192
Alternaria, 33, 50
 A. alternata, 39, 42, 47, 50, 236
Amanita, 63, 66–67
 A. rubescens, 65–67
Amphid, 96–97, 99–100, 102
Anticancer, 35
Antimicrobial, 2, 72, 120, 152, 234
Antioxidant, 37, 40–41, 196, 239–240
Aphelenchoides besseyi, 100
Aphids, 75
Apigenin, 170
Apple, 220
Arabidopsis, 101, 106, 120–122, 128–131, 171, 232
 A. thaliana, 101, 103, 109, 182–183, 192, 218, 220
Aspergillus, 11, 18–19, 42
 A. flavus, 18, 42
 A. ochraceus, 19
 A. parasiticus, 18
Attraction, 73, 76, 82–85, 87–88, 98, 101
Auxin, 77, 102–103, 123, 156, 170–172, 174, 195–196, 208, 211
 transport inhibition, 171

Avena, 1, 3–5, 8–10
 A. strigosa, 5, 6
Avenacinase, 4–5, 8–10
Avenacinase-like protein (ALP), 5, 11
Avenacins, 2–5, 6, 8, 11
 avenacin A-1, 3–5, 8, 10
Avenacosidase, 8–11
Avenacosides, 2–5, 7, 8, 10

Bacillus thuringiensis, 221
Bacterial blight, 233, 243
Banana, 35, 39
Barley, 4, 18–19, 24, 72–73, 99, 103, 190–191, 220
Benzoic acid (BA), 122–123
Benzothiadiazole, 121, 127, 129
Benzoxazinones, 73, 75–76, 79
 1,4-benzoxazin-3-ones, 73, 77
 DIMBOA, 74–90, 214–215
 GDIMBOA, 78–81, 82–87
 MBOA, 73, 75–80, 82, 84, 86
Benzoxazolinones, 73–80, 84–85
Betaines, 168, 171
Binary vector, 220
Biological activity, 75, 78, 89, 181, 199
Biosynthesis, 17, 19, 20–23, 25–27, 35–36, 41, 73, 103, 120–121, 127, 131, 139, 141–142, 144–148, 150, 153, 155–156, 169, 171, 181–182, 185–188, 192, 194–196, 198, 235, 237, 240
 cercosporin, 31, 34–48, 50
 2,4-dihydroxy-2H-1,4-benzoxazin-3(4H)-one, 73
 exopolysaccharides, 139–142, 144, 147–148, 150, 152–153, 157
 flavonoids, 26, 59, 140, 156, 167–174, 196–197
 jasmonic acid (JA), 180–183, 185–197

249

Biosynthesis (*cont.*)
  octadecanoids, 179–199
  phytoalexins, 2, 26–27, 87, 109, 120–121, 152, 155–156, 194, 197, 234–237, 240
  trichothecenes, 17–21, 23–27
Border repeats, 212
Bradyrhizobium, 59, 139, 144, 155
  B. japonicum, 144, 148, 150, 155
Bradysia impatiens, 192
Brassica napus, 221
Bryonia dioica, 190, 194

Caenorhabditis elegans, 97
Callose, 154, 234, 238
Capsular polysaccharide (KPS), 153–154, 156–157
β-Carotene, 41–42, 239–240
Carotenoids, 37, 40–42, 221
Casbene, 237
Cassava, 220, 231–244
Catalase, 37, 40–41, 124–127, 234–235, 239–240
Catechin, 63, 66–67, 236
Catechin gallate, 65–66
Catechol, 61, 121
Cell suspension culture, 235, 241–243
Cellulase, 102
Cenococcum, 62, 68
  C. geophilum, 67
  C. graniforme, 61
Cercospora, 31, 33–35, 38–39, 40–44, 46, 50–51
  C. beticola, 38, 47, 50
  C. coffeicola, 38
  C. kikuchii, 35–36, 47, 50
  C. musae, 39
  C. nicotianae, 41, 46
Cercosporin, 31–32, 34–48, 50
Cereal, 72–76, 79, 86
α-Chaconine, 98
Chalcone synthase (CHS), 120, 155–156, 169–171, 221
Chalcones, 170
Chemical defense, 26, 62
Chemoattractant, 59, 172
Chemotaxis, 73
Cinnamic acid, 122
Cladosporium fulvum, 1
Clostridium thermocellum, 147
Coffee, 35, 38
Colletotrichum lagenarium, 121, 123

Coniferyl alcohol, 213–214
Conjugation, 123, 216, 218
Corn, 18–19, 35, 75, 78, 193
Coronafacic acid, 190
Coronatine, 180, 190, 191, 193–195
Corydalis semperviren, 185
Coumarins, 19, 26, 34, 235; *see also* Furanocoumarins
Crown gall, 208, 211
Cucumber, 22, 100, 121–123
Cucurbic acid, 180, 182
Cucurbitacin A, 100
Cyanogenic glucosides, 237
Cyclic hydroxamic acids, 72–75, 77, 79, 82, 87–88
Cytochrome P 450 enzymes, 75, 183
Cytokinins, 103, 123, 196, 208, 211
Cytotoxic, 3, 33, 41

Deoxynivalenol, 19, 21, 24–25
26-Desglucoavenacosides, 3, 7
Deterrents, 75, 237
Diacetoxyscirpenol, 19, 21, 23–26
Diatraea grandiosella, 75
2,6-Dichloroisonicotinic acid (INA), 125, 129
Dihydro-JA, 180, 182, 189
DIMBOA, 74–90, 214–215
Dinor-oxo-phytodienoic acid, 198–199
Diploida maydis, 74
Disease resistance, 2, 4, 13, 119–120, 124–125, 127, 129–131
Diterpenoids, 236–237

Ectomycorrhiza, 58–63, 65–68
Elicitor, 120, 128, 152, 155–156, 186–187, 192, 194, 242–243
Epicatechin, 62, 65–66
Epicatechin gallate, 66
12,13-Epoxylinolenic acid, 182–184, 189
Ericaceae, 58
Erysiphe graminis, 191
Escherichia coli, 153, 183–184, 211, 214, 218, 220
Eschscholtzia californica, 187, 190
Esculin, 235–236
Esophageal gland, 96, 104–105
Esulentin, 235–236
Ethylene, 47, 103, 120, 127, 131, 155, 157, 170, 172, 195–196, 234
Eucalyptus, 61
Euphorbiaceae, 231

INDEX

Evolution, 169, 172
Exopolysaccharides (EPS), 139–148, 150, 152–157
  succinoglycan, 141, 144–147, 152

Fisetin, 168, 171
Flavonoids, 26, 59, 140, 156, 167–174, 196–197
  biosynthesis, 171
  chalcones, 170
  flavanols, 235
  flavanones, 170
  flavones, 168, 171
  isoflavonoids, 151, 156
Fumonisin, 18–19, 23
Fungal succession, 60
Fungus, 2–6, 9–11, 13, 17, 24–26, 33, 35–36, 38–39, 42, 44, 51, 62, 186
  pathogenic, 2, 6, 11, 33–34, 51, 75–76
Furanocoumarins, 26, 34
Fusarium, 2, 18–19, 21–27, 194
  F. avenaceum, 4
  F. culmorum, 5, 19, 21–22
  F. graminearum, 5, 24–25
  F. moniliforme, 19, 74–75
  F. pulicaris, 26
  F. sambucinum, 23
  F. sporotrichioides, 19, 21, 23, 25

Gaeumannomyces, 4–5
  G. graminis, 4–5, 8–11
Gall, 96, 102–103, 106, 208, 211–212
Gallocatechin, 235–236
GDIMBOA, 78–87
Gene disruption, 4, 8–9, 21–25, 36, 41, 47, 50
Giant cell, 96, 104–107
Globodera, 72, 97, 99, 102–104, 107–108
  G. pallida, 102, 107–108
  G. rostochiensis, 72, 97, 99, 102–104, 107–108
Glucanases, 108, 147, 234, 242
  β-1,3-glucanases, 109, 241–242
β-Glucans, 155, 242
β-Glucosidase, 73, 123, 214
Glutathione S-transferases, 75, 120
Glyceollin, 109, 155
Glycine soja, 169–170
Glycinoeclepin, 98
Glycoprotein, 97, 104, 120, 235, 241–242
Glycosyl hydrolase, 1–2, 9, 11,
Gossypol, 109

Hatching, 72, 88, 97–99, 108
  factor, 97–99
  stimulus, 72
Herbivore defense, 192, 197; *see also* Plant defense
Heterodera, 72, 82, 86, 88, 97–98, 101, 103
  H. avenae, 72, 82, 99, 103
  H. cajani, 108
  H. glycines, 72, 98–99, 104, 107
  H. oryzicola, 108
  H. sacchari, 108
  H. schachtii, 99–102
  H. sorghi, 108
Host recognition, 72, 85, 87, 89, 120
Host specificity, 4, 59, 62, 168–169
13(S)-Hydroperoxylinolenic acid, 182–184
Hydroxamic acids, 73–76, 78, 80, 83, 87–89; *see also* Cyclic hydroxamic acids
2-Hydroxy-2H-1,4-benzoxazin-3(4H)-ones, 73
p-Hydroxybenzoic acid, 61–62
1-Hydroxyindanoyl-L-isoleucine, 190, 193, 195
Hydroxyproline-rich glycoproteins (HRGP), 235, 241–242, 244
Hypericin, 34
Hypericum, 34
Hypersensitive response (HR), 33, 51, 104, 120, 152, 155–156, 194, 240, 243

Illegitimate recombination, 219
Induction, 51, 87, 96–97, 101, 108, 120–121, 123–125, 128–129, 131, 140, 150, 152, 154, 156, 167–173, 186–187, 190, 192, 194, 211, 213–214, 218, 240, 242–243
Infection, 5, 9, 10, 26–27, 36, 39, 42, 44, 59–60, 62, 72, 75, 86–87, 101–104, 106, 108–109, 120–123, 125, 127, 129, 139–140, 148, 150, 152, 154–157, 167, 170, 186, 194, 208, 212, 214, 220, 235, 237
Inulanthera calva, 182
Isoflavonoids, 151, 156

Jasmonic acid (JA), 180–183, 185–197

Kaempferol, 170

Lactarius, 67
  L. affinis, 66
Leccinum, 67

Legumes, 19, 148, 150, 154–155, 157, 167, 169, 170
Leucaena, 148, 150, 157
Leucinopine, 209–210
Lignin, 60, 121, 197, 234–235, 238, 240
Linolenic acid (LA), 181–182, 186–188, 192, 194, 198, 237
Lipo-chitin oligosaccharides (LCO), 168–174
Lipoxygenase, 182, 187
Lycopersicon, 1, 105
  L. esculentum, 105

Maize, 24, 72–75, 77, 79–82, 84, 86–89, 184, 188–189, 214, 220, 232
Manihot esculenta, 231; *see also* Cassava
MBOA, 73, 75–80, 82, 84, 86
Mechanotransduction, 190–191, 194, 197
Medicago, 150, 152–153, 155, 157, 169
Meloidogyne, 88, 97, 99, 104
  M. hapla, 108
  M. incognita, 72, 99–100, 102–109, 111
  M. javanica, 97, 102–103
Membrane damage, 38, 237–238, 244
Methyl jasmonate (JAMe), 182, 191, 192, 194, 195
Mixtures, 168–169, 186, 192
Multidrug resistance, 24–25
Mutant, 1–2, 4–5, 8–10, 19, 21–24, 36–37, 39, 41–42, 44–47, 50, 97, 103, 129–131, 146–148, 150–156, 168, 171, 188, 191–192, 216–218, 221
Mycorrhizal fungi, 57–62, 65, 67, 167
  ectomycorrhiza, 58–63, 65–68
  ericoid, 58, 60
  vesicular-arbuscular, 57–59
Mycorrhizas, 57, 58, 61; *see also* Mycorrhizal fungi
Mycotoxins, 17, 18, 19, 23–24, 26–27

Nematistatic, 109
Nematodes, 27, 72, 77, 80–84, 86–89, 95–97, 99–109
  behavior, 84, 102
  cyst nematodes, 88, 96, 98–99, 105, 107
  endoparasitic, 77, 80, 96
  root-knot, 88, 108
  root lesion, 73, 83, 87–90
Nematotoxic, 100
Neosolaniol, 21, 25
Nicotiana tabacum, 47, 171
Nod gene, 156, 167–168, 170–173

Nod factor, 140, 144, 150, 156
NodD, 168–169
Nodulation, 59, 140, 143, 147–148, 150, 152–155, 167, 169, 172
Nopaline, 128, 209–212

Oats, 1–6, 8–9
Ochratoxin, 18–19
Octadecanoids, 180–199
Octopine, 128, 209–212
Oligogalacturonides, 192
Oligosaccharides, 120, 140, 146, 168
OPDA-10,11-reductase, 179, 185
Opines, 209–212, 221
Orientation, 98, 101, 103, 222
Ostrinia, 75
  O. furnacalis, 75
  O. nubialis, 75
β-Oxidation, 182, 185, 188, 190, 198
Oxidative burst, 197, 240, 243
12-Oxo-phytodienoic acid (OPDA), 180, 182–192, 194–195, 198
Oxylipins, 199

Parsnips, 2, 23, 26
Pathogen defense, 193, 197
Pathogenesis-related proteins (PR), 108–109, 120–121, 123–126, 128–129, 131, 135
Pathogens, 2, 10, 13, 23, 26, 33–35, 38, 51, 72, 73–75, 87–88, 104, 108, 120–121, 129, 131, 156, 197, 221, 233, 242–244
Pea, 148–150
Peroxidase, 40–41, 120, 125–126, 182, 234–235, 239–240, 242
Perylenequinone, 32–34, 39–41, 50
Phanacoccus manihoti, 232
Phaseolus, 150, 190
  P. vulgaris, 190
Phenolics, 59–62, 65–67, 171, 194, 197, 213–214, 235–236, 239–240, 242
Phenylalanine ammonia lyase (PAL), 120–122, 129, 169, 234–235, 241–244
Phenylpropanoids, 121, 155–156, 169, 235, 241–242
  pathway, 121, 155–156, 169
Pheromones, 95, 107
Phospholipase, 187
Photosensitizer, 34–35, 37, 39–42, 44–48, 50
Phytoalexins, 2, 26–27, 75, 87, 109, 120–121, 152, 155–156, 170, 194, 197, 234–237, 240

Phytohormones, 102–103, 212, 242
Phytophthora, 155, 186, 194
  P. infestans, 194
  P. megasperma, 194
  P. sojae, 155, 186
Phytotoxic, 23, 74, 76
Picea abies, 61
α-Pinene, 65
β-Pinene, 65
Pinus, 61–62, 64–66
  P. radiata, 61
  P. resinosa, 65–66
Pisolithus, 67
  P. tinctorius, 66
Pisum, 140, 150, 157, 169
Plant defense, 2, 33, 47, 72, 75, 77, 87, 119–121, 125, 127, 152, 154–156, 170, 180, 187, 194, 197–198, 234–235; see also Herbivore defense
Plant disease, 21, 23–24, 27, 33, 51, 74, 119, 130
Plant–herbivore interactions, 179, 191–193
Plant hormones, 103, 185, 208, 221; see also Phytohormones
  abscisic acid, 103, 185
  auxin, 77, 102–103, 123, 156, 170–172, 174, 195–196, 208, 211
  cytokinins, 103, 123, 196, 208, 211
  ethylene, 47, 103, 120, 127, 131, 155, 157, 170, 172, 195–196, 234
Plant–pathogen interactions, 119, 155–156
Polyphenol oxidases, 58, 235
Post-harvest, 233, 234, 237–238
Post-harvest physiological deterioration (PPD), 233–244
Potato, 2, 23, 25–27, 72, 97–99, 103–104, 180, 184, 187, 194, 198, 232, 235, 238, 242, 244
Pratylenchus zeae, 73, 78, 81–83, 85–88, 90
Prostaglandins, 184
Protein transport, 218
Proteinase inhibitors, 109, 120, 188, 192, 234, 237
Protocatechuic acid, 61
Pseudomonas syringae, 51, 123, 125, 190, 212, 243
Puccinia graminis, 75

Quercetin, 59, 168, 170–171

Rapeseed, 220
Rauvolfia serpentina, 187

Reactive oxygen species (ROS), 120, 196–197, 239–240, 243; see also Active oxygen
Repellents, 72, 81–82, 84, 86, 88, 100
Resistance genes, 25, 27, 50, 104, 119–120
Rhizobia, 139–141, 144, 148, 150, 154, 156–157, 167–170, 172–173
Rhizobium, 59, 139, 150, 152, 154–157, 221
  R. etli, 140–141, 144, 156
  R. leguminosarum, 140–144, 148, 150–152, 154–155, 169, 210
  R. meliloti, 140–141, 171, 210
  R. tropici, 140, 143, 148
Ricinus communis, 237
Rishitin, 26
Roots, 3–5, 8, 23–24, 26, 58–63, 67–68, 73, 75–76, 78–83, 85–90, 97–110, 140–141, 149, 151, 153, 156–158, 168–175, 184, 189, 209, 215, 222, 232–245
  development, 170–171, 173–174
  exudate, 73, 78–79, 81, 83, 85–86, 89–90, 100–102, 108–109, 170, 215

SA β-glucoside (SAG), 122–123, 125
Saccharomyces cerevisiae, 47, 219; see also Yeast
Salicylic acid (SA), 47, 119–131, 194, 235
Salicylic acid-binding proteins (SABP), 124, 127
Sapogenin, 2
Saponins, 1–6, 8–11, 13
  avenacins, 2–5, 6, 8, 11
  avenacosides, 2–5, 7, 8, 10
  bisdesmosidic, 2
  degrading enzyme, 10
  monodesmosidic, 9
  oat, 3, 7
  steroidal, 1
  tomato, 2
  triterpenoid, 2
Scopoletin, 235–236
Scopolin, 235–236
Scrophulariaceae, 73
Semiochemical, 75, 95
Senescence, 99, 108, 127, 131, 190, 195–197, 234
Septoria, 1, 6, 9, 10, 15, 74
  S. avenae, 6, 8–11, 89
  S. lycopersici, 9–11
Sesquiterpenes, 2, 19, 23, 26–27

Signal, 59, 72, 87–89, 95–97, 99–101, 105–106, 108–109, 119–121, 123–127, 129–131, 140, 154–155–157, 167–169, 171–174, 181, 186, 188, 191–192, 194, 196, 198, 211–213, 216–217, 237–238, 244
  molecules, 59, 96, 120, 139–140, 154–157, 167–169, 171–174, 179, 191, 238
  transduction, 72, 105, 119, 125, 129–131, 167, 173, 181, 237
Sinapic acid, 213
Singlet oxygen, 32, 46, 50
Sinorhizobium, 139, 140–141, 143–148, 150, 152–157
  S. fredii, 148, 153, 156
  S. meliloti, 139–141, 143–148, 150, 152–157
Solanine, 98
Solanum tuberosum, 99
Solavetivone, 26–27
Soybean, 35–36, 39, 72, 104, 108–109, 144, 148, 150, 155–156, 171
Steroids, 2, 185, 236
  steroidal glycoalkaloids, 2, 9–10
  steroidal saponins, 1
Structure–activity relationships, 77
Stylet, 87, 96, 101, 105
Suberin, 234–235, 238
Succinamopine, 209–210
Succinoglycan, 141, 144–147, 152
Suillus, 63, 66–67
  S. intermedius, 62, 65–66
Superoxide dismutase, 37, 40–41, 239
Symbiosis, 57–59, 62, 139, 148, 150, 153–157, 168–169, 172–173
Syncytium, 96, 103–104, 108
Syringic acid, 213
Systemic acquired resistance (SAR), 33, 108, 120–121, 123–125, 129–130, 194, 239
Systemin, 192

T-2 toxin, 19, 21, 23–25
T-DNA, 208–209, 211–213, 216–221
T-DNA integration, 218–221
Tamarixetin, 170
Tannins, 58, 60–62, 234–235
  condensed, 60, 235
Taxifolin, 66
Tendril coiling, 189–190, 194
Terpenoids, 18, 25
  diterpenoids, 236–237
  sesquiterpenes, 2, 19, 23, 26–27
  triterpenoids, 2, 10, 100, 171
  trichothecenes, 17–25

α-Terthienyl, 100
Ti plasmid, 208, 210–212, 217, 220
Tobacco, 35, 37–38, 42, 45, 51, 99, 120–125, 128, 131, 155, 187–188, 192, 211, 213, 216–217, 220–221, 243
Tobacco mosaic virus (TMV), 121–125, 221
Tomatinase, 9–11
α-Tomatine, 9–10
Tomato, 1–2, 9–10, 102–104, 106, 108, 155, 186, 188, 190, 192, 194, 211, 221
Toxins, 21, 23–24, 26–27, 33–34, 36, 41, 50
Transformation, 21, 36, 38, 46–47, 75–76, 207–208, 219–220, 232
Transgenic plants, 106, 120–121, 124–125, 131, 171, 187, 216, 218, 221
Trichodiene, 19, 21–23, 26
Trichothecenes, 17–25
Trifolium, 141, 150, 157, 169, 171
  T. repens, 171
  T. subterraneum, 169
Triterpenoids, 2, 10, 100, 171
Tritrophic interactions, 192
Tuberonic acid, 180
Tumor, 35, 40, 208–213, 216, 218, 220
Tumorigenesis, 210–211, 216

Uridine, 172

Vaccinium myrtillus, 61
Vanillic acid, 107
Verticillium, 194
Vicia, 140–141, 148, 150, 152, 155, 157
  V. sativa, 155, 169–170
Virulence, 11, 24, 26, 59, 208, 210–212, 216
Volicitin, 180, 192–193

Wheat, 4, 18–19, 24, 72–75, 79, 99, 220, 233
Wheat head scab, 23–24
Wounding, 3, 59, 120, 127, 131, 183, 186–187, 191–192, 197, 213–214, 234–235, 237–239, 242, 244

Xanthomonas campestris, 146, 155, 233, 243
Xanthones, 18
Xanthotoxin, 26

Yeast, 22, 24–25, 47–48, 50, 155, 187, 194, 219, 243

Zea mays, 73, 75, 79–81, 87
Zearalenone, 18